Lecture Notes in Computer Science   14752

Founding Editors

Gerhard Goos
Juris Hartmanis

AF173006

Editorial Board Members

The series Lecture Notes in Computer Science (LNCS), including its subseries Lecture Notes in Artificial Intelligence (LNAI) and Lecture Notes in Bioinformatics (LNBI), has established itself as a medium for the publication of new developments in computer science and information technology research, teaching, and education.

LNCS enjoys close cooperation with the computer science R & D community, the series counts many renowned academics among its volume editors and paper authors, and collaborates with prestigious societies. Its mission is to serve this international community by providing an invaluable service, mainly focused on the publication of conference and workshop proceedings and postproceedings. LNCS commenced publication in 1973.

Bo Li · Minming Li · Xiaoming Sun
Editors

Frontiers
of Algorithmics

18th International Joint Conference, IJTCS-FAW 2024
Hong Kong SAR, China, July 29-31, 2024
Proceedings

 Springer

Editors
Bo Li
The Hong Kong Polytechnic University
Hong Kong, China

Minming Li
City University of Hong Kong
Hong Kong, China

Xiaoming Sun
Chinese Academy of Sciences
Beijing, China

ISSN 0302-9743   ISSN 1611-3349 (electronic)
Lecture Notes in Computer Science
ISBN 978-981-97-7751-8   ISBN 978-981-97-7752-5 (eBook)
https://doi.org/10.1007/978-981-97-7752-5

This Springer imprint is published by the registered company Springer Nature Singapore Pte Ltd.
The registered company address is: 152 Beach Road, #21-01/04 Gateway East, Singapore 189721, Singapore

If disposing of this product, please recycle the paper.

Preface

We are delighted to present the proceedings of the International Joint Conference on Theoretical Computer Science - Frontier of Algorithmic Wisdom, IJTCS-FAW 2024. This conference was a combination of the 5th International Joint Conference on Theoretical Computer Science (IJTCS) and the 18th International Conference on Frontiers of Algorithmic Wisdom (FAW), which took place in Hong Kong, China, from July 29 to July 31, 2024. The FAW conference originated as the Frontiers in Algorithmics Workshop in 2007 in Lanzhou, China, and has been held annually from 2007 to 2023 with published archival proceedings. IJTCS started in 2020 with the aim of attracting presentations on active topics in various tracks related to theoretical computer science.

To support a range of emerging research avenues, IJTCS and FAW collaborated to host an event that fostered the exchange of insights into both recent breakthroughs and enduring contributions in theoretical computer science. The conference featured contributed talks submitted to three tracks in IJTCS-FAW 2024, namely

- Track A: The 18th Conference on Frontiers of Algorithmic Wisdom
- Track B: Blockchain Theory and Technology
- Track C: Computational Economics and Algorithmic Game Theory

The Program Committee, comprising 48 active researchers from the field, reviewed 33 submissions from the three tracks and selected 17 of them as full papers and 3 as short papers. Each submission underwent a rigorous double-blind peer review process and received an average of three reviews. Following the review process and extensive discussions among the Program Committee members, two Best Paper Awards were presented:

- Clustering with a Knapsack Constraint: Parameterized Approximation Algorithms for the Knapsack Median Problem, authored by Zhen Zhang, Limei Liu, Yao Liu, Jie Chen, and Qilong Feng.
- On the Optimal Mixing Problem of Approximate Nash Equilibria in Bimatrix Games, authored by Xiaotie Deng, Dongchen Li, and Hanyu Li.

This year, we have a special track collaborating with the CCF Computational Economics Committee

- Track S: Selected Papers from 2024 CCF Annual Meeting on Computational Economics (CE)

Track S finally accepts 3 full papers from 10 submissions to CCF CE 2024.

In addition to the regular talks, IJTCS-FAW 2024 featured keynote talks by Dingzhu Du (University of Texas at Dallas), Maurice Herlihy (Brown University), and Pinyan Lu (Shanghai University of Finance and Economics). The conference also included invited talks in focused tracks on Blockchain Theory, Learning Theory, Quantum Computing, Multi-agent Learning, Multi-agent Systems, Multi-agent Games, and Mechanism Design in Digital Economy; and forums for young researchers, female participants, students, and the CSIAM Forum.

We extend our sincere gratitude to all those who contributed to the success of this conference: the authors for submitting their papers, the Program Committee members and Track Chairs for their excellent coordination of the review process and speaker invitations, and all the keynote and invited speakers. We would also like to thank the Advisory Committee, Steering Committee, and the General Chair for their valuable advice throughout the conference planning. Special thanks go to the members of the Local Organization Committee from the Hong Kong Polytechnic University for their invaluable organizational support. Finally, we express our appreciation to Springer for their constant encouragement and cooperation throughout the conference preparation.

July 2024   Bo Li
                                                                                      Minming Li
                                                                                      Xiaoming Sun

Organization

General Chair

Qing Li   Hong Kong Polytechnic University, Hong Kong, China

Program Committee Chairs

Bo Li   Hong Kong Polytechnic University, Hong Kong, China

Minming Li   City University of Hong Kong, Hong Kong, China

Xiaoming Sun   Chinese Academy of Sciences, China

Track Chairs

The 18th Conference on Frontiers of Algorithmic Wisdom

Zhihao Tang   Shanghai University of Finance and Economics, China

Prudence Wong   University of Liverpool, UK

Blockchain Theory and Technology

Man Ho Allen Au   Hong Kong Polytechnic University, Hong Kong, China

Computational Economics and Algorithmic Game Theory

Yukun Cheng   Suzhou University of Science and Technology, China

Zhengyang Liu   Beijing Institute of Technology, China

Biaoshuai Tao   Shanghai Jiao Tong University, China

Steering Committee

Xiaotie Deng	Peking University, China
Jian Li	Tsinghua University, China
Pinyan Lu	Shanghai University of Finance and Economics, China
Jianwei Huang	Chinese University of Hong Kong Shenzhen, China
Lijun Zhang	Chinese Academy of Sciences, China

Program Committee

Xiaohui Bei	Nanyang Technological University, Singapore
Paul Bell	University of Keele, UK
Hubert Chan	University of Hong Kong, China
Sisi Duan	Tsinghua University, China
Tomer Ezra	Harvard University, USA
Yiding Feng	University of Chicago, USA
Tom Friedetzky	Durham University, UK
Yuqing Kong	Peking University, China
Stanley P.Y. Fung	University of Leicester, UK
Xin Huang	Kyushu University, Japan
Dimitrios Letsios	King's College London, UK
Ya-Chun Liang	Columbia University, USA
Tao Lin	Harvard University, USA
Jinyan Liu	Beijing Institute of Technology, China
Shengxin Liu	HIT Shenzhen, China
Tianren Liu	Peking University, China
Junjie Luo	Beijing Jiaotong University, China
Weizhi Meng	Technical University of Denmark, Denmark
Shivika Narang	UNSW Sydney, Australia
Qingqin Nong	Ocean University of China, China
Pan Peng	University of Science and Technology of China, China
Alexandru Popa	University of Bucharest, Romania
Alex Psomas	Purdue University, USA
M. Sohel Rahman	Bangladesh University of Engineering and Technology, Bangladesh
Warut Suksompong	National University of Singapore, Singapore
Mashbat Suzuki	UNSW Sydney, Australia
Amitabh Trehan	Durham University, UK

Contents

On the Problem of Best Arm Retention

Houshuang Chen, Yuchen He, and Chihao Zhang$^{(\boxtimes)}$

Shanghai Jiao Tong University, Shanghai 200240, China
{chenhoushuang,yuchen_he,chihao}@sjtu.edu.cn

Abstract. This paper presents a comprehensive study on the problem of Best Arm Retention (BAR), which requires retaining m arms with the best arm included from n after some trials, in stochastic multi-armed bandit settings. We explore many perspectives of the problem.

- We begin by revisiting the lower bound for the (ε, δ)-PAC algorithm for Best Arm Identification (BAI), where we remove the previously imposed restriction of $\delta < 0.5$ in the lower bound found in the literature.
- By refining the technique above, we obtain optimal bounds for (ε, δ)-PAC algorithms for BAR.
- We further study another variant of the problem, called r-BAR, which has recently found applications in streaming algorithms for multi-armed bandits. The goal of the r-BAR problem is to ensure the expected gap between the best arm and the optimal arm retained is less than r. We prove tight sample complexity for the problem.
- We explore the regret minimization problem for r-BAR and develop algorithm beyond pure exploration. We also propose a conjecture regarding the optimal regret in this setting.

Keywords: Best arm identification · PAC learning · Multi-armed bandits

1 Introduction

The multi-armed bandit (MAB) framework, pioneered by [26], has emerged as a powerful paradigm for modeling sequential decision-making under uncertainty in various real-world applications, ranging from clinical trials to online advertising. Among myriad MAB problems, the *Best Arm Identification* (BAI), as the pure exploration version of stochastic MAB, stands out as a critical task, where the objective is to identify the best arm based on their rewards.

At the beginning of the stochastic MAB game, the player confronts n arms, each associated with an unknown distribution. At each round $t \in [T]^1$, the player chooses an arm and receives a reward. The player is trying to obtain higher accumulated rewards. Equivalently, the goal is to minimize the expected regret, which is the expected accumulated reward difference between playing the

---

[1] T can be a stopping time.

B. Li et al. (Eds.): IJTCS-FAW 2024, LNCS 14752, pp. 1–20, 2025.
https://doi.org/10.1007/978-981-97-7752-5_1

best arm with the highest mean and playing with the algorithm chosen arms. In contrast, BAI, the pure exploration version of MAB, seeks to swiftly identify the arm with the highest mean, ignoring reward considerations during the decision-making process.

While achieving a high-probability identification of the best arm remains an unsolved challenge [6,11], an alternative approach involves designing an (ε, δ)-probably approximately correct (PAC) algorithm for BAI. An (ε, δ)-PAC algorithm can find an arm whose mean reward is at most ε from the optimal one with probability at least $1 - \delta$. For this family of algorithms, [9,24] established that the sample complexity is $\Theta\left(\frac{n}{\varepsilon^2} \log \frac{1}{\delta}\right)$.

Recent research has explored streaming algorithms [1,12] employing multiple passes to retain m arms due to memory constraints. These works emphasize retaining the best arm to optimize rewards in subsequent passes. This particular setting naturally leads to the *Best Arm Retention* (BAR) problem, a pragmatic extension of BAI accommodating scenarios with limited memory or computational resources. The name of the problem was coined in [12] and was also known as "arm trapping problem" in literature [1]. However, previous study for the problem is either incomplete or suboptimal.

In BAR, the objective shifts from identifying the arm with the highest expected reward to retain a subset of size m containing the best arm for further exploration or exploitation. In practice, this subset may be subject to constraints like fixed memory capacity, making BAR an adaptable framework for addressing real-world considerations such as uncertainty, dynamic environments, and regret minimization over time. Notably, BAR reduces to the classic BAI problem when $m = 1$, and becomes easier as m increases.

Another similar extension of the BAI problem is to identify and retain the top m arms [16]. However, this extension poses greater complexity and in practice, retaining the best arm alone often suffices. For instance, if one would like to perform some regret minimization algorithm on the m arms, retaining the optimal one already yields optimal regret. Notably, (ε, δ)-PAC algorithm requires $\Omega\left(\frac{n}{\varepsilon^2} \log \frac{m}{\delta}\right)$ samples to retain the top m arms [17], which is worse than our bounds for BAR in Theorem 1.

In this work, we call an arm ε-optimal if the mean gap between the best arm and this arm is less than ε. We address the (ε, δ)-BAR problem, the PAC setting of BAR, where the objective is to ensure that the set of m retained arms contains an ε-optimal arm with at least $1 - \delta$ probability after observing as few samples as possible. The least number of samples to fulfill the requirement is called the sample complexity of (ε, δ)-BAR.

Theorem 1. *For any (ε, δ)-PAC algorithm for BAR satisfying $\varepsilon \le \frac{1}{8}$ and $\delta \le \frac{n-m}{n}(1 - \beta)$, where $\beta \in (0, 1)$ is a universal constant, the sample complexity is*

$$\Theta\left(\frac{n - m}{\varepsilon^2} \log \frac{n - m}{n\delta}\right).$$

It is trivial that the sample complexity is zero when $\delta \geq \frac{n-m}{n}$ because we can choose m arms uniformly at random. In fact, the lower bound in Theorem 1 addresses almost all feasible δ except $\frac{n-m}{n} - \delta = o\left(\frac{1}{n}\right)$, as explained in Remark 1.

If the subsequent exploitation only requires obtaining a low regret, as in [12], then it suffices for the expected gap between the mean of the best and the optimal in the retained arm to be small. That is a weaker requirement than (ε, δ)-PAC learnability. To capture the complexity of this requirement, we define the problem r-BAR, where the goal is to guarantee the expected gap is less than r. As before, when $m = 1$, it is equivalent to identifying an arm whose mean is at most r from the optimal one in expectation. We call it r-BAI problem, which has been investigated in [3]. We determine the sample complexity of this problem.

Theorem 2. *The sample complexity of r-BAR is* $\Theta\left(\frac{(n-m)^3}{(nr)^2}\right)$.

We further consider the decision-making process beyond pure exploration. Like the classic MAB, we prove both upper bounds and lower bounds for regret minimization. To this end, we introduce a new complexity measure called *regret complexity*, which intuitively measures how much regret one has to pay to retain an arm whose expected mean reward is at most r from the best. The formal definition of the regret complexity is in Sect. 3.3.

Theorem 3. *There exists an algorithm for r-BAR such that the regret complexity is no more than*

$$O\left(\frac{(n-m)^2}{nr}\left(1 + \sqrt{\frac{m}{n-m}}\right)\right),$$

and for any algorithm, the regret complexity is no less than

$$\Omega\left(\frac{(n-m)^2}{nr}\right).$$

The gap between the upper and the lower bounds is $\left(1 + \sqrt{\frac{m}{n-m}}\right)$. Thus our bounds are tight for $n - m = \Omega(n)$. When m is very close to n, the gap is $\sqrt{n}$, and we will explain in Sect. 5.3 that eliminating this gap is not easy because different instances requires different sample size and therefore a more sophisticated adaptive strategy is required for an optimal algorithm.

2   Related Work

Pure exploration in stochastic MAB (see [22]) has garnered significant attention and has been explored in various settings, with the most prominent being BAI. Research in this area has investigated sample complexity under fixed confidence [6,9,11,14,27,32], success probability of identifying the best arm with a fixed budget [2,18,34], and (ε, δ)-PAC algorithms for BAI, aiming to identify an ε-optimal arm with fixed confidence [13,15,24]. These settings naturally extend to finding the top m arms [4,16,17,30,33], or other structured arm groups [7,8,10, 20,23,25,29].

The concept of *Best Arm Retention* (BAR) was introduced by [12], with a similar idea of "trapping the best arm" first appearing in [1] in the context of stream algorithms. The notion of r-BAI (as a special case of r-BAR when $m = 1$), also referred to as "simple regret", was discussed by [3]. To the best of our knowledge, this paper is the first systematic investigation of the BAR problem.

3   Notation and Preliminaries

For any integer $n > 0$, let $[n]$ denote the set $\{1, 2, \ldots, n\}$. If $x, y \in \mathbb{R}$ and $x \leq y$, $[x, y]$ denotes the closed interval $\{z : x \leq z \leq y\}$. Let $Ber(x)$ denote the Bernoulli distribution with mean $x \in [0, 1]$. The Kullback-Leibler (KL) divergence between two Bernoulli distributions with means x and y is given by $\mathfrak{d}(x, y) := x \log \frac{x}{y} + (1 - x) \log \frac{1-x}{1-y}$ for brevity. There are some properties of the KL divergence:

Fact 1.

(a) $\mathfrak{d}(\cdot, y)$ (or $\mathfrak{d}(x, \cdot)$) is convex for any fixed y (or x);
(b) for any $0 \leq a \leq x \leq y \leq b \leq 1$, $\mathfrak{d}(a, b) \geq \mathfrak{d}(x, y)$;

Proof. (a) Directly calculating $\frac{\partial^2 \mathfrak{d}(x,y)}{\partial x^2} \geq 0$ implies $\mathfrak{d}(\cdot, y)$ is convex and $\mathfrak{d}(x, \cdot)$ is similar.
(b) Since $\frac{\partial \mathfrak{d}(x,y)}{\partial x}\Big|_{x=y} = 0$, $\mathfrak{d}(\cdot, y)$ achieves the minimum in $x = y$. Similarly, $\mathfrak{d}(x, y)$ is the minimum of $\mathfrak{d}(x, \cdot)$. Therefore, $\mathfrak{d}(a, b) \geq \mathfrak{d}(x, b) \geq \mathfrak{d}(x, y)$.

3.1   Mutil-armed Bandits

In this paper, we exclusively consider the stochastic *Multi-Armed Bandit* (MAB) problem, which can be represented by an n-dimensional product distribution $\nu = (\nu_1, \nu_2, \ldots, \nu_n)$. Each distribution corresponds to an arm. At each round/day $t \in [T]$, the player selects an arm $a_t \in [n]$ and receives a reward $r_t \sim \nu_{a_t}$ independently.

Let $\mu = (\mu_1, \mu_2, \ldots, \mu_n)$ denote the mean vector of ν. Define $i^* = \arg\max_{i \in [n]} \mu_i$ as the best arm with the highest mean, and let $\Delta_i = \mu_{i^*} - \mu_i$ represent the mean gap between the best arm and arm i. Additionally, let $T_i = \sum_{t=1}^{T} \mathbb{K}_{a_t=i}$ denote the number of times arm i is pulled. The player's objective is to maximize the accumulated reward $\sum_{t=1}^{T} r_t$, or equivalently, to minimize the regret, defined as $R(n, T) = T\mu_{i^*} - \mathbf{E}\left[\sum_{t=1}^{T} r_t\right] = \mathbf{E}\left[\sum_{i=1}^{n} \Delta_i T_i\right]$, which measures the difference between the accumulated expected reward of the best arm and that of the algorithm. We abbreviate $R(n, T)$ as R when the context is clear and refer to a product distribution ν as an MAB instance. If each ν_i is a Bernoulli distribution, we also use μ to denote an MAB instance.

For the regret of the MAB problem, there exist several algorithms that achieve tight bounds of $\Theta\left(\sqrt{nT}\right)$ up to a constant factor, such as the *Online Stochastic Mirror Descent* (OSMD) or the *Follow the Regularized Leader* (FTRL) algorithm. The following result from [21] provides a refined constant factor and refer to Appendix C for the detail of the algorithm.

Proposition 1 ([21], Theorem 11). *Running the procedure* MIRRORDESCENT *on any MAB instance with specific parameters, the regret is at most* $R(n,T) \leq \sqrt{2nT}$.

3.2 Best Arm Identification

Given an MAB instance ν, the *Best Arm Identification* (BAI) problem aims to identify the arm i^* with the highest mean based on as few samples as possible. Different from the fixed T in MAB, the sample size T here is a stopping time with respect to the filtration $(\mathcal{F}_t)_{t\in\mathbb{N}}$ where $\mathcal{F}_t = \sigma(a_1, r_1, a_2, r_2, \ldots, a_t, r_t)$. In our paper, we focus on the (ε, δ)-PAC setting of BAI, denoted as (ε, δ)-BAI, which requires the algorithm to output an ε-optimal arm with probability at least $1 - \delta$ for any MAB instance. Here, an arm i is considered ε-optimal if $\mu_{i^*} - \mu_i < \varepsilon$.

The first tight bound for (ε, δ)-BAI was achieved by a median elimination algorithm in [9]. We will provide the details of this algorithm in Appendix D for the completeness.

Proposition 2 ([9], Theorem 10). *Given* ε *and* δ, *and an arm set* S, *the median elimination algorithm* $\hat{i} = $ MEDIANELIMINATION(ε, δ, S), *taking* ε, δ, *and* S *as input and outputting an* ε-*optimal arm, is an* (ε, δ)-*BAI algorithm with sample size* $T = O\left(\frac{n}{\varepsilon^2} \log \frac{1}{\delta}\right)$.

3.3 Best Arm Retention

Given an MAB instance ν, the *Best Arm Retention* (BAR) problem involves retaining m arms S_T out of n after T samples, ensuring that the best arm is included in the retained set $i^* \in S_T$. Similar to BAI, in our paper the sample size T is a stopping time with respect to $(\mathcal{F}_t)_{t\in\mathbb{N}}$. Correspondingly, the (ε, δ)-PAC version of BAR, denoted as (ε, δ)-BAR, requires that the retained m arms contain at least one ε-optimal arm with probability at least $1 - \delta$.

In practice, to achieve low regret in a multiple-pass streaming algorithm, it suffices for the *expected gap*[2] between the mean of the best arm and the optimal arm in the retained m arms to be small. We define r-BAR as the problem to guarantee that the expected gap is at most r, namely $\mathbf{E}\left[\mu_{i^*} - \max_{i \in S_T} \mu_i\right] < r$.

---

[2] We use *mean gap* to refer to the mean difference between i^* and the other fixed arm, and *expected gap* to denote the expected difference in means between i^* and the optimal arm of an arm subset, where the randomness of the expectation arises from the arm subset.

The regret for a r-BAR algorithm, $R(n) = T\mu_{i^*} - \mathbf{E}\left[\sum_{t=1}^{T} r_t\right]$, is similar to the regret defined in Sect. 3.1 except that T is a stopping time. When referring to the sample (or regret) complexity of r-BAR, we denote the minimum samples (or regret) required by any algorithm capable of solving r-BAR for any MAB instance.

4   An (ε, δ)-PAC Algorithm for BAR

In this section, we will design a simple algorithm with the assistance of the median elimination algorithm to establish an upper bound, followed by a lower bound based on likelihood ratio.

4.1   Upper Bound for (ε, δ)-BAR

Our algorithm presented in Algorithm 1 is straightforward. We first uniformly at random choose $n - m + 1$ arms from the set of n arms. Next, we execute the MEDIANELIMINATION algorithm (Algorithm 6) to obtain an ε-optimal arm (with respect to the chosen arms) with probability $1 - \frac{n}{n-m+1} \cdot \delta$. Finally, we output this arm along with the remaining $m - 1$ arms that were not chosen in the first stage.

Algorithm 1 (ε, δ)-PAC Algorithm for BAR

    Input: ε, δ, m, arm set S
    Output: m arms
1: Choose $n - m + 1$ arms denoted as S' from n arms uniformly at random.
2: Run the median elimination algorithm $i' = \text{MEDIANELIMINATION}(\varepsilon, \frac{n}{n-m+1}\delta, S')$.
3: **return** $S \setminus S' \cup \{i'\}$.

Theorem 4 (Part of Theorem 1). *Algorithm 1 is an (ε, δ)-BAR algorithm with a sample size of $O\left(\frac{n-m+1}{\varepsilon^2} \log \frac{n-m+1}{n\delta}\right)$.*

Proof. Our algorithm fails if and only if:

(1) the optimal arm was chosen in the first stage, and
(2) the procedure MEDIANELIMINATION failed to return a nearly optimal arm.

Therefore, the failure probability is given by $\frac{n-m+1}{n} \cdot \frac{n}{n-m+1} \cdot \delta = \delta$. Furthermore, the sample complexity due to the MEDIANELIMINATION procedure is $O\left(\frac{n-m+1}{\varepsilon^2} \log \frac{n-m+1}{n\delta}\right)$ by Proposition 2.

4.2   Lower Bounds

Recall the mean vector $\mu = (\mu_1, \mu_2, \cdots, \mu_n)$ denotes a MAB instance, where each arm i follows a Bernoulli distribution with mean μ_i. Consider the following n instances: $\mathscr{H}_1 = (\frac{1}{2} + \varepsilon, \frac{1}{2}, \ldots, \frac{1}{2})$, and for $j \neq 1$, $\mathscr{H}_j$ differs from $\mathscr{H}_1$ only in $\mathscr{H}_j(j) = \frac{1}{2} + 2\varepsilon$. We use $\mathbf{Pr}_i[\cdot]$ and $\mathbf{E}_i[\cdot]$ to denote probability and expectation of the algorithm running on instance $\mathscr{H}_i$.

At a high level, if an algorithm outputs arm j as the best arm with a higher probability in $\mathscr{H}_j$ than in $\mathscr{H}_1$, then this algorithm, to some extent, distinguishes the two instances, indicating that arm j should be pulled enough times.

The well-known lower bound for (ε, δ)-BAI, $\Omega\left(\frac{n}{\varepsilon^2} \log \frac{1}{\delta}\right)$, is tight but with the restriction of $\delta < 0.5$ [24]. The lower bound proof techniques from previous literature (e.g. [24]) are mainly based on the following observation: given two instances $\mathscr{H}_1$ and $\mathscr{H}_j$, any proper algorithm should retain arm j with probability at least $1 - \delta$ on $\mathscr{H}_j$, but at most δ on $\mathscr{H}_1$. When $1 - \delta > \delta$, enough pulls for arm j are required to distinguish between these two instances, which necessitates $\delta < \frac{1}{2}$. However, when $\delta \geq \frac{1}{2}$, a more refined argument is required.

We will begin with a warm-up lower bound for BAR with $m = 1$ (or equivalently, BAI) to demonstrate how to eliminate this restriction in Theorem 5. For $\delta \geq 0.5$, we know that $\Theta\left(\frac{n}{\varepsilon^2} \log \frac{1}{\delta}\right) = \Theta\left(\frac{(1-\delta)n}{\varepsilon^2}\right)$. Therefore our result shows that the lower bound $\Omega\left(\frac{n}{\varepsilon^2} \log \frac{1}{\delta}\right)$ holds for almost all feasible δ. Subsequently, we will extend this method to BAR with general m.

BAI: Warm-up

Theorem 5. *For any (ε, δ)-BAI algorithm satisfying $1 - \delta = \frac{1+\Omega(1)}{n}$ and $\varepsilon \leq \frac{1}{8}$, the sample size when running on $\mathscr{H}_1$ is $\mathbf{E}_1[T] = \Omega\left(\frac{(1-\delta)n-1}{\varepsilon^2}\right)$.*

We will prove Theorem 5 in this part. Given any algorithm $\mathcal{A}$, we define B as $\left\{ i \in \{2, \ldots, n\} : \mathbf{Pr}_1[\mathcal{A} \text{ outputs arm } i] \leq \frac{\delta}{n-k} \right\}$, where k is an integer to be determined later such that $\frac{\delta}{n-k} \leq 1 - \delta$. It is evident that $\|B\| \geq k$. Otherwise, there are $n - k$ arms $j \in \{2, \ldots, n-1\}$ satisfying $\mathbf{Pr}_1[\mathcal{A} \text{ outputs arm } j] > \frac{\delta}{n-k}$ and this contradicts that $\mathcal{A}$ is an (ε, δ)-PAC algorithm. However, for each $i \in B$, we have $\mathbf{Pr}_i[\mathcal{A} \text{ outputs arm } i] \geq 1 - \delta$. The following lemma obtained by likelihood ratio shows that arm i must be pulled enough times. The proof is given in Appendix A for completeness.

Lemma 1 ([19], Lemma 1). *For any two MAB instances μ, μ' with n arms, and for any algorithm with almost-surely finite stopping time T and event $\mathcal{E} \in \mathcal{F}_T$, $\sum_{i=1}^{n}\left(\mathbf{E}_\mu[T_i] \cdot \mathfrak{d}(\mu_i, \mu_i')\right) \geq \mathfrak{d}(\mathbf{Pr}_\mu[\mathcal{E}], \mathbf{Pr}_{\mu'}[\mathcal{E}])$.*

Here we only consider the algorithm with almost-surely finite stopping time. Otherwise the sample complexity is infinite and the theorem obviously holds. Therefore, for any $i \in B$, we can apply Lemma 1 to instances $\mathscr{H}_1$ and $\mathscr{H}_i$ with $\mathcal{E}_i = \{\mathcal{A} \text{ outputs arm } i\}$ to obtain $\mathbf{E}_1[T_i] \cdot \mathfrak{d}(0.5, 0.5 + 2\varepsilon) \geq$

$\mathfrak{d}\left(\mathbf{Pr}_1\left[\mathcal{E}_i\right],\mathbf{Pr}_i\left[\mathcal{E}_i\right]\right)$. Since $\mathfrak{d}(0.5,0.5+2\varepsilon)\le 12\varepsilon^2$ and $\mathfrak{d}\left(\mathbf{Pr}_1\left[\mathcal{E}_i\right],\mathbf{Pr}_i\left[\mathcal{E}_i\right]\right)\ge \mathfrak{d}\left(\frac{\delta}{n-k},1-\delta\right)$ by Fact 1, then $12\varepsilon^2\cdot\mathbf{E}_1\left[T_i\right]\ge\mathfrak{d}\left(\frac{\delta}{n-k},1-\delta\right)$. Summing up all $i\in B$, we have $12\varepsilon^2\mathbf{E}_1\left[T\right]\ge k\cdot\mathfrak{d}\left(\frac{\delta}{n-k},1-\delta\right)$. Here we choose $k=n-\frac{\delta}{\frac{1-\delta}{2}+\frac{1}{2n}}=$ $\Omega(n)$, and thus $\frac{\delta}{n-k}=\frac{1-\delta}{2}+\frac{1}{2n}$. The following lemma, which will be proven in Appendix B, assists us in bounding the KL divergence:

Lemma 2. $\mathfrak{d}(\frac{1-\delta}{2}+\frac{1}{2n},1-\delta)=\Omega\left(\frac{1-\delta}{2}-\frac{1}{2n}\right)$ if $1-\delta=\frac{1+\Omega(1)}{n}$.

Therefore sample complexity $\mathbf{E}_1\left[T\right]=\Omega\left(\frac{(1-\delta)n-1}{\varepsilon^2}\right)$ if $1-\delta=\frac{1+\Omega(1)}{n}$.

Lower Bound for (ε,δ)-BAR. In this part, we establish a more general lower bound for the (ε,δ)-PAC algorithms for best arm retention (BAR) by refining arguments in the proof of Theorem 5. Similar to the proof for BAI, we only need to consider the algorithm with the almost-surely finite stopping time for the sample complexity. The following theorem is stronger than the lower bound in Theorem 1.

Theorem 6. *For any (ε,δ)-BAR algorithm with almost-surely finite stopping time such that $\varepsilon\le\frac{1}{8}$ and $\delta\le\frac{n-m}{n}(1-\beta)$, where β is a constant, its sample complexity on the input $\mathcal{H}_1$ satisfies $\mathbf{E}_1\left[T-T_1\right]=\Omega\left(\frac{n-m-\delta}{\varepsilon^2}\log\frac{n-m-\delta}{(n-1)\delta}\right)$.*

We reserve the notations introduced in the previous parts except $\mathcal{E}_i=\{\mathcal{A}$ retains arm $i\}$. For any algorithm, we have $m=\mathbf{E}_1\left[\sum_{i=1}^n\mathbb{1}_{\mathcal{E}_i}\right]=\sum_{i=1}^n\mathbf{Pr}_1\left[\mathcal{E}_i\right]$.

If we directly apply an argument similar to that in the proof of Theorem 5 here, then there exists at least k arms retained with probability at most $\frac{m-(1-\delta)}{n-k}$. Therefore the lower bound is $12\varepsilon^2\mathbf{E}_1\left[T-T_1\right]\ge k\cdot\mathfrak{d}\left(\frac{m-(1-\delta)}{n-k},1-\delta\right)$. We should choose a k satisfying $\frac{m-(1-\delta)}{n-k}<1-\delta$ to maximize $k\cdot\mathfrak{d}\left(\frac{m-(1-\delta)}{n-k},1-\delta\right)$. Consider a special case with $m=n-1$ and $\delta=\frac{1}{2n}$, where we can only choose $k=1$. Thus $k\cdot\mathfrak{d}(\frac{m-(1-\delta)}{n-k},1-\delta)=\mathfrak{d}(1-\frac{1-\frac{1}{2n}}{n-1},1-\frac{1}{2n})=\Theta\left(\frac{1}{n}\right)$, which leads to $\mathbf{E}_1\left[T\right]=\Omega\left(\frac{1}{\varepsilon^2 n}\right)$. However, the upper bound is $T\le O\left(\frac{1}{\varepsilon^2}\right)$ in this case.

The above analysis is too pessimistic as we classify suboptimal arms (those in B and those not in B) via a single threshold. The following lemma proved in Appendix B allows us to argue about their sum directly.

Lemma 3. *For any $x_1,x_2\ldots,x_n\in[0,1]$ with average $a:=\frac{\sum_i x_i}{n}<b\in[0,1]$, then $\sum_{i:x_i<b}\mathfrak{d}(x_i,b)\ge n\cdot\mathfrak{d}(a,b)$.*

Armed with Lemma 3, we can sum up all i such that $\mathbf{Pr}_1\left[\mathcal{E}_i\right]\le 1-\delta$ to which Lemma 1 can be applied.

$$12\varepsilon^2 \mathbf{E}_1\left[T - T_1\right] \geq \sum_{i:\mathbf{Pr}_1[\mathcal{E}_i]<1-\delta} \mathfrak{d}\left(\mathbf{Pr}_1\left[\mathcal{E}_i\right], \mathbf{Pr}_i\left[\mathcal{E}_i\right]\right) \qquad \text{(By Lemma 1)}$$

$$\geq \sum_{i:\mathbf{Pr}_1[\mathcal{E}_i]<1-\delta} \mathfrak{d}\left(\mathbf{Pr}_1\left[\mathcal{E}_i\right], 1 - \delta\right) \qquad \text{(By Fact 1)}$$

$$\geq (n-1)\cdot\mathfrak{d}\left(\frac{m-(1-\delta)}{n-1}, 1-\delta\right). \qquad \text{(By Lemma 3)}$$

Finally, we use a lemma proved in Appendix B to help analyze the KL divergence:

Lemma 4. *For any* $0 < a < b < 1$, *if* $\frac{b-a}{a} = \Omega(1)$, *then* $\mathfrak{d}(b,a) = \Omega\left(b\cdot\log\frac{b}{a}\right)$.

Now we are ready to bound $\mathfrak{d}(\frac{m-(1-\delta)}{n-1}, 1-\delta)$. Let $\delta = \frac{n-m}{n}(1-\beta)$ where β is a universal constant.

$$\mathfrak{d}\left(\frac{m-(1-\delta)}{n-1}, 1-\delta\right) = \mathfrak{d}\left(\frac{n-m-\delta}{n-1}, \delta\right) \qquad (\mathfrak{d}(x,y) = \mathfrak{d}(1-x, 1-y))$$

$$= \mathfrak{d}\left(\frac{n-m}{n}\left(1 + \frac{\beta}{n-1}\right), \frac{n-m}{n}(1-\beta)\right)$$

$$\triangleq \mathfrak{d}(B, A).$$

Here $\frac{B-A}{A} = \frac{\beta/(n-1)+\beta}{1-\beta} = \Omega(1)$, thereby $\mathfrak{d}\left(\frac{m-(1-\delta)}{n-1}, 1-\delta\right) =$ $\Omega\left(\frac{n-m-\delta}{n-1}\log\frac{n-m-\delta}{(n-1)\delta}\right)$. Thus, $\mathbf{E}_1\left[T - T_1\right] = \Omega\left(\frac{n-m-\delta}{\varepsilon^2}\log\frac{n-m-\delta}{(n-1)\delta}\right)$.

Remark 1. This approach encounters limitations when δ approaches the boundary $\frac{n-m}{n}$, specifically when $\frac{n-m}{n} - \delta = o\left(\frac{1}{n}\right)$. For instance, consider the scenario where $m = n-1$ and $\delta = \frac{1}{n} - \frac{1}{n^2}$. In this case, $\mathfrak{d}(\frac{m-(1-\delta)}{n-1}, 1-\delta) = \mathfrak{d}(\frac{n-1}{n} - \frac{1}{n^2(n-1)}, \frac{n-1}{n} + \frac{1}{n^2}) = \Theta\left(\frac{1}{n^2}\right)$. Consequently, the resulting lower bound is $\Omega\left(\frac{1}{\varepsilon^2 n}\right)$.

Suppose there exists an algorithm that achieves this lower bound, making it an $(\varepsilon, \frac{1}{n} - \frac{1}{n^2})$-BAR algorithm with a sample complexity of $c\frac{1}{\varepsilon^2 n}$, where c is a universal constant independent of n. However, as n grows sufficiently large, such that $c\frac{1}{\varepsilon^2 n} \leq 0$, this algorithm is paradoxical. It allows for retaining an ε-optimal arm with a higher probability than $\frac{n-1}{n}$ but without exploration, which is logically impossible.

5   *r*-BAR

Recall that r-BAR requires the mean difference between the best arm from n arms and the optimal arm from the retained pool of size $m < n$ is less than r. In this section, we study both the sample complexity and the minimum regret of this problem. Our results reveal some connections and distinctions between these two optimization objectives.

5.1   Sample Complexity for r-BAR

Exploration Algorithm for r-BAR. Directly adapting the (ε, δ)-PAC algorithm to a r-BAR setting would imply an expected gap bounded by $\delta + (1-\delta)\varepsilon \leq r$. This translates to $\delta \leq r$ and $\varepsilon \leq 2r$ for $\delta \leq 0.5$. Consequently, the sample complexity of this algorithm becomes $O\left(\frac{n-m}{r^2}\log\frac{n-m}{nr}\right)$. However, when r is small, this bound is not tight compared to the optimal bound in Theorem 2. To address this, we leverage the insight that a lower expected gap suggests lower regret. Thus, we employ the procedure MIRRORDESCENT and choose arms with probabilities proportional to their pull counts. A similar approach has been explored in [3,5,12]. Let us restate the algorithm for completeness:

Algorithm 2 Find best arm with online stochastic mirror descent
---
Input: arm set S of size n and time horizon T
Output: a good arm
1: **procedure** FINDBEST(S, T)
2:     Run MIRRORDESCENT on S with T rounds
3:     Compute $T_i, \forall i \in [n]$: the number of times arm i is pulled during T rounds
4:     Choose arm i' from S with probability $\frac{T_{i'}}{T}$
5:     **return** arm i'

Lemma 5. *Let i^* be the best arm among S, then for n arms with any mean μ as input, Algorithm 2 satisfies $\mathbf{E}\left[\mu_{i^*} - \mu_{i'}\right] \leq \sqrt{\frac{2n}{T}}$.*

Proof. Direct calculation yields

$$\mathbf{E}\left[\mu_{i^*} - \mu_{i'}\right] = \mathbf{E}\left[\mathbf{E}\left[\mu_{i^*} - \mu_{i'}|T_1, T_2, \ldots, T_n\right]\right]$$

$$= \mathbf{E}\left[\sum_{\text{arm } j \in S} \Delta_j \cdot \mathbf{Pr}\left[i' = j|T_1, T_2, \ldots, T_n\right]\right] = \mathbf{E}\left[\sum_{\text{arm } j \in S} \Delta_j \cdot \frac{T_j}{T}\right] \leq \sqrt{\frac{2n}{T}},$$

where the last inequality follows from Proposition 1.

We can employ the previously described subroutine to devise our final algorithm. Firstly, we randomly select $n - m + 1$ arms from the set of n arms, denoted as S'. We then run Algorithm 2 with a sufficient number of rounds. Finally, we add the output arm to the remaining unchosen arms to form the output set.

Theorem 7. (Part of Theorem *2*). *Algorithm 3 is an algorithm for r-BAR with sample complexity $O\left(\frac{(n-m)^3}{(nr)^2}\right)$.*

Algorithm 3 Optimal sampling for r-BAR

Input: arm set S of size $n \geq m$ and expectation gap r
Output: m arms
1: Sample $n - m + 1$ arms, denoted as S', uniformly at random from S
2: $i' = $ FINDBEST(S', T^*) where $T^* = \frac{2(n-m+2)^3}{(nr)^2}$
3: **return** $\{i'\} \cup S \backslash S'$

Proof. The sample complexity is straightforward. To demonstrate that the expected gap between the best arm in $\{\,i'\,\} \cup S \backslash S'$, denoted by $\hat{i}$, and the best arm i^* among all n arms is less than r, note that i^* is excluded only if $i^* \in S'$. Thus,

$$\mathbf{E}\left[\mu_{i^*} - \mu_{\hat{i}}\right] = \mathbf{Pr}\left[i^* \notin S'\right]\mathbf{E}\left[\mu_{i^*} - \mu_{\hat{i}}|i^* \notin S'\right] + \mathbf{Pr}\left[i^* \in S'\right]\mathbf{E}\left[\mu_{i^*} - \mu_{\hat{i}}|i^* \in S'\right]$$

$$\leq 0 + \frac{n - m + 1}{n}\sqrt{\frac{2(n - m + 1)}{T^*}} < r,$$

where the inequality follows from Lemma 5.

Lower Bound of Theorem 2. Let $\hat{i}$ be the best arm among the retained m arms. For any r-BAR algorithm, we have $\mathbf{E}\left[\mu_{i^*} - \mu_{\hat{i}}\right] \leq r$. By Markov inequality, we have $\mathbf{Pr}\left[\mu_{i^*} - \mu_{\hat{i}} \geq cr\right] \leq \frac{1}{c}$ for any $c > 0$. This implies that an r-BAR algorithm is also a $(cr, \frac{1}{c})$-PAC algorithm for BAR. Thus then sample complexity is bounded below by $\Omega\left(\frac{n-m}{(cr)^2}\log\frac{c(n-m)}{n}\right)$, as per Theorem 6. Choosing $c = \frac{2n}{n-m}$ completes the proof for the lower bound of Theorem 2.

5.2 Regret Minimization for r-BAR

Upper Bound. While a pure exploration algorithm suffices for r-BAR, it may yield large regret. For instance, if i^* is not chosen by Algorithm 3 initially, a low-regret algorithm like MIRRORDESCENT running on the suboptimal arm can still result in significant regret. To address this issue, we first run FINDBEST on all n arms for a few rounds. We then add the output arm to the randomly chosen $n - m + 1$ arms. Subsequently, we run FINDBEST on these $n - m + 2$ arms again, with the subsequent process identical to that of Algorithm 4, except for retaining the optimal arm from the initial process. Define $L_2 = \frac{2(n-m+2)^3}{(n-1)^2 r^2}$ and $L_1 = \frac{m-2}{n-1}L_2$. Our algorithm outlined in Algorithm 4 follows a similar approach as proposed in [12], albeit with distinct objectives.

Theorem 8. *Algorithm 4 is an algorithm for r-BAR with regret $O\left(\sqrt{\frac{(n-m)^3}{nr^2}}\right)$.*

Algorithm 4 Low-regret sampling for BAR

    Input: arm set S of size $n \geq m$ and expectation gap r
    Output: m arms
1: $i_1 = $ FINDBEST(S, L_1)
2: Sample $n - m + 1$ arms, denoted as S', uniformly at random from $S \setminus i_1$
3: $i_2 = $ FINDBEST$(S' \cup \{\,i_1\,\}, L_2)$
4: Uniformly at random choose $n - m$ arms from $S' \setminus \{\,i_1, i_2\,\}$ to drop
5: **return** the remaining arms

Proof. The proof follows a similar structure to that of Theorem 7. Let $\hat{i}$ denote the best arm among the retained arms. Since i^* will be dropped only if $i^* \in S'$, thus

$$
\begin{aligned}
\mathbf{E}\left[\mu_{i^*} - \mu_{\hat{i}}\right] &= \mathbf{Pr}\left[i^* \in S'\right] \mathbf{E}\left[\mu_{i^*} - \mu_{\hat{i}}|i^* \in S'\right] \\
&= \mathbf{Pr}\left[i^* \neq i_1\right] \mathbf{Pr}\left[i^* \in S'|i^* \neq i_1\right] \mathbf{E}\left[\mu_{i^*} - \mu_{\hat{i}}|i^* \in S'\right] \\
&\leq \frac{n-m+1}{n-1} \mathbf{E}\left[\mu_{i^*} - \mu_{\hat{i}}|i^* \in S'\right] \\
&\leq \frac{n-m+1}{n-1}\sqrt{\frac{2(n-m+2)}{L_2}} < r,
\end{aligned}
$$

where the last inequality follows from Lemma 5.

Regarding regret, the initial FINDBEST procedure incurs a regret of $\sqrt{2nL_1}$. In the subsequent step, let i' denote the best arm in $S' \cup i_1$. The regret between playing i' and the algorithm over L_2 rounds amounts to $\sqrt{2(n-m+2)L_2}$. If i' is not i^*, then

$$
\begin{aligned}
\mathbf{E}\left[\mu_{i^*} - \mu_{i'}\right] &= \mathbf{Pr}\left[i^* \notin S' \cup \{i_1\}\right]\mathbf{E}\left[\mu_{i^*} - \mu_{i'}|i^* \notin S' \cup \{i_1\}\right] \\
&= \mathbf{Pr}\left[i_1 \neq i^*\right]\mathbf{Pr}\left[i^* \notin S'|i^* \neq i_1\right]\mathbf{E}\left[\mu_{i^*} - \mu_{i'}|i^* \notin S' \cup \{i_1\}\right] \\
&\leq \frac{m-2}{n-1}\mathbf{Pr}\left[i_1 \neq i^*\right]\mathbf{E}\left[\mu_{i^*} - \mu_{i_1}|i^* \notin S' \cup \{i_1\}\right] \\
&\stackrel{\heartsuit}{=} \frac{m-2}{n-1}\mathbf{Pr}\left[i_1 \neq i^*\right]\mathbf{E}\left[\mu_{i^*} - \mu_{i_1}|i_1 \neq i^*\right] \\
&= \frac{m-2}{n-1}\mathbf{E}\left[\mu_{i^*} - \mu_{i_1}\right] \stackrel{\clubsuit}{\leq} \frac{m-2}{n-1}\sqrt{\frac{2n}{L_1}},
\end{aligned}
$$

where $\heartsuit$ follows that $\mu_{i^*} - \mu_{i_1}$ is independent of $\mathbb{K}_{i^* \notin S'}$ conditioned on $i^* \neq i_1$, and $\clubsuit$ is because of Lemma 5. Therefore the regret is

$$
\begin{aligned}
\sqrt{2nL_1} + \sqrt{2(n-m+2)L_2} + \frac{m-2}{n-1}\sqrt{\frac{2n}{L_1}}L_2 &\leq O\left(\frac{\sqrt{m(n-m)^3}}{nr} + \frac{(n-m)^2}{nr}\right) \\
&\leq O\left(\frac{(n-m)^2}{nr}\left(1 + \sqrt{\frac{m}{n-m}}\right)\right).
\end{aligned}
$$

When $n - m = \Omega(n)$, our regret bound is $O\left(\frac{(n-m)^2}{nr}\right)$.

Proof of the Lower Bound of Theorem 3. For the scenario where the algorithm does not almost surely stop within finite time, achieving a large regret lower bound requires more effort. In such cases, we cannot deduce a large regret by infinite sample complexity because the algorithm may continually pull the best arm. To tackle this, we first establish a lower bound for algorithms with an almost-surely finite stopping time, and then reduce any algorithm to this case.

For the algorithm with almost-surely finite stopping time, similar to the proof of the lower bound in Sect. 5.1, an r-BAR algorithm also acts as a $(\frac{2nr}{n-m}, \frac{n-m}{2n})$-PAC algorithm for BAR. Consequently, it must play the suboptimal arms

$\mathbf{E}_1 [T - T_1] = \Omega \left(\frac{(n-m)^3}{(nr)^2} \right)$ times on $\mathscr{H}_1$ with $\varepsilon = \frac{2nr}{n-m}$. Therefore, the regret by Wald's equation (see e.g. [28]) is $\Omega \left(\frac{2nr}{n-m} \mathbf{E}_1 [T - T_1] \right) = \Omega \left(\frac{(n-m)^2}{nr} \right)$.

Now assume there exists a $\frac{r}{2}$-BAR algorithm $\mathcal{A}$ with regret $o \left(\frac{(n-m)^2}{nr} \right)$, and let $T' = \omega \left(\frac{(n-m)^2}{nr^2} \right)$ be a fixed number. We can construct an algorithm $\mathcal{A}'$ with finite stopping time as follows: If $\mathcal{A}$ stops with $T < T'$ and outputs S_T, then $\mathcal{A}'$ simulates it. Otherwise, $\mathcal{A}'$ stops in the T'-th round, chooses an arm i' proportional to the pull times of each arm in T' rounds, similar to the procedure FINDBEST, and outputs it with $m - 1$ randomly chosen arms as $S_{T'}$.

We use $\mathbf{E}_{\mathcal{A}} [\cdot]$ and $\mathbf{E}_{\mathcal{A}'} [\cdot]$ to denote the expectation of the corresponding algorithms running on some MAB instance and let $\hat{i}$ denote the optimal arm among the retained subset arms, similarly for $\mathbf{Pr}_{\mathcal{A}} [\cdot]$ and $\mathbf{Pr}_{\mathcal{A}'} [\cdot]$. We use T_i to denote the number of times that i is pulled and $\mathcal{R} = \sum_{i=1}^{n} \Delta_i T_i$.

It is evident that the regret of $\mathcal{A}'$ is less than that of $\mathcal{A}$ because it may stop earlier. Now we claim that $\mathcal{A}'$ is an r-BAR algorithm with regret $o \left(\frac{(n-m)^2}{nr} \right)$:

$$
\begin{aligned}
& \mathbf{E}_{\mathcal{A}'} [\mu_{i^*} - \mu_{\hat{i}}] \\
& = \mathbf{Pr}_{\mathcal{A}'} [T \geq T'] \mathbf{E}_{\mathcal{A}'} [\mu_{i^*} - \mu_{\hat{i}} | T \geq T'] + \mathbf{Pr}_{\mathcal{A}'} [T < T'] \mathbf{E}_{\mathcal{A}'} [\mu_{i^*} - \mu_{\hat{i}} | T < T'] \\
& \leq \mathbf{Pr}_{\mathcal{A}} [T \geq T'] \mathbf{E}_{\mathcal{A}'} [\mu_{i^*} - \mu_{i'} | T \geq T'] + \mathbf{Pr}_{\mathcal{A}} [T < T'] \mathbf{E}_{\mathcal{A}} [\mu_{i^*} - \mu_{\hat{i}} | T < T'] \\
& \leq \mathbf{Pr}_{\mathcal{A}} [T \geq T'] \mathbf{E}_{\mathcal{A}'} \left[\frac{\mathcal{R}}{T'} \Big| T \geq T' \right] + \mathbf{E}_{\mathcal{A}} [\mu_{i^*} - \mu_{\hat{i}}] \\
& \leq \frac{1}{T'} \mathbf{Pr}_{\mathcal{A}} [T \geq T'] \mathbf{E}_{\mathcal{A}} [\mathcal{R} | T \geq T'] + \frac{r}{2} \leq \frac{1}{T'} \mathbf{E}_{\mathcal{A}} [\mathcal{R}] + \frac{r}{2} \leq r,
\end{aligned}
$$

which leads to a contradiction. Hence, the regret of any r-BAR algorithm is $\Omega \left(\frac{(n-m)^2}{nr} \right)$.

5.3   Difference Between Sample Complexity and Regret Minimization

The proof of the lower bound in Sect. 5.1 reveals that the challenging scenario for r-BAR with optimal regret occurs in $\mathscr{H}_1$ with $\varepsilon = \Theta \left(\frac{nr}{n-m} \right)$. Our analysis in Sect. 4.2 shows that, on this instance, the requisite number of rounds T is $\Theta \left(\frac{(n-m)^3}{(nr)^2} \right)$ in expectation.

If we consider an MAB game with fixed rounds $T = \Theta \left(\frac{(n-m)^3}{(nr)^2} \right)$, it is well known that the optimal regret is $\Theta \left(\sqrt{nT} \right) = O \left(\frac{(n-m)^2}{nr} \left(1 + \sqrt{\frac{m}{n-m}} \right) \right)$, which matches our upper bound in Theorem 3. Previous works have shown that this regret lower bound for MAB problem is achieved on the hard instance $\mathscr{H}_1$ but with mean gap parameter $\varepsilon' = \Theta \left(\sqrt{\frac{n}{T}} \right) = \Theta \left(\sqrt{\frac{n}{n-m}} \cdot \varepsilon \right)$ (see [22]). This indicates a regret lower bound for Algorithm 4: if an r-BAR algorithm runs for

$T = \Theta\left(\frac{(n-m)^3}{(nr)^2}\right)$ rounds on any instances (as our Algorithm 4 does), then it has to suffer a regret of $\Omega\left(\frac{(n-m)^2}{nr}\left(1 + \sqrt{\frac{m}{n-m}}\right)\right)$ on some instances.

This discrepancy between our regret upper and lower bounds indicates a natural idea to improve the algorithm. Note that $\mathcal{H}_1$ with ε' is not the hardest instance for r-BAR. An optimal algorithm need not play $\Theta\left(\frac{(n-m)^3}{(nr)^2}\right)$ rounds on this instance. It should be more adaptive, sampling for different number of times on different instances, rather than treating all instances equally by handling them as the hardest one.

In a nutshell, we conjecture that our lower bound is tight, and a more sophisticated algorithm is required to obtain optimal regret upper bounds.

Conjecture 1. The regret complexity of the r-BAR is $\Theta\left(\frac{(n-m)^2}{nr}\right)$.

A   Proof of Lemma 1

Let $T_i(s)$ denote the index of the s-th pull of arm i for $s \leq T_i$. Define the log-likelihood $L_T(a_1, r_1, a_2, r_2, \ldots, a_T, r_T) = \log \frac{\mathbf{Pr}_\mu[a_1, r_1, a_2, r_2, \ldots, a_T, r_T]}{\mathbf{Pr}_{\mu'}[a_1, r_1, a_2, r_2, \ldots, a_T, r_T]}$, abbreviated as L_T when the context is clear. By applying the chain rule to L_T, we have

$$L_T = \log \frac{\prod_{t=1}^T \mathbf{Pr}_\mu[a_t|\mathcal{F}_{t-1}] \cdot \mathbf{Pr}_\mu[r_t|\mathcal{F}_{t-1}, a_t]}{\prod_{t=1}^T \mathbf{Pr}_{\mu'}[a_t|\mathcal{F}_{t-1}] \cdot \mathbf{Pr}_{\mu'}[r_t|\mathcal{F}_{t-1}, a_t]}$$

$$= \sum_{t=1}^T \log \frac{\mathbf{Pr}_\mu[r_t|a_t]}{\mathbf{Pr}_{\mu'}[r_t|a_t]} = \sum_{i=1}^n \sum_{s=1}^{T_i} \log \frac{\mathbf{Pr}_\mu\left[r_{T_i(s)}|a_{T_i(s)}\right]}{\mathbf{Pr}_{\mu'}\left[r_{T_i(s)}|a_{T_i(s)}\right]},$$

where the second equality follows from $\mathbf{Pr}_\mu[a_t|\mathcal{F}_{t-1}] = \mathbf{Pr}_{\mu'}[a_t|\mathcal{F}_{t-1}]$ and that r_t is independent of $\mathcal{F}_{t-1}$ conditioned on a_t. With $\mathbf{E}_\mu\left[\log \frac{\mathbf{Pr}_\mu\left[r_{T_i(s)}|a_{T_i(s)}\right]}{\mathbf{Pr}_{\mu'}\left[r_{T_i(s)}|a_{T_i(s)}\right]}\right] = \mathfrak{d}(\mu_i, \mu_i')$, we apply Wald's Lemma (see e.g. [28]) to $\sum_{i=1}^n \sum_{s=1}^{T_i} \log \frac{\mathbf{Pr}_\mu\left[r_{T_i(s)}|a_{T_i(s)}\right]}{\mathbf{Pr}_{\mu'}\left[r_{T_i(s)}|a_{T_i(s)}\right]}$ to obtain:

$$\mathbf{E}_\mu[L_T] = \sum_{i=1}^n \mathbf{E}_\mu[T_i]\,\mathfrak{d}(\mu_i, \mu_i'). \tag{1}$$

The remaining task is to prove $\mathbf{E}_\mu[L_T] \geq \mathfrak{d}(\mathbf{Pr}_\mu[\mathcal{E}], \mathbf{Pr}_{\mu'}[\mathcal{E}])$ for any event $\mathcal{E} \in \mathcal{F}_T$, we reformulate the definition of L_T as

$$\mathbf{Pr}_{\mu'}[a_1, r_1, a_2, r_2, \ldots, a_T, r_T] = \exp{-L} \cdot \mathbf{Pr}_\mu[a_1, r_1, a_2, r_2, \ldots, a_T, r_T]$$

Summing over all $\omega \in \mathcal{E}$, we obtain

$$\mathbf{Pr}_{\mu'}[\mathcal{E}] = \mathbf{E}_\mu[\mathbb{1}_\mathcal{E} \cdot \exp{-L_T}]. \tag{2}$$

Continuing to lower bound Eq. (2), we have

$$\mathbf{Pr}_{\mu'}\left[\mathcal{E}\right] = \mathbf{E}_{\mu}\left[\mathbf{E}_{\mu}\left[\mathbb{1}_{\mathcal{E}} \cdot \exp -L \| \mathbb{1}_{\mathcal{E}}\right]\right]$$
$$\geq \mathbf{E}_{\mu}\left[\mathbb{1}_{\mathcal{E}} \cdot \exp -\mathbf{E}_{\mu}\left[L \| \mathbb{1}_{\mathcal{E}}\right]\right]$$
$$= \mathbf{Pr}_{\mu}\left[\mathcal{E}\right]\mathbf{E}_{\mu}\left[\mathbb{1}_{\mathcal{E}} \cdot \exp -\mathbf{E}_{\mu}\left[L \| \mathbb{1}_{\mathcal{E}}\right] | \mathcal{E}\right] + \mathbf{Pr}_{\mu}\left[\bar{\mathcal{E}}\right] \cdot 0$$
$$= \mathbf{Pr}_{\mu}\left[\mathcal{E}\right]\exp -\mathbf{E}_{\mu}\left[L_T | \mathcal{E}\right],$$

where the inequality follows from the Jensen inequality. Rearranging, we get $\mathbf{E}_{\mu}\left[L_T | \mathcal{E}\right] \geq \log \frac{\mathbf{Pr}_{\mu}[\mathcal{E}]}{\mathbf{Pr}_{\mu'}[\mathcal{E}]}$. Similarly, $\mathbf{E}_{\mu}\left[L_T | \bar{\mathcal{E}}\right] \geq \log \frac{\mathbf{Pr}_{\mu}[\bar{\mathcal{E}}]}{\mathbf{Pr}_{\mu'}[\bar{\mathcal{E}}]}$. Hence, we conclude

$$\mathbf{E}_{\mu}\left[L_T\right] = \mathbf{Pr}_{\mu}\left[\mathcal{E}\right]\mathbf{E}_{\mu}\left[L_T | \mathcal{E}\right] + \mathbf{Pr}_{\mu}\left[\bar{\mathcal{E}}\right]\mathbf{E}_{\mu}\left[L_T | \bar{\mathcal{E}}\right]$$
$$\geq \mathbf{Pr}_{\mu}\left[\mathcal{E}\right]\log\frac{\mathbf{Pr}_{\mu}\left[\mathcal{E}\right]}{\mathbf{Pr}_{\mu'}\left[\mathcal{E}\right]} + \mathbf{Pr}_{\mu}\left[\bar{\mathcal{E}}\right]\log\frac{\mathbf{Pr}_{\mu}\left[\bar{\mathcal{E}}\right]}{\mathbf{Pr}_{\mu'}\left[\bar{\mathcal{E}}\right]}$$
$$= \mathfrak{d}(\mathbf{Pr}_{\mu}\left[\mathcal{E}\right], \mathbf{Pr}_{\mu'}\left[\mathcal{E}\right]),$$

which completes our proof in conjunction with Eq. (1).

B   Bounds of KL Divergence

We will utilize the following inequalities from [31] to bound the KL divergence.

Fact 2. *The following inequalities hold.*

(a) $\log(1+x) \geq \frac{x}{1+x}, \forall x > -1$;
(b) $\log(1+x) \geq \frac{x}{1+x}(1+\frac{x}{2+x}) = \frac{2x}{2+x}, \forall x > 0$;
(c) $\log(1+x) \geq \frac{x}{1+x}\frac{2+x}{2}$, *if* $-1 < x \leq 0$.

Lemma 6 (Restate 2). $\mathfrak{d}(\frac{1-\delta}{2}+\frac{1}{2n}, 1-\delta) = \Omega\left(\frac{1-\delta}{2}-\frac{1}{2n}\right)$ *if* $1-\delta = \frac{1+\Omega(1)}{n}$.

Proof. By definition,

$$\mathfrak{d}(\frac{1-\delta}{2}+\frac{1}{2n}, 1-\delta)$$
$$= \left(\frac{1-\delta}{2}+\frac{1}{2n}\right)\log\frac{\frac{1-\delta}{2}+\frac{1}{2n}}{1-\delta} + \left(1-\frac{1-\delta}{2}-\frac{1}{2n}\right)\log\frac{1-\frac{1-\delta}{2}-\frac{1}{2n}}{\delta}$$
$$= \log\left(1+\frac{\frac{1-\delta}{2}-\frac{1}{2n}}{\delta}\right) + \left(\frac{1-\delta}{2}+\frac{1}{2n}\right)\left(\log\left(1-\frac{\frac{1-\delta}{2}-\frac{1}{2n}}{1-\delta}\right) + \log\left(1-\frac{\frac{1-\delta}{2}-\frac{1}{2n}}{\frac{1+\delta}{2}-\frac{1}{2n}}\right)\right)$$
$$\geq \left(\frac{1-\delta}{2}-\frac{1}{2n}\right)\left(\frac{1}{\frac{1+\delta}{2}-\frac{1}{2n}} - \left(\frac{1-\delta}{2}+\frac{1}{2n}\right)\left(\frac{1}{\frac{1-\delta}{2}+\frac{1}{2n}} \cdot \left(1-\frac{\frac{1-\delta}{2}-\frac{1}{2n}}{2(1-\delta)}\right) + \frac{1}{\delta}\right)\right)$$
$$= \left(\frac{1-\delta}{2}-\frac{1}{2n}\right)\left(\frac{1}{\frac{1+\delta}{2}-\frac{1}{2n}} - \frac{3}{4} - \frac{1}{4(1-\delta)n} - \frac{\frac{1-\delta}{2}+\frac{1}{2n}}{\delta}\right)$$
$$= \Omega\left(\frac{1-\delta}{2}-\frac{1}{2n}\right),$$

where the inequality follows from (a) & (b) of Fact 2.

Lemma 7 (Restate Lemma 3). *For any* $x_1, x_2 \ldots, x_n \in [0,1]$ *with average* $a := \frac{\sum_i x_i}{n} < b \in [0,1]$, *then* $\sum_{i:x_i<b} \eth(x_i, b) \geq n \cdot \eth(a, b)$.

Proof. Recall that $\eth(\cdot, y)$ is convex for any fixed y in Fact 1. Let $S = \{i : x_i < b\}$ and $k = \|S\|$. By the convexity of $\eth(\cdot, b)$, we have $\frac{1}{k} \sum_{i \in S} \eth(x_i, b) \geq \eth\left(\frac{\sum_{i \in S} x_i}{k}, b\right)$. Since $\eth(x, b) > \eth(y, b)$ if $x < y < b$ in Fact 1,

$$\sum_{i \in S} \eth(x_i, b) \geq k \cdot \eth\left(\frac{\sum_{i \in S} x_i}{k}, b\right) \geq k \cdot \eth\left(\frac{an - (n-k)b}{k}, b\right).$$

Using the convexity of $\eth(\cdot, b)$ again, we get

$$\frac{k}{n} \cdot \eth\left(\frac{an - (n-k)b}{k}, b\right) + \frac{n-k}{n} \cdot \eth(b, b) \geq \eth(a, b),$$

which implies $k \cdot \eth\left(\frac{an-(n-k)b}{k}, b\right) \geq n \cdot \eth(a, b)$ since $\eth(b, b) = 0$.

Lemma 8 (Restate Lemma 4). *For any* $0 < a < b < 1$, *if* $\frac{b-a}{a} = \Omega(1)$, *then* $\eth(b, a) = \Omega\left(b \cdot \log \frac{b}{a}\right)$.

Proof. By definition of the KL divergence, $\eth(b, a) = b \log \frac{b}{a} + (1 - b) \log \frac{1-b}{1-a}$. By Fact 2 (b) & (c),

$$b \log \frac{b}{a} = b \log \left(1 + \frac{b-a}{a}\right) \geq (b - a)\left(1 + \frac{(b-a)/a}{2 + (b-a)/a}\right)$$

and

$$(1 - b) \log \frac{1-b}{1-a} = (1 - b) \log \left(1 + \frac{a-b}{1-a}\right) \geq -(b - a)\left(1 - \frac{b-a}{2(1-a)}\right).$$

Therefore if $r := \frac{b-a}{a} = \Omega(1)$,

$$\eth(b, a) = \left(1 - \frac{1}{1 + r/(2+r)}\right) b \log \frac{b}{a} + \frac{1}{1 + r/(2+r)} b \log \frac{b}{a} + (1 - b) \log \frac{1-b}{1-a}$$

$$\geq \left(1 - \frac{1}{1 + r/(2+r)}\right) b \log \frac{b}{a} + (b - a) - (b - a)\left(1 - \frac{b-a}{2(1-a)}\right)$$

$$\geq \left(1 - \frac{1}{1 + r/(2+r)}\right) b \log \frac{b}{a}.$$

C Details of the OSMD Algorithm Corresponding to Proposition 1

For completeness, we provide a description of the OSMD algorithm used in Algorithm 1. For more detailed information, please refer to the work of [21].

Let $\Delta_{(n-1)}$ denote the probability simplex with $n - 1$ dimensions, defined as $\Delta_{(n-1)} = \{ \mathbf{q} \in \mathbb{R}_{\geq 0} : \sum_{i=1}^{n} \mathbf{q}(i) = 1 \}$. Here, $\mathbf{q}(i)$ represents the value at the i-th position of vector $\mathbf{q}$. Consider a function $F : \mathbb{R}^n \to \mathbb{R} \cup \{ \infty \}$. The Bregman divergence with respect to F is defined as $B_F(\mathbf{q}, \mathbf{p}) = F(\mathbf{q}) - F(\mathbf{p}) - \langle \nabla F(\mathbf{p}), \mathbf{q} - \mathbf{p} \rangle$ for any $\mathbf{q}, \mathbf{p} \in \mathbb{R}^n$.

The algorithm proposed in [21] is designed for loss cases, where each pull results in a loss associated with the corresponding arm instead of a reward. To adapt their algorithm to our setting, we can perform a simple reduction by constructing the loss of each arm $\ell_t(i)$ as $1 - r_t(i)$, where $r_t(i)$ is the reward of arm arm_i. It is straightforward to verify that the results in [21] also hold for the reward setting. Let η be the learning rate and $F : \mathbb{R}^{\|S\|} \to \mathbb{R} \cup \{ \infty \}$ be the potential function, where S is the arm set. Without loss of generality, we index the arms in S by $[\|S\|]$.

Algorithm 5 Online Stochastic Mirror Descent([21])

Input: a set of arms S and the number of rounds L

1: **procedure** MIRRORDESCENT(S, L)
2:      $Q_1 \leftarrow \arg\min_{\mathbf{q} \in \Delta_{(|S|-1)}} F(\mathbf{q})$
3:      **for** $t = 1, 2 \ldots, L$ **do**
4:          Sample arm $a_t \sim Q_t$, observe reward $r_t(a_t)$ and let $\ell_t(a_t) = 1 - r_t(A_t)$
5:          Compute reward estimator $\hat{\ell}_t$ as

$$\hat{\ell}_t(i) = \mathbb{1}\left[A_t = i\right]\left(\ell_t(i) - \frac{1}{2} + \frac{\eta}{8}\left(1 + \frac{1}{Q_t(i) + \sqrt{Q_t(i)}}\right)\right) - \frac{\eta Q_t(A_t)}{8\left(Q_t(i) + \sqrt{Q_t(i)}\right)}$$

6:          Set $Q_{t+1} = \arg\min_{\mathbf{q} \in \Delta_{(|S|-1)}} \langle \mathbf{q}, \hat{\ell}_t \rangle + \frac{1}{\eta} \cdot B_F(\mathbf{q}, Q_t)$

By choosing $\eta = \sqrt{\frac{8}{L}}$ and $F(\mathbf{q}) = -2 \sum_{i=1}^{\|S\|} \sqrt{\mathbf{q}(i)}$, the conclusion in Proposition 1 can be directly derived from Theorem 11 in [21].

D Details of the MEDIANELIMINATION Algorithm Corresponding to Proposition 2

For completeness, we present the description of the MEDIANELIMINATION algorithm we used in Algorithm 1. For more detailed information, please refer to Theorem 10 of [9].

Algorithm 6 Median Elimination([9])

Input: a set of arms S of size n and (ε, δ)
Output: an arm

1: **procedure** MEDIANELIMINATION(ε, δ, S)
2:     Set $S_1 = S, \varepsilon_1 = \varepsilon/4, \delta_1 = \delta/2, \ell = 1$
3:     **while** $|S| > 1$ **do**
4:         Sample every arm $a \in S$ for $\frac{4}{\varepsilon_\ell^2} \log \frac{3}{\delta_\ell}$ times, and let $\hat{p}_a^\ell$ denote its empirical mean
5:         Find the median of $\hat{p}_a^\ell$, denoted by m_ℓ
6:         Update: $S_{\ell+1} = S_\ell \setminus \left\{ a : \hat{p}_a^\ell < m_\ell \right\}$
7:         Update: $\varepsilon_{\ell+1} = \frac{3}{4}\varepsilon_\ell, \delta_{\ell+1} = \frac{\delta_\ell}{2}, \ell = \ell + 1$

References

1. Assadi, S., Wang, C.: Single-pass streaming lower bounds for multi-armed bandits exploration with instance-sensitive sample complexity. In: Proceedings of the 35th Annual Conference on Neural Information Processing Systems (NeurIPS) (2022)
2. Audibert, J., Bubeck, S., Munos, R.: Best arm identification in multi-armed bandits. In: Proceedings of the 23th Conference on Learning Theory (COLT) (2010)
3. Bubeck, S., Munos, R., Stoltz, G.: Pure exploration in multi-armed bandits problems. In: Proceedings of the 20th International Conference on Algorithmic Learning Theory (ALT) (2009)
4. Chen, C.H., He, D., Fu, M., Lee, L.H.: Efficient simulation budget allocation for selecting an optimal subset. INFORMS J. Comput. **20**(4), 579–595 (2008). https://doi.org/10.1287/IJOC.1080.0268
5. Chen, H., He, Y., Zhang, C.: On interpolating experts and multi-armed bandits. arXiv preprint arXiv:2307.07264 (2023). https://doi.org/10.48550/ARXIV.2307.07264
6. Chen, L., Li, J., Qiao, M.: Towards instance optimal bounds for best arm identification. In: Proceedings of the 30th Conference on Learning Theory (COLT) (2017)
7. Chen, S., Lin, T., King, I., Lyu, M.R., Chen, W.: Combinatorial pure exploration of multi-armed bandits. In: Proceedings of the 27th Annual Conference on Neural Information Processing Systems (NeurIPS) (2014)
8. Degenne, R., Ménard, P., Shang, X., Valko, M.: Gamification of pure exploration for linear bandits. In: Proceedings of the 37th International Conference on Machine Learning (ICML) (2020)
9. Even-Dar, E., Mannor, S., Mansour, Y.: Action elimination and stopping conditions for the multi-armed bandit and reinforcement learning problems. J. Mach. Learn. Res. **7**, 1079–1105 (2006)
10. Fiez, T., Jain, L., Jamieson, K.G., Ratliff, L.: Sequential experimental design for transductive linear bandits. In: Proceedings of the 32th Annual Conference on Neural Information Processing Systems (NeurIPS) (2019)
11. Garivier, A., Kaufmann, E.: Optimal best arm identification with fixed confidence. In: Proceedings of the 29th Conference on Learning Theory (COLT) (2016)
12. He, Y., Ye, Z., Zhang, C.: Understanding memory-regret trade-off for streaming stochastic multi-armed bandits. arXiv preprint arXiv:2405.19752 (2024)

13. Howard, S.R., Ramdas, A.: Sequential estimation of quantiles with applications to A/B testing and best-arm identification. Bernoulli **28**(3), 1704–1728 (2022)
14. Jamieson, K., Malloy, M., Nowak, R., Bubeck, S.: lil'UCB: an optimal exploration algorithm for multi-armed bandits. In: Proceedings of the 27th Conference on Learning Theory (COLT) (2014)
15. Jourdan, M., Degenne, R., Kaufmann, E.: An ε-best-arm identification algorithm for fixed-confidence and beyond. In: Proceedings of the 36th Annual Conference on Neural Information Processing Systems (NeurIPS) (2023)
16. Kalyanakrishnan, S., Stone, P.: Efficient selection of multiple bandit arms: theory and practice. In: Proceedings of the 27th International Conference on Machine Learning (ICML) (2010)
17. Kalyanakrishnan, S., Tewari, A., Auer, P., Stone, P.: Pac subset selection in stochastic multi-armed bandits. In: Proceedings of the 29th International Conference on Machine Learning (ICML) (2012)
18. Karnin, Z., Koren, T., Somekh, O.: Almost optimal exploration in multi-armed bandits. In: Proceedings of the 30th International Conference on Machine Learning (ICML) (2013)
19. Kaufmann, E., Cappé, O., Garivier, A.: On the complexity of best arm identification in multi-armed bandit models. J. Mach. Learn. Res. **17**, 1–42 (2016)
20. Kone, C., Kaufmann, E., Richert, L.: Bandit pareto set identification: the fixed budget setting. arXiv preprint arXiv:2311.03992 (2023). https://doi.org/10.48550/ARXIV.2311.03992
21. Lattimore, T., Gyorgy, A.: Mirror descent and the information ratio. In: Proceedings of the 34th Conference on Learning Theory (COLT) (2021)
22. Lattimore, T., Szepesvári, C.: Bandit Algorithms. Cambridge University Press, Cambridge (2020)
23. Locatelli, A., Gutzeit, M., Carpentier, A.: An optimal algorithm for the thresholding bandit problem. In: Proceedings of the 33th International Conference on Machine Learning (ICML) (2016)
24. Mannor, S., Tsitsiklis, J.N.: The sample complexity of exploration in the multi-armed bandit problem. J. Mach. Learn. Res. **5**, 623–648 (2004)
25. Mason, B., Jain, L., Tripathy, A., Nowak, R.: Finding all ϵ-good arms in stochastic bandits. In: Proceedings of the 33th Annual Conference on Neural Information Processing Systems (NeurIPS) (2020)
26. Robbins, H.: Some aspects of the sequential design of experiments. Bull. Am. Math. Soc. **58**(5), 527–535 (1952)
27. Russo, D.: Simple Bayesian algorithms for best arm identification. In: Proceedings of the 29th Conference on Learning Theory (COLT) (2016)
28. Siegmund, D.: Sequential Analysis: Tests and Confidence Intervals. Springer, Heidelberg (2013)
29. Simchi-Levi, D., Wang, C., Xu, J.: On experimentation with heterogeneous subgroups: An asymptotic optimal δ-weighted-PAC design. Available at SSRN 4721755 (2024)
30. Simchowitz, M., Jamieson, K., Recht, B.: The simulator: understanding adaptive sampling in the moderate-confidence regime. In: Proceedings of the 30th Conference on Learning Theory (COLT) (2017)
31. Topsøe, F.: Some bounds for the logarithmic function. Inequality Theory Appl. **4**, 137 (2007)
32. Wang, P.A., Tzeng, R.C., Proutiere, A.: Fast pure exploration via frank-wolfe. In: Proceedings of the 34th Annual Conference on Neural Information Processing Systems (NeurIPS) (2021)

33. You, W., Qin, C., Wang, Z., Yang, S.: Information-directed selection for top-two algorithms. In: Proceedings of the 36th Conference on Learning Theory (COLT) (2023)
34. Zhao, Y., Stephens, C., Szepesvári, C., Jun, K.S.: Revisiting simple regret: fast rates for returning a good arm. In: Proceedings of the 40th International Conference on Machine Learning (ICML) (2023)

Clustering with a Knapsack Constraint: Parameterized Approximation Algorithms for the Knapsack Median Problem

Zhen Zhang[1,2], Limei Liu[1,2], Yao Liu[1,2], Jie Chen[1,2(✉)], and Qilong Feng[2,3(✉)]

[1] School of Advanced Interdisciplinary Studies, Hunan University of Technology and Business, Changsha 410205, People's Republic of China
`chemjay@hnu.edu.cn`
[2] Xiangjiang Laboratory, Changsha 410205, People's Republic of China
`csufeng@mail.csu.edu.cn`
[3] School of Computer Science and Engineering, Central South University, Changsha 410083, People's Republic of China

Abstract. The KNAPSACK MEDIAN problem was known to be W[2]-hard if parameterized by the maximal number of opened facilities in feasible solutions (denoted by k), implying that exactly solving this problem in $\text{FPT}(k)$ time is unlikely. We focus on parameterized approximation algorithms for the KNAPSACK MEDIAN problem. We give a sampling-based approach for reducing the solution search space, which yields a $(3 + \varepsilon)$-approximation algorithm that runs in $(k\varepsilon^{-1})^{O(k)} n^{O(1)}$ time in general metric spaces and a $(1 + \varepsilon)$-approximation algorithm with similar running time in d-dimensional Euclidean space.

Keywords: Approximation algorithms · Fixed-parameter tractability · Knapsack median

1 Introduction

Facility location-type problems are frequently encountered in many fields related to operational research, which aim to locate facilities to serve a given set of clients. Most of these problems involve the following trade-off: We would like to connect each client to a nearby facility, while the opened facilities should be as few as possible.

We focus on an extensively studied facility location-type problem, called KNAPSACK MEDIAN. An instance $(\mathcal{C}, \mathcal{F}, B, w)$ of the KNAPSACK MEDIAN problem considers a set $\mathcal{C}$ of clients and a set $\mathcal{F}$ of facilities located in a metric

This work was supported by National Natural Science Foundation of China (62202161, 71991465, 62202160), Open Project of Xiangjiang Laboratory (22XJ03013), Natural Science Foundation of Hunan Province (2023JJ40240, 2021JJ40158), Scientific Research Fund of Hunan Provincial Education Department (23B0597, 21A0376), and Project of Hunan Social Science Achievement Appraisal Committe (XSP21YBZ113).

B. Li et al. (Eds.): IJTCS-FAW 2024, LNCS 14752, pp. 21–32, 2025.
https://doi.org/10.1007/978-981-97-7752-5_2

space, and a budget $B > 0$, where each facility $f \in \mathcal{F}$ is associated with a positive opening cost $w(f)$, and connecting a client $c \in \mathcal{C}$ to f incurs a connection cost $\Delta(c, f)$ equaling the distance from c to f. A feasible solution $(\mathcal{S}, \tau)$ to this instance opens a subset $\mathcal{S} \subseteq \mathcal{F}$ of facilities satisfying $\sum_{f \in \mathcal{S}} w(f) \le B$ and connects each client $c \in \mathcal{C}$ to an opened facility $\tau(c) \in \mathcal{S}$. The cost of such a solution is $\sum_{c \in \mathcal{C}} \Delta(c, \tau(c))$, and the goal of the KNAPSACK MEDIAN problem is to find a feasible solution with minimal cost.

Designing algorithms for the KNAPSACK MEDIAN problem remains an active area of research. The first algorithm with provable guarantees was given by Lin and Vitter [1]. They showed that a greedy set covering-based approach yields a solution violating the budget by a factor of $(1 + \varepsilon^{-1})(\ln n + 1)$ (n denotes the size of the considered instance and ε is a positive constant), whose cost is within $1 + \varepsilon$ times the cost of an optimal solution. Krishnaswamy et al. [2] latter gave a constant-factor approximation algorithm that violates the budget by the maximal facility-opening cost based on an iterative rounding method. Moreover, Charikar and Guha [3] and Byrka et al. [4] introduced constant-factor approximation algorithms with arbitrarily small budget violations. For the case where the constraints posed on the instances need to be strictly satisfied, Kumar [5] strengthened the natural linear programming relaxation of the problem and gave a 2700-approximation algorithm. This approximation guarantee was improved by a series of work [4,6–8] to the state-of-the-art ratio of $6.387 + \varepsilon$ [9]. There has also been work focusing on constrained versions of the KNAPSACK MEDIAN problem. Han et al. [7] generalized the KNAPSACK MEDIAN problem by posing a lower bound constraint on the number of clients connected to each facility, and gave a 1608-approximation algorithm for this generalization. Grover et al. [10] considered the CAPACITATED KNAPSACK MEDIAN problem where no facility can be connected with too many clients, and gave a constant-factor approximation algorithm violating the budget by a $1 + \varepsilon$ factor and the capacities by a $2 + \varepsilon$ factor.

1.1 Our Results

A commonly used way for relaxing facility location-type problems is to assume that the numbers of facilities opened by feasible solutions are small, which is reasonable since the locations servicing demands are much fewer than the clients in most real-world applications of these problems. We study the KNAPSACK MEDIAN problem under such an assumption, that is, the maximal number of opened facilities among all the feasible solutions to our considered instance, denoted by k, is limited. Given an instance $(\mathcal{C}, \mathcal{F}, B, w)$ of the KNAPSACK MEDIAN problem, we can determine the value of k by sorting the facilities according to their opening costs, and it is clear that an optimal solution can be found by brute-force enumeration in $(|\mathcal{C}||\mathcal{F}|)^{O(k)}$ time, but we ask: What can be done in FPT(k) time (i.e., $(|\mathcal{C}||\mathcal{F}|)^{O(1)} g(k)$ time for a positive function g)?

Unfortunately, a reduction to the SET COVER problem implies that the KNAPSACK MEDIAN problem, even for the case where the facility-opening costs are uniform, is W[2]-hard if parameterized by k [11], and the approximation ratio of any $\mathrm{FPT}(k)$-time algorithm for it cannot be better than $1 + 2e^{-1}$ if the Gap-Exponential Time Hypothesis is true [12]. These negative results say that exactly solving the KNAPSACK MEDIAN problem in $\mathrm{FPT}(k)$ time is unlikely, but the possibility of constructing better approximation solutions in $\mathrm{FPT}(k)$ time is not ruled out. Indeed, FPT approximation algorithms parameterized by the upper bound on the number of opened facilities have been developed for many facility location-type problems, including UNCAPACITATED FACILITY LOCATION [12], k-MEDIAN [12–14], MATROID MEDIAN [12], and SUM-OF-RADII CLUSTERING [15,16]. These algorithms lead to significantly improved approximation guarantees compared with polynomial-time algorithms. In this paper we get similar improvements for the KNAPSACK MEDIAN problem.

Theorem 1. *Given a real number $\varepsilon \in (0,1)$ and an instance $\mathcal{I} = (\mathcal{C}, \mathcal{F}, B, w)$ of the* KNAPSACK MEDIAN *problem, there exists a $(3 + O(\varepsilon))$-approximation algorithm that runs in $(k\varepsilon^{-1})^{O(k)} n^{O(1)}$ time for $\mathcal{I}$, where $n = |\mathcal{C} \cup \mathcal{F}|$ and k denotes the maximal number of opened facilities among all the feasible solutions to $\mathcal{I}$ (i.e., $k = \max_{\mathcal{S} \subseteq \mathcal{F} \wedge \sum_{f \in \mathcal{S}} w(f) \leq B} |\mathcal{S}|$).*

We also consider the instances of the KNAPSACK MEDIAN problem lying in Euclidean spaces. We can hope to obtain near-optimal solutions to these instances within $\mathrm{FPT}(k)$ time since the lower bound of $1 + 2e^{-1}$ given in [12] only holds in general metrics. In fact, for the case where the facilities have the same opening cost and can be opened at any location of the space, exploring the properties of Euclidean metrics has yielded many $\mathrm{FPT}(k)$-time $(1 + \varepsilon)$-approximation algorithms for the KNAPSACK MEDIAN problem [17–19]. However, whether similar results can be obtained without such limitations on the instances remains elusive. In this paper we give an affirmative answer to this question.

Theorem 2. *Given a real number $\varepsilon \in (0,1)$ and an instance $\mathcal{I} = (\mathcal{C}, \mathcal{F}, B, w)$ of the* KNAPSACK MEDIAN *problem satisfying $\mathcal{C} \cup \mathcal{F} \subset \mathbb{R}^d$, there exists a $(1 + O(\varepsilon))$-approximation algorithm that runs in $(nd\varepsilon^{-1})^{O(1)} + (k\varepsilon^{-1})^{k\varepsilon^{-O(1)}} n^{O(1)}$ time for $\mathcal{I}$, where $n = |\mathcal{C} \cup \mathcal{F}|$ and $k = \max_{\mathcal{S} \subseteq \mathcal{F} \wedge \sum_{f \in \mathcal{S}} w(f) \leq B} |\mathcal{S}|$.*

1.2 Our Techniques

Our algorithms for the KNAPSACK MEDIAN problem use a sampling-based approach to construct small solution search spaces. Denote by $\mathcal{I} = (\mathcal{C}, \mathcal{F}, B, w)$ an instance of the KNAPSACK MEDIAN problem, and let $(\mathcal{S}^*, \tau^*)$ be an optimal solution to it. Our approach for constructing solution search spaces finds a subset $\mathcal{H} \subset \mathcal{C}$ that contains a neighbouring client of each facility from $\mathcal{S}^*$. It seems intuitively that identifying such clients is quite difficult when optimal solutions are unclear: There are $|\mathcal{F}|^k$ choices of the members of $\mathcal{S}^*$. Counter-intuitively, we show that a simple sampling method yields the desired set of neighbouring clients of the facilities from $\mathcal{S}^*$.

The client set $\mathcal{H}$ is quite valuable in selecting opened facilities and constructing candidate solutions to the instance. For example, if we open a neighbouring facility of each member of $\mathcal{H}$ and connect each client to the nearest opened facility, then triangle inequality[1] implies that the sum of the incurred connection costs is close to the cost of $(\mathcal{S}^*, \tau^*)$. However, it should be pointed out that constructing low-cost feasible solutions based on $\mathcal{H}$ is not so trivial, since the facilities are associated with non-uniform opening costs and the trade-off between facility-opening costs and client-connection costs needs to be considered when picking opened facilities. To achieve the trade-off goal, we estimate the locations of the facilities from $\mathcal{S}^*$ using the clients from $\mathcal{H}$, based on which we limit the search range of the opened facilities. We show that this yields a set of candidate solutions containing an $O(1)$-approximation one. For the instances lying in Euclidean spaces, we further make use of the client set $\mathcal{H}$ and carefully narrow the range where the opened facilities are selected based on the properties of Euclidean metrics, which yields the desired $(1 + \varepsilon)$-approximation solution.

2 Preliminaries

From now on we consider an instance $\mathcal{I} = (\mathcal{C}, \mathcal{F}, B, w)$ of the KNAPSACK MEDIAN problem satisfying $|\mathcal{C} \cup \mathcal{F}| = n$ and a constant $\varepsilon \in (0, 1)$. Given an integer $a > 1$, define $[a] = \{1, \ldots, a\}$. Define $k = \max_{\mathcal{S} \subseteq \mathcal{F} \wedge \sum_{f \in \mathcal{S}} w(f) \leq B} |\mathcal{S}|$. As mentioned above, the value of k can be determined by sorting the facilities from $\mathcal{F}$ according to their opening costs. Let $(\mathcal{S}^*, \tau^*)$ denote an optimal solution to $\mathcal{I}$ satisfying $|\mathcal{S}^*| = k^*$, where $\mathcal{S}^* = \{f_1^*, \ldots, f_{k^*}^*\}$. Given a real number $i \in [k^*]$, define $\mathcal{C}_i^* = \{c \in \mathcal{C} : \tau^*(c) = f_i^*\}$. Without loss of generality, we can assume that $|\mathcal{C}_i^*| \geq |\mathcal{C}_j^*|$ for each $\{i, j\} \subseteq [k^*]$ satisfying $i < j$. For each $x, y \in \mathcal{C} \cup \mathcal{F}$ and $\mathcal{A} \subseteq \mathcal{C} \cup \mathcal{F}$, denote by $\Delta(x, y)$ the distance from x to y, and define $\Delta(\mathcal{A}, x) = \sum_{z \in \mathcal{A}} \Delta(z, x)$ and $\Delta(x, \mathcal{A}) = \min_{z \in \mathcal{A}} \Delta(z, x)$. Define $opt_i = \Delta(\mathcal{C}_i^*, f_i^*)$ for each $i \in [k^*]$, and denote by $opt = \sum_{i \in [k^*]} opt_i$ the cost of $(\mathcal{S}^*, \tau^*)$. Given a multi-set $\mathcal{B}$, let $|\mathcal{B}|$ denote its size and $||\mathcal{B}||$ denote the number of its distinct members.

The following result provides a way of estimating the probability of finding a neighboring client of an unknown facility by randomly sampling.

Lemma 1. *Given a point x and a set $\mathcal{A}$ located in a metric space and a real number $\alpha > 1$, we have $|\{y \in \mathcal{A} : \Delta(x, y) \leq \Delta(\mathcal{A}, x)|\mathcal{A}|^{-1}\alpha\}| > (1 - \alpha^{-1})|\mathcal{A}|$.*

Proof. Define $\mathcal{A}' = \{y \in \mathcal{A} : \Delta(x, y) \leq \Delta(\mathcal{A}, x)|\mathcal{A}|^{-1}\alpha\}$ for brevity. This definition implies that $\Delta(x, y) > \Delta(\mathcal{A}, x)|\mathcal{A}|^{-1}\alpha$ for each $y \in \mathcal{A} \backslash \mathcal{A}'$, and thus

$$\Delta(\mathcal{A} \backslash \mathcal{A}', x) > \frac{1}{|\mathcal{A}|} \Delta(\mathcal{A}, x) \alpha |\mathcal{A} \backslash \mathcal{A}'|. \tag{1}$$

Moreover, the fact that $\mathcal{A} \backslash \mathcal{A}' \subseteq \mathcal{A}$ implies that

$$\Delta(\mathcal{A} \backslash \mathcal{A}', x) \leq \Delta(\mathcal{A}, x). \tag{2}$$

---

[1] Given three points x, y, and z located in a metric space, we have $\Delta(x, z) \leq \Delta(x, y) + \Delta(y, z)$, where Δ is the distance function.

Algorithm 1: Finding neighboring clients of opened facilities

---

Input: A constant $\varepsilon \in (0,1)$ and an instance $\mathcal{I} = (\mathcal{C}, \mathcal{F}, B, w)$ of the KNAPSACK MEDIAN problem;

Output: A set $\mathbb{H}$ of client sets;

1 $\mathbb{H} \Leftarrow \emptyset$;

2 $k \Leftarrow \max_{\mathcal{S} \subseteq \mathcal{F} \wedge \sum_{f \in \mathcal{S}} w(f) \leq B} |\mathcal{S}|$;

3 **for** $t \Leftarrow 1$ *to* $(k\varepsilon^{-1})^{O(k)}$ **do**

4      $\lfloor$ **Sampling**$(k, \emptyset, \mathcal{C}, \mathbb{H})$;

5 **return** $\mathbb{H}$.

Using inequality (1) and inequality (2), we have $|\mathcal{A} \backslash \mathcal{A}'| < \alpha^{-1}|\mathcal{A}|$. Consequently, it is the case that

$$|\mathcal{A}'| = |\mathcal{A}| - |\mathcal{A} \backslash \mathcal{A}'|$$
$$> (1 - \frac{1}{\alpha})|\mathcal{A}|. \qquad (3)$$

This completes the proof of Lemma 1. $\square$

The following algebraic fact is used to analyze the running times of our algorithms.

Lemma 2. *Given two real numbers i and j larger than 1, we have satisfy $\log^j i \leq ij^{O(j)}$.*

Proof. For the case where $j < \frac{\log i}{\log \log i}$, we have $\log^j i < \log^{\frac{\log i}{\log \log i}} i = i$, as desired. For the case where $j \geq \frac{\log i}{\log \log i}$, we have $\log i \leq j^{O(1)}$, which implies that $\log^j i \leq j^{O(j)}$. Thus, Lemma 2 is true. $\square$

3 Finding Clients Close to Opened Facilities

In this section we give a sampling-based approach to identify the neighboring clients of the facilities from $\mathcal{S}^*$, which is described in Algorithm 1. This algorithm iteratively invokes Algorithm 2 to find the desired client set. Algorithm 2 takes as input a real number k and three sets $\mathcal{H}$, $\mathcal{C}^\dagger$, and $\mathbb{H}$, where k denotes an upper bound on the numbers of facilities opened by feasible solutions to $\mathcal{I}$, $\mathcal{H}$ denotes the set of clients that have been picked, $\mathcal{C}^\dagger$ is the sampling range, and $\mathbb{H}$ contains the client sets that have been constructed.

Given a real number $i \in [k^*]$, define (informally) $\mathcal{G}_i$ as a set of clients from $\mathcal{C}_i^*$ close to f_i^*. Algorithm 2 aims to find a client from $\mathcal{G}_i$ for each $i \in [k^*]$. Using Lemma 1, we know that a certain number of clients from $\mathcal{C}_i^*$ lie in $\mathcal{G}_i$, which implies that the members of $\mathcal{G}_i$ have high probabilities of being selected by randomly sampling for the case where $\mathcal{C}_i^*$ is a subset of the sampling range and occupies a relatively large proportion of the latter. This case can indeed

---

Algorithm 2: Sampling$(k, \mathcal{H}, \mathcal{C}^\dagger, \mathbb{H})$

---

if $|\mathcal{H}| = k$ then
  $\lfloor\ \mathbb{H} \Leftarrow \mathbb{H} \cup \{\mathcal{H}\};$
else if $|\mathcal{C}^\dagger| = 1$ then
  $\lfloor\ \mathbb{H} \Leftarrow \mathbb{H} \cup \{\mathcal{H} \cup \mathcal{C}^\dagger\};$
else
  Randomly and uniformly pick a client $c \in \mathcal{C}^\dagger$;
  Sampling$(k, \mathcal{H} \cup \{c\}, \mathcal{C}^\dagger, \mathbb{H})$;
  if $\mathcal{H} \neq \emptyset$ then
    Find the median value φ of the real numbers from $\{\Delta(c, \mathcal{H}) : c \in \mathcal{C}^\dagger\}$;
    $\mathcal{C}^\ddagger \Leftarrow \{c \in \mathcal{C}^\dagger : \Delta(c, \mathcal{H}) \geq \varphi\};$
    Sampling$(k, \mathcal{H}, \mathcal{C}^\ddagger, \mathbb{H})$.

---

be encountered: We show that Algorithm 2 can recursively reduce the sampling range to make it satisfy the desired properties. This leads to the proof of the following guarantee of Algorithm 1, where $\mathbb{H}$ is the set returned by it.

Lemma 3. *The following event occurs with a constant probability: There is a client set $\mathcal{H} \in \mathbb{H}$ satisfying $\sum_{i=1}^{k^*} |\mathcal{C}_i^*| \Delta(f_i^*, \mathcal{H}) \leq (1 + 2\varepsilon)opt$.*

The following lemma gives upper bounds on the running time of Algorithm 1 and the size of $\mathbb{H}$.

Lemma 4. *The running time of Algorithm 1 is $(k\varepsilon^{-1})^{O(k)}|\mathcal{C}|$ in general metrics and $(k\varepsilon^{-1})^{O(k)}d|\mathcal{C}|$ in $\mathbb{R}^d$, and $|\mathbb{H}| \leq (k\varepsilon^{-1})^{O(k)}|\mathcal{C}|$.*

4   The Approximation Algorithms

Let $\mathbb{H}$ denote the set constructed by Algorithm 1, and let $\mathcal{H}$ denote a set of no more than k clients satisfying

$$\sum_{i=1}^{k^*} |\mathcal{C}_i^*| \Delta(f_i^*, \mathcal{H}) \leq (1 + 2\varepsilon)opt. \tag{4}$$

Lemma 3 implies that such a set $\mathcal{H}$ exists in $\mathbb{H}$ with a constant probability.

In this section we show how to construct approximation solutions to $\mathcal{I}$ based on $\mathcal{H}$. For each $i \in [k^*]$, let h_i denote the facility from $\mathcal{H}$ nearest to f_i^*. Define $\gamma = n \cdot \max_{c \in \mathcal{C}} \Delta(c, \tau^*(c))$. We have $\gamma \in [opt, n \cdot opt]$, combining which with inequality (4) yields

$$\max_{i \in [k^*]} \Delta(h_i, f_i^*) \leq \sum_{i=1}^{k^*} |\mathcal{C}_i^*| \Delta(f_i^*, \mathcal{H})$$
$$\leq (1 + 2\varepsilon)opt$$
$$\leq (1 + 2\varepsilon)\gamma \tag{5}$$

Algorithm 3: The algorithm in general metrics

Input: An instance $\mathcal{I} = (\mathcal{C}, \mathcal{F}, B, w)$ of the KNAPSACK MEDIAN problem, a positive integer k^*, and k^* sets $\mathcal{K}_1^*, \ldots, \mathcal{K}_{k^*}^*$;

Output: A solution $(\mathcal{S}_1, \tau_1)$ to $\mathcal{I}$;

1   $\mathcal{S}_1 \Leftarrow \emptyset$;

2   **for** $i \Leftarrow 1$ *to* k^* **do**

3     $\lfloor \; \mathcal{S}_1 \Leftarrow \mathcal{S}_1 \cup \{\arg\min_{f \in \mathcal{K}_i^*} w(f)\};$

4   **for** *each* $c \in \mathcal{C}$ **do**

5     $\lfloor \; \tau_1(c) = \arg\min_{f \in \mathcal{S}_1} \Delta(c, f);$

6   **return** $(\mathcal{S}_1, \tau_1)$.

---

and

$$\frac{\varepsilon}{n^2}\gamma \leq \frac{\varepsilon}{n} opt. \tag{6}$$

Given a real number $\alpha > 0$ and a client $c \in \mathcal{C}$, define $\mathcal{G}(c, \alpha) = \{f \in \mathcal{F} : \Delta(c, f) \leq \alpha\}$. Define $\mathcal{K}(i, j) = \mathcal{G}(h_i, \varepsilon(1 + \varepsilon)^j \gamma n^{-2}) \backslash \mathcal{G}(h_i, \varepsilon(1 + \varepsilon)^{j-1} \gamma n^{-2})$ and $\mathcal{K}(i, 0) = \mathcal{G}(h_i, \varepsilon\gamma n^{-2})$ for each $i \in [k^*]$ and $j \in [\lceil 4\varepsilon^{-1} \log n \rceil]$. For the case where $j = \lceil 4\varepsilon^{-1} \log n \rceil$, inequality (5) implies that

$$\varepsilon(1 + \varepsilon)^j \frac{\gamma}{n^2} > (1 + 2\varepsilon)\gamma$$
$$\geq \max_{i \in k^*} \Delta(h_i, f_i^*).$$

This inequality and the definitions of $\mathcal{K}(i, 0)$ and $\mathcal{K}(i, j)$ imply that for each $i \in [k^*]$, there exists an integer $j \in [0, \lceil 4\varepsilon^{-1} \log n \rceil]$ satisfying $f_i^* \in \mathcal{K}(i, j)$. Let $\mathcal{K}_i^*$ denote such a set $\mathcal{K}(i, j)$ containing f_i^*. The following lemma implies that the sets $\mathcal{K}_1^*, \ldots, \mathcal{K}_{k^*}^*$ can be guessed with a $(k\varepsilon^{-1})^{O(k)} n^{O(1)}$ multiplicative overhead in the running times of our algorithms.

Lemma 5. *Given the set* $\mathbb{H}$, *the sets* $\mathcal{K}_1^*, \ldots, \mathcal{K}_{k^*}^*$ *can be guessed by enumerating over at most* $(k\varepsilon^{-1})^{O(k)} n^{O(1)}$ *choices.*

To reduce the search space of the solutions to $\mathcal{I}$, we guess the sets $\mathcal{K}_1^*, \ldots, \mathcal{K}_{k^*}^*$ and only select opened facilities from $\bigcup_{i=1}^{k^*} \mathcal{K}_i^*$. In Sect. 4.1, we show that opening the facility associated with the minimal opening cost from $\mathcal{K}_i^*$ for each $i \in [k^*]$ and connecting each client to the nearest opened facility yields a $(3 + O(\varepsilon))$-approximation solution to $\mathcal{I}$. In Sect. 4.2, we further narrow the solution search space using the properties of Euclidean metrics: We identify a low-cost facility close to f_i^* from $\mathcal{K}_i^*$ for each $i \in [k^*]$, based on which a $(1 + O(\varepsilon))$-approximation solution to $\mathcal{I}$ is constructed.

4.1   A $(3 + O(\varepsilon))$-Approximation Algorithm in General Metrics

In this section we consider the KNAPSACK MEDIAN problem in general metrics. Our algorithm opens the facility with smallest opening cost from $\mathcal{K}_i^*$ for each

$i \in [k^*]$ and connects each client to the nearest opened facility, as described in Algorithm 3.

Let $(\mathcal{S}_1, \tau_1)$ be the solution constructed by Algorithm 3. Lemma 4 implies that constructing $\mathbb{H}$ takes $(k\varepsilon^{-1})^{O(k)}n$ time. Given the set $\mathbb{H}$, there are k choices of the value of k^* and $(k\varepsilon^{-1})^{O(k)}n^{O(1)}$ choices of the set $\{\mathcal{K}_1^*, \ldots, \mathcal{K}_{k^*}^*\}$ (due to Lemma 5). Consequently, we can assume that all inputs of Algorithm 3 are known with a $(k\varepsilon^{-1})^{O(k)}n^{O(1)}$ multiplicative overhead in the running time of the algorithm. Moreover, the fact that Algorithm 3 runs in $O(nk^*)$ time implies that $(\mathcal{S}_1, \tau_1)$ can be constructed in $(k\varepsilon^{-1})^{O(k)}n^{O(1)}$ time. Combining this with the following lemma that states the approximation ratio of $(\mathcal{S}_1, \tau_1)$, we complete the proof of Theorem 1.

Lemma 6. $(\mathcal{S}_1, \tau_1)$ *can be constructed in* $(k\varepsilon^{-1})^{O(k)}n^{O(1)}$ *time, and it satisfies* $\sum_{f \in \mathcal{S}_1} w(f) \leq B$ *and* $\sum_{c \in C} \Delta(c, \tau_1(c)) \leq (3 + O(\varepsilon))opt$.

4.2   A $(1 + O(\varepsilon))$-Approximation Algorithm in Euclidean Metrics

In this section we consider the case where instance $\mathcal{I}$ lies in $\mathbb{R}^d$. We select opened facilities by constructing nets for the instance. Such a net can be formally defined as follows.

Definition 1 (σ-net[20]). *Given a real number* $\sigma > 0$ *and a set* $\mathcal{P} \subset \mathbb{R}^d$, *a set* $\mathcal{N} \subset \mathcal{P}$ *is a* σ-net *of* $\mathcal{P}$ *if it satisfies the following two properties:*

(1) $\Delta(x, \mathcal{N}) < \sigma$ *for each* $x \in \mathcal{P}$;
(2) $\Delta(x, \mathcal{N} \setminus \{x\}) > \sigma$ *for each* $x \in \mathcal{N}$.

Har-Peled and Mendel [20] showed that in low-dimensional Euclidean spaces, one can construct a net of a given set in near-linear time.

Lemma 7 ([20]). *Given a real number* $\sigma > 0$ *and a set* $\mathcal{P} \subset \mathbb{R}^d$, *a* σ-net *of* $\mathcal{P}$ *of size at most* $\min\{|\mathcal{P}|, (\sigma^{-1} \max_{x,y \in \mathcal{P}} \Delta(x, y))^d\}$ *can be constructed in* $|\mathcal{P}| \log |\mathcal{P}| 2^{O(d)}$ *time.*

The algorithm for constructing nets given in [20] does not work in high-dimensional Euclidean spaces. Thus, we combine the method for constructing coresets given by Chen [13] with the stronger version of Johnson-Lindenstrauss transform given by Mahabadi et al. [21] to convert high-dimensional instances to low-dimensional ones. Lemma 8 and Lemma 9 are the guarantees of the methods given in [13] and [21], respectively.

Lemma 8 ([13]). *Given an integer* $t > 0$, *a real number* $\lambda \in (0, 1)$, *and a set* $\mathcal{P} \subset \mathbb{R}^d$ *of more than* t *points, one can construct a multi-set* $\mathcal{P}' \subset \mathbb{R}^d$ *satisfying* $|\mathcal{P}'| = |\mathcal{P}|$ *and* $||\mathcal{P}'|| \leq d(\log |\mathcal{P}|t\lambda^{-1})^{O(1)}$ *in* $O(|\mathcal{P}|dt)$ *time, such that* $\sum_{x \in \mathcal{P}'} \Delta(x, \mathcal{A}) \in [1 - \lambda, 1 + \lambda] \sum_{x \in \mathcal{P}} \Delta(x, \mathcal{A})$ *for each* $\mathcal{A} \subset \mathbb{R}^d$ *with* $|\mathcal{A}| = t$.

Lemma 9 ([21]). *Given a real number* $\lambda \in (0, 1)$ *and a set* $\mathcal{P} \subset \mathbb{R}^d$, *one can construct a mapping* $g : \mathbb{R}^d \to \mathbb{R}^{d'}$ *satisfying* $d' = O(\lambda^{-4} \log |\mathcal{P}|)$ *in* $(|\mathcal{P}|d)^{O(1)}$ *time, such that* $\Delta(g(x), g(y)) \in [1, 1 + \lambda]\Delta(x, y)$ *for each* $x \in \mathcal{P}$ *and* $y \in \mathbb{R}^d$.

Algorithm 4: The algorithm in Euclidean metrics

Input: A constant $\varepsilon \in (0,1)$, an instance $\mathcal{I} = (\mathcal{C}, \mathcal{F}, B, w)$ of the KNAPSACK MEDIAN problem satisfying $\mathcal{C} \cup \mathcal{F} \subset \mathbb{R}^d$, a positive integer k^*, and k^* sets $\mathcal{K}_1^*, \ldots, \mathcal{K}_{k^*}^*$;

Output: A solution $(\mathcal{S}_2, \tau_2)$ to $\mathcal{I}$;

1 $\mathcal{V} \Leftarrow \emptyset$;

2 **for** $i \Leftarrow 1$ *to* k^* **do**

3   **if** $|\mathcal{K}_i^*| = 1$ **then**

4           $\mathcal{V} \Leftarrow \mathcal{V} \cup \mathcal{K}_i^*$;

5   **else**

6   Construct a $\max_{x,y \in \mathcal{K}_i^*} \varepsilon\Delta(x,y)$-net $\mathcal{N}_i$ of $\mathcal{K}_i^*$;

7   **for** *each* $f \in \mathcal{N}_i$ **do**

8               $\mathcal{V}_i(f) \Leftarrow \{f^\dagger \in \mathcal{K}_i^* : \Delta(f^\dagger, f) = \min_{f' \in \mathcal{N}_i} \Delta(f^\dagger, f')\}$;

9               $\mathcal{V} \Leftarrow \mathcal{V} \cup \{\arg\min_{f' \in \mathcal{V}_i(f)} w(f')\}$;

10 $\mathcal{S}_2 \Leftarrow \arg\min_{\mathcal{S}' \subseteq \mathcal{V} \wedge \sum_{f \in \mathcal{S}'} w(f) \leq B} \sum_{c \in \mathcal{C}} \Delta(c, \mathcal{S}')$;

11 **for** *each* $c \in \mathcal{C}$ **do**

12       $\tau_2(c) \Leftarrow \arg\min_{f \in \mathcal{S}_2} \Delta(c, f)$;

13 **return** $(\mathcal{S}_2, \tau_2)$.

As a corollary of Lemma 8 and Lemma 9, we have the following result about dimensionality reduction.

Corollary 1. *Given an integer $t > 0$, a real number $\lambda \in (0,1)$, and a set $\mathcal{P} \subset \mathbb{R}^d$ of more than t points, one can construct a mapping $g : \mathbb{R}^d \to \mathbb{R}^{d'}$ satisfying $d' = \lambda^{-O(1)}(\log t + \log\log |\mathcal{P}|)$ and a multi-set $\mathcal{P}' \subset \mathbb{R}^{d'}$ satisfying $|\mathcal{P}'| = |\mathcal{P}|$ in $(|\mathcal{P}| d \lambda^{-1})^{O(1)}$ time, such that $\sum_{x \in \mathcal{P}'} \Delta(x, \{g(y) : y \in \mathcal{A}\}) \in [1 - O(\lambda), 1 + O(\lambda)] \sum_{x \in \mathcal{P}} \Delta(x, \mathcal{A})$ for each $\mathcal{A} \subset \mathbb{R}^d$ with $|\mathcal{A}| = t$.*

For each $i \in [k^*]$ satisfying $|\mathcal{K}_i^*| > 1$, our algorithm constructs a net of $\mathcal{K}_i^*$, and adds the facility associated with smallest opening cost from each Voronoi cell defined by the members of the net to the candidate set of opened facilities, as described in Algorithm 4.

Denote by $(\mathcal{S}_2, \tau_2)$ the solution returned by Algorithm 4. Let $I = \{i \in [k^*] : |\mathcal{K}_i^*| \neq 1\}$. We define $\mathcal{N}_i$, $\mathcal{V}_i(f)$, and $\mathcal{V}$ in the same way as Algorithm 4, where $i \in I$ and $f \in \mathcal{N}_i$. We have

$$|\mathcal{N}_i| \leq \left(\frac{\max_{x,y \in \mathcal{K}_i^*} \Delta(x,y)}{\max_{x,y \in \mathcal{K}_i^*} \varepsilon\Delta(x,y)}\right)^d = \varepsilon^{-d} \tag{7}$$

for each $i \in I$, where the first step is due to the definition of $\mathcal{N}_i$ and Lemma 7. Moreover, the definition of $\mathcal{V}$ implies that

$$
\begin{aligned}
|\mathcal{V}| = \sum_{i \in I} |\mathcal{N}_i| + k^* - |I| \\
\leq \varepsilon^{-d} k^* \\
\leq \varepsilon^{-d} k,
\end{aligned}
\tag{8}
$$

where the second step is due to inequality (7).

Algorithm 4 constructs a net of $\mathcal{K}_i^*$ based on the maximal distance between the members of $\mathcal{K}_i^*$ and selects a set of low-cost facilities from $\mathcal{K}_i^*$ for each $i \in I$, which takes

$$
\sum_{i \in I} 2^{O(d)} |\mathcal{K}_i^*|^{O(1)} \leq 2^{O(d)} n^{O(1)}
$$

time in total due to Lemma 7. After this, the algorithm enumerates all subsets $\mathcal{S} \subseteq \mathcal{V}$ satisfying $\sum_{f \in \mathcal{S}} w(f) \leq B$ and selects the one corresponding to the solution with smallest cost. The fact that $k = \max_{\mathcal{S} \subseteq \mathcal{F} \wedge \sum_{f \in \mathcal{S}} w(f) \leq B} |\mathcal{S}|$ implies that this enumeration can be done in $O(|\mathcal{V}|^k ndk)$ time, which is no more than $O((\varepsilon^{-d} k)^k ndk)$ time due to inequality (8). Summing up, Algorithm 4 runs in $(\varepsilon^{-d} k)^{O(k)} n^{O(1)}$ time. Given the set $\mathbb{H}$ (which can be constructed in $(k\varepsilon^{-1})^{O(k)} dn$ time due to Lemma 4), Lemma 5 implies that we have k choices of the value of k^* and $(k\varepsilon^{-1})^{O(k)} n^{O(1)}$ choices of the set $\{\mathcal{K}_1^*, \ldots, \mathcal{K}_{k^*}^*\}$. Consequently, all inputs of Algorithm 4 can be assumed to be known with a $(k\varepsilon^{-1})^{O(k)} n^{O(1)}$ multiplicative overhead and a $(k\varepsilon^{-1})^{O(k)} dn$ additive overhead in the running time of the algorithm. Putting everything together, we know that $(\mathcal{S}_2, \tau_2)$ can be constructed in $(\varepsilon^{-d} k)^{O(k)} n^{O(1)}$ time. Based on the method given in Corollary 1 that runs in $(nd\varepsilon^{-1})^{O(1)}$ time, we can assume that $d = \varepsilon^{-O(1)}(\log k + \log \log n)$, which incurs an arbitrarily small loss in the approximation guarantee. Thus, the time for constructing $(\mathcal{S}_2, \tau_2)$ can be reduced to $(nd\varepsilon^{-1})^{O(1)} + (k \log n)^{k\varepsilon^{-O(1)}}$, which is upper-bounded by $(nd\varepsilon^{-1})^{O(1)} + (k\varepsilon^{-1})^{k\varepsilon^{-O(1)}} n^{O(1)}$ due to Lemma 2.

By the argument above, we know that $(\mathcal{S}_2, \tau_2)$ can be constructed in FPT(k) time based on the dimension-reduction method given in Corollary 1. Moreover, the following lemma suggests that $(\mathcal{S}_2, \tau_2)$ is a $(1 + O(\varepsilon))$-approximation solution to $\mathcal{I}$. This completes the proof of Theorem 2.

Lemma 10. $(\mathcal{S}_2, \tau_2)$ can be constructed in $(nd\varepsilon^{-1})^{O(1)} + (k\varepsilon^{-1})^{k\varepsilon^{-O(1)}} n^{O(1)}$ time, and it satisfies $\sum_{f \in \mathcal{S}_2} w(f) \leq B$ and $\sum_{c \in \mathcal{C}} \Delta(c, \tau_2(c)) \leq (1 + O(\varepsilon)) opt$.

5   Conclusions

In this paper we consider the KNAPSACK MEDIAN problem for the case where the maximal number of facilities opened by feasible solutions is a fixed parameter. Based on a simple sampling-based approach for limiting the search space of solutions, we give FPT-time $(3 + \varepsilon)$-approximation and $(1 + \varepsilon)$-approximation algorithms for the problem in general metrics and Euclidean metrics, respectively.

An interesting direction for future work is to extend our approach to the constrained versions of the KNAPSACK MEDIAN problem, such as the problems of LOWER-BOUNDED KNAPSACK MEDIAN [7] and CAPACITATED KNAPSACK MEDIAN [10]. The additional constraints posed on the instances deepen the conflict between the requirements of feasibility and cost of solutions, which makes these problems more challenging.

References

1. Lin, J., Vitter, J.S.: ε-approximations with minimum packing constraint violation. In: Proceedings of the 24th Annual ACM Symposium on Theory of Computing (STOC), pp. 771–782 (1992)
2. Krishnaswamy, R., Kumar, A., Nagarajan, V., Sabharwal, Y., Saha, B.: The matroid median problem. In: Proceedings of the 22nd Annual ACM-SIAM Symposium on Discrete Algorithms (SODA), pp. 1117–1130 (2011)
3. Charikar, M., Guha, S.: Improved combinatorial algorithms for facility location problems. SIAM J. Comput. **34**(4), 803–824 (2005)
4. Byrka, J., Pensyl, T.W., Rybicki, B., Spoerhase, J., Srinivasan, A., Trinh, K.: An improved approximation algorithm for knapsack median using sparsification. Algorithmica **80**(4), 1093–1114 (2018)
5. Kumar, A.: Constant factor approximation algorithm for the knapsack median problem. In: Proceedings of the 21st Annual ACM-SIAM Symposium on Discrete Algorithms (SODA), pp. 824–832 (2012)
6. Swamy, C.: Improved approximation algorithms for matroid and knapsack median problems and applications. ACM Trans. Algorithms **12**(4), 49:1-49:22 (2016)
7. Han, L., Hao, C., Wu, C., Zhang, Z.: Approximation algorithms for the lower-bounded knapsack median problem. In: Proceedings of the 14th International Conference on Algorithmic Aspects in Information and Management (AAIM), pp. 119–130 (2020)
8. Krishnaswamy, R., Li, S., Sandeep, S.: Constant approximation for k-median and k-means with outliers via iterative rounding. In: Proceedings of the 50th Annual ACM SIGACT Symposium on Theory of Computing (STOC), pp. 646–659 (2018)
9. Gupta, A., Moseley, B., Zhou, R.: Structural iterative rounding for generalized k-median problems. In: Proceedings of the 48th International Colloquium on Automata, Languages, and Programming (ICALP), pp. 77:1–77:18 (2021)
10. Grover, S., Gupta, N., Khuller, S., Pancholi, A.: Constant factor approximation algorithm for uniform hard capacitated knapsack median problem. In: Proceedings of the 38th IARCS Annual Conference on Foundations of Software Technology and Theoretical Computer Science (FSTTCS), pp. 23:1–23:22 (2018)
11. Guha, S., Khuller, S.: Greedy strikes back: improved facility location algorithms. J. Algorithms **31**(1), 228–248 (1999)
12. Cohen-Addad, V., Gupta, A., Kumar, A., Lee, E., Li, J.: Tight FPT approximations for k-median and k-means. In: Proceedings of the 46th International Colloquium on Automata, Languages, and Programming (ICALP), pp. 42:1–42:14 (2019)

13. Adamczyk, M., Byrka, J., Marcinkowski, J., Meesum, S.M., Wlodarczyk, M.: Constant-factor FPT approximation for capacitated k-median. In: Proceedings of the 27th Annual European Symposium on Algorithms (ESA), pp. 1:1–1:14 (2019)
14. Cohen-Addad, V., Li, J.: On the fixed-parameter tractability of capacitated clustering. In: Proceedings of the 46th International Colloquium on Automata, Languages, and Programming (ICALP), pp. 41:1–41:14 (2019)
15. Inamdar, T., Varadarajan, K.R.: Capacitated sum-of-radii clustering: an FPT approximation. In: Proceedings of the 28th Annual European Symposium on Algorithms (ESA), pp. 62:1–62:17 (2020)
16. Bandyapadhyay, S., Lochet, W., Saurabh, S.: FPT constant-approximations for capacitated clustering to minimize the sum of cluster radii. In: Proceedings of the 39th International Symposium on Computational Geometry (SoCG), pp. 12:1–12:14 (2023)
17. Kumar, A., Sabharwal, Y., Sen, S.: Linear-time approximation schemes for clustering problems in any dimensions. J. ACM **57**(2), 5:1-5:32 (2010)
18. Chen, K.: On k-median clustering in high dimensions. In: Proceedings of the 17th Annual ACM-SIAM Symposium on Discrete Algorithms (SODA), pp. 1177–1185 (2006)
19. Jaiswal, R., Kumar, A., Sen, S.: A simple D^2-sampling based PTAS for k-means and other clustering problems. Algorithmica **70**(1), 22–46 (2014)
20. Har-Peled, S., Mendel, M.: Fast construction of nets in low-dimensional metrics and their applications. SIAM J. Comput. **35**(5), 1148–1184 (2006)
21. Mahabadi, S., Makarychev, K., Makarychev, Y., Razenshteyn, I.P.: Nonlinear dimension reduction via outer Bi-Lipschitz extensions. In: Proceedings of the 50th Annual ACM SIGACT Symposium on Theory of Computing (STOC), pp. 1088–1101 (2018)

On the Existence of EFX (and Pareto-Optimal) Allocations for Binary Chores

Biaoshuai Tao[1], Xiaowei Wu[2], Ziqi Yu[1], and Shengwei Zhou[2(✉)]

[1] Shanghai Jiao Tong University, Shanghai, China
{bstao,yzq.111}@sjtu.edu.cn
[2] University of Macau, Taipa, Macau SAR
{xiaoweiwu,yc17423}@um.edu.mo

Abstract. We study the problem of allocating a set of indivisible chores among agents while each chore has a binary marginal. We focus on the fairness criteria of envy-freeness up to any item (EFX) and investigate the existence of EFX allocations. We show that when agents have additive binary cost functions, there exist EFX and Pareto-optimal (PO) allocations that can be computed in polynomial time. To the best of our knowledge, this is the first setting with a general number of agents that admits EFX and PO allocations, before which EFX and PO allocations have only been shown to exist for three bivalued agents. We further consider more general cost functions: cancelable and general monotone (both with binary marginal). We show that EFX allocations exist and can be computed for binary cancelable chores, but EFX is incompatible with PO. For general binary marginal functions, we propose an algorithm that computes (partial) envy-free (EF) allocations with at most $n - 1$ unallocated items.

1 Introduction

The *fair division* problem mainly focuses on allocating heterogeneous resources fairly to a group of agents. It is a classic resource allocation problem that has been widely studied by mathematicians, economists, and computer scientists. In this research direction, a popular question is to focus on allocating *indivisible* items. An ideal fairness notion is *envy-freeness* (EF) [23], which requires that every agent values her own bundle not lower than any other bundle held by others. However, EF allocations are not guaranteed to exist when allocating indivisible items, e.g. allocating one item to two agents. Caragiannis et al. [17] proposed a relaxed notion called *envy-freeness up to any item* (EFX), which requires that for any pair of agents i and j, i does not envy j after removing an arbitrary item

---

The authors are ordered alphabetically. Biaoshuai Tao is funded by the National Natural Science Foundation of China (Grant No. 62102252). Xiaowei Wu is funded by the Science and Technology Development Fund (FDCT), Macau SAR (file no. 0014/2022/AFJ, 0085/2022/A, 0143/2020/A3 and SKL-IOTSC-2024-2026).

B. Li et al. (Eds.): IJTCS-FAW 2024, LNCS 14752, pp. 33–52, 2025.
https://doi.org/10.1007/978-981-97-7752-5_3

from j's bundle. The existence of EFX allocations still remains open for general instances and is only known for some special cases, e.g., two agents with general valuations [33], three agents with some special valuations [1, 19] and *bivalued* instances for additive functions [3].

In the traditional setting of fair division problems, items are assumed to have non-negative effects on agents. Roughly speaking, giving an item to any agent will make her happier. This kind of question is called fair division of *goods*. On the contrary, the complement setting is called fair division of *chores*. In this case, each item is of negative value to agents, and we use *cost* functions to describe the preference of each agent. Some definitions of fairness criteria (such as EFX) can be naturally extended to chores division [5]. However, most of the known results do not directly extend to the chores setting. The minor differences between goods and chores settings nullify some existing techniques, such as Cycle-Elimination, when considering the allocation of chores [37]. Another example can be given by the comparison of goods and chores on fair and efficient allocations. As one of the influential results in the world of goods, Caragiannis et al. [17] showed that maximizing *Nash social welfare* guarantees the fairness of *envy-freeness up to one item* (EF1) (a fairness notion weaker than EFX) and the efficiency of Pareto-optimality (PO). However, the existence of EF1 and PO allocations remains a major open problem in the context of chores, except for a limited class of bivalued additive cost functions [22, 25, 36] and lexicographic preference [28].

Major of our results focus on the settings when the agents' costs for chores have *binary marginals*. To be exact, for a binary cost function c, $c(S \cup \{g\}) - c(S) \in \{0, 1\}$, where S is any set of chores and g is a single chore. Binary marginals have received much attention in the literature for both goods [7, 9, 30] and chores [8], due to various real-world applications such as page cache, reviewer assignment, bilateral matching [14] and housing allocations [10]. The binary marginal gives us extra properties to develop new techniques.

1.1 Our Results

In this paper, we study the fair (and efficient) allocations of indivisible chores when each chore has a binary marginal cost. We consider the fairness notion of envy-freeness up to any item (EFX) and investigate the existence of EFX allocations for different cost functions. For additive cost functions with binary marginal, we show that EFX and Pareto-optimal (PO) allocations exist and can be computed in polynomial time. Prior to our result, it has only been shown that EFX and PO allocations exist for restricted instances with three bivalued agents [26] and lexicographic preferences [28].

Result 1 (Theorem 3). There exists a polynomial time algorithm that computes EFX and PO allocations for additive cost functions with binary marginals.

We further consider the *cancelable* cost function (with binary marginals) that was introduced by [12]. Cancelable cost functions generalize several cost functions including *additive, budget additive,* and more. We show that EFX is no longer compatible with PO for the more general cancelable cost functions. We propose a polynomial-time algorithm that computes EFX allocations for binary cancelable cost functions. To the best of our knowledge, this is the first non-trivial result regarding the existence and computation of EFX allocations for non-identical cost functions that beyond additive.

Result 2 (Theorem 4). There exists a polynomial time algorithm that computes EFX allocations for cancelable cost functions with binary marginals.

Result 3 (Theorem 5). For any number of agents $n \geq 2$, there exist instances with binary cancelable cost functions where all EFX allocations are not Pareto-optimal.

We further consider submodular and general cost functions with binary marginals. We show that under the binary marginal property, the set of submodular cost functions is a superset of the set of cancelable cost functions. We propose a polynomial-time algorithm that computes 2-approximately EFX allocations for binary submodular cost functions, and (partial) EF allocations with at most $n-1$ unallocated items for general cost functions with binary marginals.

Result 4 (Theorem 6). There exists a polynomial time algorithm that computes a partial allocation that is envy-free and leaves at most $n-1$ items unallocated, for general cost functions with binary marginals.

Result 5 (Theorem 7, Corollary 1). There exists a polynomial time algorithm that computes 2-approximately EFX allocations for submodular cost functions with binary marginals.

We summarize our results in Fig. 1 which also presents the relationship between different cost functions with binary marginals.

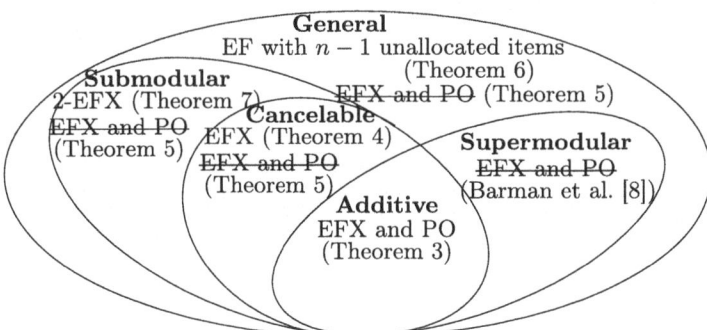

Fig. 1. Illustration of our results and the relationship between different set functions with binary marginals

1.2   Other Related Work

The setting of binary marginal cases has been widely considered in recent years. In the literature of goods, Babaioff et al. [7] considered the class of binary submodular (or matroid-rank) valuations and showed that EFX and PO allocations always exist. They designed a mechanism that maximizes *Nash social welfare* which generates EFX and PO properties under the matroid-rank valuations. Similar mechanism has also been studied in [27]. Viswanathan and Zick [35] improved the reuslt by proposing the "Yankee-Swap algorithm" that runs significantly faster. However, maximizing Nash social welfare might fail to guarantee EFX allocations when it comes to more general valuations, which makes it impossible to extend Babaioff et al.'s and Halpern et al.'s mechanisms directly. Very recently, Bu et al. [15] developed a polynomial-time algorithm, based on cycle-elimination techniques, that computes EFX allocations for general valuations with binary marginals. However, these results can not be easily extended to that of chores. In the setting of chores division, the cycle-elimination process will sometimes break the EFX property, which will not happen in the world of goods division. When it comes to binary chores division, Barman et al. [8] considered binary supermodular costs and showed that: 1) EF1 and PO allocations exist and can be computed in polynomial time; 2) EFX and PO are incompatible, under the binary supermodular valuations. However, the existence of EFX allocations for non-identical cost functions and the compatibility of EFX and PO for other binary functions, e.g., *additive* and *submodular*, are still open.

Though the existence of EFX allocations remains open, much partial progress has been made over the past years. A line of the literature focuses on exploring under which restriction EFX allocations exist. For the allocation of goods, Plaut and Roughgarden [33] showed that EFX allocations exist when all agents have identical valuations. On the restriction of the number of agents, EFX allocations have been shown to exist for two agents with general valuations [33] and three agents with additive valuations [19]. The latter result was recently extended to more general valuations by Akrami et al. [1]. In the context of chores, it has been shown that EFX allocations exist for some special cases, e.g., agents have monotone identical functions [11], identical ordering (IDO) instances [32], three agents with bivalued additive functions [37], two types of chores [6]. Another branch of the literature is to explore the relaxations of EFX, e.g., approximately EFX allocations [4,18,33,37] and partial EFX allocations with reasonable guarantees [1,12,16,20,21,29]. For a more comprehensive overview of the fair division problem, we refer to the recent surveys [2] and [5].

2   Preliminaries

We consider how to fairly allocate a set of m indivisible chores/items M to a group of n agents N. We call a subset of items, e.g., $S \subseteq M$, a *bundle*. For ease of notation we use $X + e$ and $X - e$ to denote $X \cup \{e\}$ and $X \setminus \{e\}$, respectively, for any $X \subseteq M$ and $e \in M$. A complete allocation $\mathbf{X} = (X_1, \ldots, X_n)$ is an n-partition of the items M such that $X_i \cap X_j = \emptyset$ for all $i \neq j$ and $\cup_{i \in N} X_i = M$, where agent i receives bundle X_i. When $\cup_{i \in N} X_i \subsetneq M$, we call it a partial allocation. Given an instance $I = (N, M, \mathbf{c})$ where $\mathbf{c} = (c_1, \ldots, c_n)$ is the set of *cost functions* (to be defined immediately), our goal is to find an allocation $\mathbf{X}$ that is *fair* to all agents.

Cost Functions. Each agent $i \in N$ has a cost function $c_i : 2^M \rightarrow \mathbb{R}^+ \cup \{0\}$ that assigns a cost to every bundle of items. More specifically, for any subset of items $S \subseteq M$, the cost of S to agent i is denoted as $c_i(S)$. For convenience we use $c_i(e)$ to denote $c_i(\{e\})$, the cost of agent $i \in N$ on item $e \in M$. We use $\mathbf{c} = (c_1, \ldots, c_n)$ to denote the cost functions of agents. For any subset of items $S \subseteq M$, any item $e \in M$, we use $c_i(e \mid S)$ to denote the marginal cost of item e to subset S, under agent i's cost function, i.e., $c_i(S + e) - c_i(S)$. Similarly, for $S, T \subseteq M$, we denote $c_i(T \mid S) = c_i(T \cup S) - c_i(S)$.

- *Binary Marginal.* We focus on the instances in which each item gives *binary* marginal cost to any subset of items, that is, for any $i \in N, e \in M, S \subseteq M$, we have $c_i(e \mid S) \in \{0, 1\}$. Note that, any binary marginal cost function is also *monotone*, i.e., $c_i(S + e) \geq c_i(S)$ for any $i \in N, e \in M, S \subseteq M$.
- *Additive Function.* A cost function $c_i(\cdot)$ is said to be additive if $c_i(S) = \sum_{e \in S} c_i(e)$ for any $S \subseteq M$.
- *Cancelable Function.* As a generalization of additive cost functions, cancelable functions are first introduced by Berger et al. [12] in the fair allocation of goods. A cost function $c_i(\cdot)$ is said to be cancelable if for any two bundles $S, T \subseteq M$ and item $e \in M \setminus (S \cup T)$, we have

$$c_i(S + e) > c_i(T + e) \Rightarrow c_i(S) > c_i(T). \tag{1}$$

  As [12] pointed out, cancelable functions describe many natural meaningful valuation functions in economics, including additive, unit-demand, and budget-additive cost functions.
- *Submodular Function.* A cost function $c_i(\cdot)$ is submodular if and only if for any $S \subseteq T, e \in M \setminus T$, we have $c_i(e \mid S) \geq c_i(e \mid T)$.

Fairness. We first introduce the *envy-freeness* (EF) for the allocation of chores. An allocation $\mathbf{X}$ is EF if for any agents $i, j \in N$, $c_i(X_i) \leq c_i(X_j)$. Since EF allocations are not guaranteed to exist for indivisible chores, we focus on a relaxation of EF: *envy-freeness up to any item* (EFX). An allocation $\mathbf{X}$ is EFX if for any agents $i, j \in N$, either $X_i = \emptyset$, or $c_i(X_i - e) \leq c_i(X_j)$ for any item $e \in X_i$. For $\alpha \geq 1$, an allocation is α-approximately envy-free, denoted by α-EF, if $c_i(X_i) \leq \alpha \cdot c_i(X_j)$; an allocation is α-approximately EFX, denoted by α-EFX, if either $X_i = \emptyset$ or $c_i(X_i - e) \leq \alpha \cdot c_i(X_j)$ for any $i, j \in N$ and any $e \in X_i$.

Efficiency. An allocation $\mathbf{X}'$ *Pareto-dominates* another allocation $\mathbf{X}$ if $c_i(X_i') \leq c_i(X_i)$ for all $i \in N$ and the inequality is strict for at least one agent. An allocation $\mathbf{X}$ is said to be *Pareto-optimal* (PO) if $\mathbf{X}$ is not Pareto-dominated by any other allocation. For any allocation $\mathbf{X} = (X_1, \cdots, X_n)$, the social cost $\mathrm{sc}(\mathbf{X})$ is defined as the sum of the cost of each agent, i.e., $\mathrm{sc}(\mathbf{X}) = \sum_{i \in N} c_i(X_i)$. An allocation that minimizes the social cost is Pareto-optimal since any Pareto-improvement strictly decreases the social cost.

Remark. We remark that several works [10,11,31] differ in the definitions of EFX_+ and EFX_0, by considering the cost of the removal item is whether or not strictly positive. Their definition of EFX_0 is consistent with our definition of EFX, while EFX_+ is weaker, requiring removing any chore with positive (marginal) cost. However, in the binary setting, the EFX_+ allocation coincides with EF1 allocation, which can be found in the polynomial time, even for general cost functions [13]. In the remaining of this paper, we focus on the stronger version of EFX_0 without additional instructions.

2.1   Relationships Between Set Functions

In Appendix A, we prove the following two theorems, which state that additive functions form a special case of cancelable functions, and cancelable functions form a special case of submodular functions, when we are considering functions with binary marginals. Notice that the latter statement only holds for functions with binary marginals.

Theorem 1. *The set of binary additive cost functions is a proper subset of the set of cancelable cost functions with binary marginals.*

Next, we show that, by applying binary marginal property to cancelable functions, any cancelable function with binary marginals is also submodular.

Theorem 2. *The set of cancelable cost functions with binary marginals is a proper subset of the set of submodular cost functions with binary marginals.*

We also list some properties for cancelable and submodular functions.

Proposition 1. *Let $\phi : M \to \mathbb{Z}_{\geq 0}$ be a cancelable function.*

1. *For any $S, T \subseteq M$ and any $e \in M \setminus (S \cup T)$, if $\phi(S) = \phi(T)$, then $\phi(S + e) = \phi(T + e)$.*
2. *For any $S, T \subseteq M$ and any $U \subseteq M$ with $U \cap (S \cup T) = \emptyset$, if $\phi(S) = \phi(T)$, then $\phi(S \cup U) = \phi(T \cup U)$.*

Proposition 2. *Let $\phi : M \to \mathbb{Z}_{\geq 0}$ be a submodular function with binary marginals.*

1. *For any $S, T \subseteq M$, $\phi(S) + \phi(T) \geq \phi(S \cup T) + \phi(S \cap T)$. In particular, $\phi(S) + \phi(T) \geq \phi(S \cup T)$.*
2. *If $\phi(S) = 1$ holds for some S with $|S| \geq 2$, there exists $e \in S$ s.t. $\phi(S - e) = 1$.*

3   EFX and PO Allocations for Bianry Additive Chores

In this section, we explore the existence of EFX and PO allocations for additive cost functions. We call an instance *binary* if each agent has an additive cost function with binary marginals. We propose an algorithm that computes EFX and PO allocations for binary instances. This is the first class with a general number of agents, for which EFX and PO allocations exist and can be computed in polynomial time. In the world of allocation of goods, it has been shown that maximizing Nash social welfare (product of agents' utilities) leads to EFX and PO allocations for bivalued instances (which is a superset of binary instances) [3, 24]. However, in the opposite of goods, minimizing Nash social welfare results in zero Nash social welfare which has no guarantee of fairness and efficiency. Before our result, it has only been shown that EFX and PO allocations exist for three bivalued agents [26] and lexicographic preference [28].

Given a binary instance, we divide the set of chores into M^0 and M^+, while M^+ includes the items that cost 1 to all agents, i.e., $c_i(e) = 1$ for all $i \in N$.

$$M^0 = \{e \in M : \exists i \in N, c_i(e) = 0\}, \quad M^+ = \{e \in M : \forall i \in N, c_i(e) = 1\}.$$

In this section, we prove the following main result.

Theorem 3. *For additive cost functions with binary marginals, EFX and PO allocations always exist and can be computed in polynomial time.*

Algorithm 1: Finding an EFX and PO allocation for binary additive cost

---

Input: A binary instance $< M, N, \mathbf{c} >$

1 initialize $X_i \leftarrow \emptyset$ for all $i \in N$, $P \leftarrow M$;

   // Phase 1: compute a partial allocation $\mathbf{X}^0$ with $\mathrm{sc}(\mathbf{X}^0) = 0$.

2 **for** *each item $e \in M^0$* **do**

3 $\quad$ pick agent $i \in N$ s.t. $c_i(e) = 0$. break tie arbitrary;

4 $\quad$ update $X_i \leftarrow X_i + e$;

   // Phase 2: allocate items in M^+.

5 initialize $P \leftarrow M^+$;

6 **while** $P \neq \emptyset$ **do**

7 $\quad$ let $i^* \leftarrow \mathrm{argmin}_{i \in N}\{c_i(X_i)\}$, break tie arbitrary;

8 $\quad$ pick arbitrary item $e \in P$;

9 $\quad$ update $X_{i^*} \leftarrow X_{i^*} + e$, $P \leftarrow P - e$;

10 $\quad$ **if** *EFX is violated between i and j for some $j \neq i$* **then**

11 $\quad\quad$ update $X_{i^*} \leftarrow X_{i^*} - e$, $X_j \leftarrow X_j + e$;

12 $\quad\quad$ **while** *there exists an item $e' \in X_j$ such that $c_{i^*}(e') = 0$* **do**

13 $\quad\quad\quad$ update $X_{i^*} \leftarrow X_{i^*} + e'$, $X_j \leftarrow X_j - e'$;

Output: $\mathbf{X} = \{X_1, \cdots, X_n\}$.

Our algorithm starts from a partial allocation $\mathbf{X}^0$ in which every agent only receives items that have zero cost to her. In the second phase, we allocate each item $e \in M^+$ to an agent if it maintains EFX property after allocating e, or we reallocate some items until an item $e \in M^+$ can be allocated without breaking EFX. The steps of the full algorithm are summarized in Algorithm 1.

Lemma 1. *Algorithm 1 returns an allocation $\mathbf{X}$ with minimum social cost.*

Proof. We prove the lemma by showing that $sc(\mathbf{X}) = |M^+|$, while no complete allocation has a strictly less social cost. In Phase 1 we compute a partial allocation X^0 with zero social cost and remaining all items in M^+ unallocated. Then in Phase 2 we allocate items in M^+ and reallocate some items, while each allocation of item $e \in M^+$ increases the total social cost by 1. Note that item $e \in M$ would be reallocated to agent i only under the case that $c_i(e) = 0$. Hence any reallocation would not change the social cost of the allocation. In conclusion, we have $sc(\mathbf{X}) = |M^+|$.

Lemma 2. *The allocation* $\mathbf{X}$ *returned by Algorithm 1 is EFX to all agents.*

Proof. We show that the allocation $\mathbf{X}$ is EFX by mathematic induction. At the beginning of Phase 2, we must have that the partial allocation is EFX since each agent only receives items that cost 0 to her. In the following, we assume that the (partial) allocation is EFX at the beginning of some round t. We show that the (partial) allocation is still EFX for all agents at the end of round t.

According to the algorithm, we only have to consider the case under the if condition (in line 9) that we reallocate some items. Let X_{i^*}, X_j be the bundles that agents i^*, j hold at the beginning of round t, respectively. During the round, we reallocate a subset of items $S \subseteq X_j$ to agent i^* and assign an item $e \in M^+$ to agent j. We must have $c_{i^*}(e) = 1$ otherwise the if-condition does not hold. We show that at the end of the round, the new allocation $\mathbf{X}'$ is EFX for all agents, while $X'_{i^*} = X_{i^*} \cup S, X'_j = X_j \setminus S + e$ and $X'_k = X_k$ for all $k \in N \setminus \{i^*, j\}$. Before we give the proof, we first show the following claim.

Claim. If the if-condition from Line 9 to Line 12 is executed, we must have $c_{i^*}(X_*) = c_j(X_j)$. After the execution, we have $X_j \setminus S \subseteq M^+$.

Proof. For the first statement, we assume otherwise that $c_j(X_j) \geq c_{i^*}(X_{i^*}) + 1$. Then we have $c_{i^*}(X_{i^*} + e) = c_{i^*}(X_{i^*}) + 1 \leq c_j(X_j) \leq c_{i^*}(X_j)$, where the last inequality holds since the (partial) allocation minimizes social cost. In other words, agent i is envy-free towards j after allocating e to i^*, which is a contradiction. Hence the only case that agent i^* is not EFX towards j after assigning item e is that $c_{i^*}(X_{i^*}) = c_j(X_j) = c_{i^*}(X_j)$. Following the minimum social cost property, for any item $e' \in X_j$, we have $c_{i^*}(e') = c_j(e')$. Hence we have $S \subseteq M^0$ and $X_j \setminus S \subseteq M^+$.

Given Claim 13, we are ready to show the allocation is EFX for all agents at the end of round t. We show the property holds for agents i^*, j and any $k \neq i^*, j$ individually.

- For any $k \neq i^*, j$, agent i^* is EFX towards j and k. Note that during the reallocation, we only reallocate those items that cost 0 to agent i^*, i.e., $c_{i^*}(S) = 0$. Hence we have $c_{i^*}(X'_{i^*}) = c_{i^*}(X_{i^*} \cup S) = c_{i^*}(X_{i^*})$. Note that the allocation $\mathbf{X}$ is EFX for agent i^* and $X_k = X'_k$ for any $k \neq i^*, j$. We have agent i^* is EFX towards any agent k at the end of the round. As for the envy between i^* and j, we have $c_{i^*}(X'_j) = c_{i^*}(X_j \setminus S + e) = c_{i^*}(X_j) + 1$, agent i is EFX towards agent j at the end of the round.

- For any $k \neq i^*, j$, agent j is EFX towards i^* and k. Following Claim 13, agent j has the minimum bundle cost among all agents at the beginning of the round. In other words, agent j is envy-free towards any agent $k \neq j$ since $c_j(X_j) \leq c_k(X_k) \leq c_j(X_k)$, where the second inequality holds since the (partial) allocation minimizes social cost. Combining $X_j \setminus M^+$ and $e \in M^+$, we have $c_j(X'_j - e') = c_j(X_j) \leq c_j(X_k)$ for any $e' \in X'_j$ and any $k \neq j$.
- For any $k \neq i^*, j$, agent k is EFX towards i and j. Due to the monotonicity of c_k, we have $c_k(X'_{i^*}) \geq c_k(X_{i^*})$, agent k is EFX towards i at the end of the round. Next, we show that k is EFX towards j. We first claim that $c_k(X_k) \leq c_{i^*}(X_{i^*}) + 1$. Assume otherwise $c_k(X_k) \geq c_i(X_i) + 2$, we consider the last time that we assign an item $e' \in M^+$ to agent k. We must have $c_k(X_k - e') \geq c_{i^*}(X_{i^*}) + 1$, which contradicts the fact that k holds a bundle with minimum cost. Hence we consider the cases that $c_k(X_k) = c_{i^*}(X_{i^*})$ and $c_k(X_k) = c_{i^*}(X_{i^*}) + 1$. For both cases k is EFX towards j since $c_k(X_k) \leq c_{i^*}(X_{i^*}) + 1 = c_j(X'_j) \leq c_k(X'_j)$, where the equality follows from Claim 13 and the fact that $e \in M^+$.

In conclusion, we show that the (partial) allocation is EFX for all agents, at the end of the round. Note that in each round we allocate one item in M^+ and the algorithm terminates Phase 2 after $|M^+|$ rounds. Hence upon the running of Algorithm 1, it computes an EFX allocation.

Furthermore, it can be verified that the algorithm runs in polynomial time.

Lemma 3. *Algorithm 1 runs in $O(nm^2)$ time.*

Proof. Algorithm 1 starts from a partial allocation $\mathbf{X}^0$ with zero social cost, which can be computed in $O(mn)$ times (there are at most m items in M^0 and each item can be allocated in $O(n)$ time). The algorithm runs in rounds in phase 2 and there are at most m rounds to allocate items in M^+. In each round, determining the agent i who receives item e can be done in $O(\log n)$ time, and determining whether i is EFX towards other agents can be done in $O(mn)$ time. The reallocation in each round can be done in $O(m)$ time since there at most m items in X_j. In conclusion, Algorithm 1 runs in $O(nm^2)$ time.

We further complete our result by exploring the non-existence of EFX and PO allocations in the additive setting. We show that extending our result to ternary instances[1] is impossible (see full version).

4   EFX Allocations for Binary Cancelable Chores

In this section, we show that, for cancelable cost functions with binary marginals, we can always find EFX allocations in polynomial time, but there exist instances for which all EFX allocations are not Pareto-optimal.

---

[1] An instance is called ternary if all cost functions are additive and $c_i(e) \in \{0, 1, 2\}$ for all $i \in N, e \in M$.

Theorem 4. *For cancelable cost functions with binary marginals, EFX allocations always exist and can be computed in polynomial time.*

Algorithm 2: Finding an EFX allocation for cancelable cost functions with binary marginals

---

Input: A binary instance $(M, N, \mathbf{c})$

// Phase 1: allocate items with marginal cost 1 from all
    agents' perspective

1 initialize $(A_1, \ldots, A_n)$ with $A_i = \emptyset$ for all i;

2 **while** *there exist n items s.t. each item e satisfies $c_i(e \mid A_i) = 1$ for all i* **do**

3   add those n items to $(A_1, \ldots, A_n)$ s.t. each A_i is added one item;

4 let $w = |A_1| = \cdots = |A_n|$;

// Phase 2: allocate the remaining items

5 initialize $(B_1, \ldots, B_n)$ where $B_i = \emptyset$ for each i;

6 consider the cost function $\mathbf{d} = (d_1, \ldots, d_n)$ with $d_i(S) = c_i(S \mid A_i)$;

7 let $M_1 = \{e \in M : d_i(e) = 1 \text{ for all } i\}$;

// We have $|M_1| < n$, for otherwise Phase 1 should not have
    been terminated.

8 for each $i = 1, \ldots, |M_1|$, let B_i be the set containing one item of M_1;

9 **while** *there is an unallocated item e* **do**

     // We maintain that $(B_1, \ldots, B_n)$ is EFX with respect to $\mathbf{d}$
       and $d_i(B_i) \leq 1$ for any i.

10   **if** $d_i(e \mid B_i) = 0$ *for some agent i and $(B_1, \ldots, B_n)$ is EFX after updating $B_i \leftarrow B_i + e$* **then**

11   update $B_i \leftarrow B_i + e$;

12   **else**

       // If we reach here, there must exist i and j, with the
         possibility $i = j$, such that $d_i(B_j) = 0$
         (Proposition 5).

13   **if** *there exists $i \in N$ with $d_i(B_i) = 0$* **then**

14   if there exists $j \neq i$ with $d_i(B_j) = 0$, update $B_i \leftarrow B_i \cup B_j$ and $B_j \leftarrow \{e\}$; otherwise, update $B_i \leftarrow B_i + e$;

15   **else**

16   find j such that $d_i(B_j) = 0$, and swap B_i and B_j;

Output: $\mathbf{X} = (X_1, \ldots, X_n)$ where $X_i = A_i \cup B_i$ for each i.

The algorithm consists of two phases. In the first phase, we iteratively allocate n items, where each agent is believed to have a marginal cost of 1, such that each agent is allocated exactly one item. Let w be the number of items allocated to each agent, and we have allocated wn items. After Phase 1, the number of items with marginal costs 1 to all agents is less than n. Let M_1 be the set of these items. Let $(A_1, \ldots, A_n)$ be the allocation of Phase 1. After the first phase, each agent believes that her own bundle costs w and each other agent's bundle also costs w, as the lemma below states.

Lemma 4. *After Phase 1, we have $c_i(A_j) = w$ for each pair of i and j, with the possibility $i = j$.*

In Phase 2, we compute an allocation $(B_1, \ldots, B_n)$ of the remaining items. We update the cost function c_i such that, for each unallocated set S of items, the cost of S is given by $c_i(S \mid A_i)$. Let $\mathbf{d}$ be the set of the updated cost functions. We will show that the allocation $(B_1, \ldots, B_n)$ output in Phase 2 satisfies some properties (see Lemma 5).

Lastly, the final allocation $\mathbf{X} = (X_1, \ldots, X_n)$ is given by $X_i = A_i \cup B_i$ for each $i \in N$. We will show that $\mathbf{X}$ is EFX with respect to $\mathbf{c}$. The cancelable property plays an important role in guaranteeing EFX property when combining the allocations in the two phases. Specifically, the cancelable property and Lemma 4 ensure a good interpolation between the two cost function sets $\mathbf{d}$ and $\mathbf{c}$. Before proving the key properties of Phase 2 allocation, we first state the following proposition whose proof is straightforward.

Proposition 3. *Each d_i is a cancelable and submodular set function with binary marginals.*

Now we are ready to show our main lemma for Phase 2. It states that the algorithm will always be terminated and the allocation output satisfies our two key properties.

Lemma 5. *The while-loop in Phase 2 is executed for at most $2m$ iterations, and the output allocation $(B_1, \ldots, B_n)$ satisfies that*

1. *$d_i(B_j) \leq 1$ for each agent i, and*
2. *$(B_1, \ldots, B_n)$ is EFX with respect to $\mathbf{d}$.*

We show that Property 1 and 2 above are always satisfied (in Appendix B).

Proposition 4. *After any number of while-loop iterations of Phase 2, the two properties in Lemma 5 are satisfied.*

To show the algorithm terminates, we prove the following observation which is also stated between Line 12 and Line 13 of the algorithm.

Proposition 5. *If the "else" part from Line 12 to Line 16 is executed, there must exist i and j, with the possibility $i = j$, such that $d_i(B_j) = 0$.*

Proof. Suppose $d_i(B_j) \geq 1$ for every pair of i and j. We will show that the if-condition at Line 10 is always satisfied. Proposition 4 indicates that Property 1 in Lemma 5 holds before the if-condition at Line 10. Therefore, the current allocation $(B_1, \ldots, B_n)$ is envy-free. Adding an item with a zero marginal cost to an agent will not destroy the envy-free property. On the other hand, for each $e \notin M_1$, there exists an agent i with $d_i(e) = 0$, and, by submodularity, $d_i(e \mid B_i) = 0$. Since all the items in M_1 have been allocated before the while-loop and the algorithm never returns an item back to the pool of unallocated items, the if-condition at Line 10 will be satisfied.

Now we are ready to prove Lemma 5.

Proof of Lemma 5: Proposition 4 ensures that the two properties always hold. To show that the while-loop will be executed for at most $2m$ iterations, it suffices to prove that at least one item is allocated in every two iterations. The only case where no item is allocated is when Line 16 is executed. In this case, Proposition 5 ensures the existence of j. After the swapping, we have $d_i(B_i) = 0$. Therefore, in the next iteration, either Line 11 or Line 14 will be executed, in which case an item is allocated. ∎

Remark 1. For the proof of Lemma 5, we have only exploited the submodularity of d_i, not the cancelability. In particular, if no item is allocated in Phase 1 and the algorithm starts with Phase 2, we have $\mathbf{d} = \mathbf{c}$. The algorithm computes an EFX allocation even if each c_i is only known to be submodular (while not necessarily cancelable).

Finally, we combine Lemma 4 and Lemma 5 to conclude Theorem 4.

Proof of Theorem 4: We first show that the allocation output by Algorithm 2 is EFX. We check the EFX condition for any agents i and j. By Lemma 5, $(B_1, \ldots, B_n)$ is EFX w.r.t. $\mathbf{d}$, and we consider two cases: 1) $d_i(B_i) \leq d_i(B_j)$ and 2) $d_i(B_i) > d_i(B_j)$.

For the first case, we have

$$
\begin{aligned}
c_i(X_i) = c_i(A_i) + d_i(B_i) &\leq c_i(A_i) + d_i(B_j) && \text{(case assumption)}\\
&= c_i(A_i) + (c_i(A_i \cup B_j) - c_i(A_i))\\
&= c_i(A_i \cup B_j) && \text{(definition of } d_i)\\
&= c_i(A_j \cup B_j) = c_i(X_j), && \text{(Property 2 of Proposition 1 and Lemma 4)}
\end{aligned}
$$

which indicates that i does not envy j.

For the second case, it must be that $|B_i| = 1$. To see this, Property 1 of Lemma 5 indicates that $d_i(B_i) = 1$ and $d_i(B_j) = 0$. Property 2 of Proposition 2 indicates that $|B_i| \geq 2$ will destroy the EFX property (Property 2 of Lemma 5) between i and j. Since we have seen $|B_i| = 1$, $|X_i| = w + 1$. Therefore, agent i's cost will be at most w after removing any item from X_i. On the other hand, by Lemma 4 and the monotonicity, $c_i(X_j) \geq c_i(A_j) = w$. Thus, the EFX condition between i and j is satisfied.

Lastly, the algorithm runs in polynomial time straightforwardly. ∎

We have seen that EFX allocations always exist for binary cancelable chores. However, unlike the case with additive cost functions, EFX is no longer compatible with PO (see Appendix B). This is in sharp contrast to the setting with goods, for which we know that EFX is compatible with PO even for the more general submodular binary cost functions: Babaioff et al. [7] shown that the allocation maximizing the Nash social welfare (product of agents' utilities) is EFX.

Theorem 5. *For any number of agents $n \geq 2$, there exist instances with binary cancelable cost functions where all EFX allocations are not Pareto-optimal.*

5   General Cost Functions with Binary Marginals

For general cost functions with binary marginals, we show that there exists a partial allocation that leaves at most $n - 1$ items unallocated and is envy-free. We will also show that the algorithm can be used to find a complete 2-EFX allocation for submodular binary cost functions. We remark that we do not know if complete EFX allocations always exist, even for the more special case with submodular binary cost functions.

Theorem 6. *For general cost functions with binary marginals, there exists a partial allocation that is envy-free and leaves at most $n - 1$ items unallocated. Moreover, such an allocation can be computed in polynomial time.*

The algorithm makes use of the *envy graph*, a technique that has been widely used in the fair division literature [13, 21, 32].

Definition 1. *Given a partial allocation $(X_1, \ldots, X_n)$ and the cost function profile* **c***, the* envy graph $G = (V = [n], E)$ *is a directed graph where each vertex in V represents an agent and $(i, j) \in E$ if and only if $c_i(A_i) = c_i(A_j)$.*

As a remark, unlike it is in most of the previous literature where an edge represents "envy", we always maintain an envy-free allocation and an edge in our envy graph represents equality, or, "about to envy".

Algorithm Description and proof of Theorem 6. Our algorithm is presented in Algorithm 3. It initializes the allocation with n empty bundles, and the initial envy graph is a complete graph. At each iteration, it attempts to allocate one or more items by using one of the three update rules:

1. if there is an item with marginal cost 0 to some agent, allocate it;
2. if an edge (i, j) is on a cycle C and i thinks adding some item e to j's bundle does not increase its cost, rotate the bundles on the cycle so that i receive j's bundle, and add e to i's bundle (which is previously j's) which does not increase the cost for i;
3. if the first two update rules do not apply, find a tail strongly connected component S (a strongly connected component with no outgoing edges) and allocate each agent in S an arbitrary item; if the number of items is insufficient for this, the algorithm is terminated with the partial allocation outputted.

We will show that the EF property is satisfied throughout the algorithm. The initial allocation is clearly envy-free. It suffices to show that, given a partial allocation, any of the three update rules does not destroy envy-freeness.

It is straightforward to check the first and the second update rules do not violate envy-freeness. For the third update rule, first notice that the envy-freeness

---

Algorithm 3: Finding a partial EF allocation for cost functions with binary marginals

---

Input: A binary instance $(M, N, \mathbf{c})$

1 initialize $(X_1, \ldots, X_n)$ with n empty bundles, and initialize the envy graph $G = (V, E)$;

2 **while** *there exist unallocated items* **do**

3   update the envy graph $G = (V, E)$;

4   **if** *there exist an unallocated item e and i with $c_i(e \mid X_i) = 0$* **then**

5   update $X_i \leftarrow X_i + e$;

6   **continue**;

7   **if** *there exist an unallocated item e and an edge $(i, j) \in E$ such that (i, j) is on a directed cycle C and $c_i(e \mid X_j) = 0$* **then**

8   rotate bundles on the cycle: for each edge $(u, v) \in C$, allocate X_v to agent u;
        // In particular, i receives X_j.

9   add e to agent i's bundle;

10   **continue**;

    // Handling the case where the above two kinds of updates fail.

11   find a tail strongly connected component S in G;
    // A tail strongly connected component S has no outgoing edge from S; in particular, a sink is a tail strongly connected component.

12   **if** *there are at least $|S|$ unallocated items* **then**

13   allocate each agent S an arbitrary unallocated item;

14   **else**

15   **break**;

Output: $\mathbf{X} = (X_1, \ldots, X_n)$

---

between an agent i in S and an agent k in $V \setminus S$ is preserved. Agent k does not envy agent i as agent i's cost can only be increased from k's perspective. Since S is a tail strongly connected component, $(i, k) \notin E$ before the update, so $c_i(X_i) \leq c_i(X_k) - 1$. Adding an item to agent i, which increases $c_i(X_i)$ by at most 1 (in fact, it is exactly 1, for otherwise the first update rule should be applied), does not make i envy k.

Now, consider any two agents $i, j \in S$. If $(i, j) \notin E$ before the update, i will not envy j after the update even in the worst case where $c_i(X_i)$ is increased by 1 and $c_i(X_j)$ is unchanged. If $(i, j) \in E$ before the update, (i, j) is on a cycle by the property of strongly connected components. Since the first two update rules do not apply, it must be that $c_i(e \mid X_i) = c_i(e \mid X_j) = 1$ for any unallocated item e. Adding an item to i and an item to j increases both $c_i(X_i)$ and $c_i(X_j)$ by 1. Envy-freeness is again preserved.

Finally, checking that the algorithm runs in polynomial time is straightforward. We can only reach a partial allocation when the if-condition at Line 12 fails, in which case the number of unallocated items is less than $|S|$, which is at most $n - 1$.

6   Submodular Cost Functions with Binary Marginals

Theorem 7. *For submodular cost functions with binary marginals, there exists an allocation* $\mathbf{X}$ *that is either EFX or 2-EF. In addition, such an allocation can be computed in polynomial time.*

Notice that a 2-EF allocation is always 2-EFX. We have the following corollary.

Corollary 1. *For submodular cost functions with binary marginals, there exists a 2-EFX allocation and it can be computed in polynomial time.*

The proof of Theorem 7 is mostly based on algorithms in the previous sections.

Proof (Proof of Theorem 7). Let $M_1 = \{e \in M \mid c_i(e) = 1 \text{ for all } i\}$. We consider two cases.

Case 1: $|M_1| < n$. In this case, we will run Algorithm 2. Note that Phase 1 will be skipped. As a result, each d_i in the algorithm will be c_i, which is submodular. By Lemma 5 and Remark 1, an EFX allocation will be output.

Case 2: $|M_1| \geq n$. In this case, we will initialize the allocation $(X_1, \ldots, X_n)$ such that each X_i contains exactly one item in M_1. The current partial allocation is clearly EF. Next, we will implement the while-loop in Algorithm 3. Lastly, for the unallocated items, we allocate them arbitrarily such that each agent receives at most one extra item.

Since we have $c_i(X_j) = 1$ for any i and j before the while-loop and the algorithm never removes items from a bundle, it holds that $c_i(X_j) \geq 1$ for any i and j for the final allocation. The allocation remains EF during the while-loop, and the envy-freeness can only be broken at the last step. By the binary marginal property, if i envies j, it must be that $c_i(X_i) = w + 1$ and $c_i(X_j) = w$ for some w. We have $c_i(X_i) \leq 2 \cdot c_i(X_j)$ since we have seen that $w \geq 1$.

7   Conclusion

In this paper, we study the existence of EFX allocations for indivisible chores. We focus on the setting with binary marginal, which has received much attention in the past years [8,9,34,35]. We first consider the additive cost function and propose a polynomial-time algorithm to compute EFX and PO allocations. We further show that EFX allocations exist and can be computed in polynomial time for cancelable functions. Moreover, part of our algorithm only depends on submodularity, which might shed light on generalizing our result to the submodular functions. Surprising and in contrast with the case of goods where EFX and PO allocations exist for binary submodular functions [7], we show that EFX and PO are non-compatible for binary cancelable functions (and hence submodular funtions). Finally, we show that there exists (partial) EF allocation with at most $n - 1$ unallocated items for general cost functions and 2-EFX allocations for submodular cost functions. We believe that it would be interesting and non-trivial to investigate the existence of EFX allocations for submodular or general functions.

A   Relationship Between Set Functions with Binary Marginals

In this section, we explore the relationship between additive, cancelable, and submodular set functions, and we focus only on set functions with binary marginals. Many results will be used in later sections, and they are interesting observations that are independent of our applications to fair division.

Firstly, it is obvious from definitions that all additive functions are cancelable. Moreover, there exist cancelable set functions with binary marginals that are not additive. An example is given below.

$$\phi(X) = \begin{cases} 5 & \text{when } |X| \geq 5 \\ |X| & \text{otherwise} \end{cases}. \tag{2}$$

Next, we show that, by applying binary marginal property to cancelable functions, any cancelable function with binary marginals is also submodular.

Theorem 1. *The set of binary additive cost functions is a proper subset of the set of cancelable cost functions with binary marginals.*

Proof. To show the set containment, we will show that any set function $\phi : M \to \mathbb{Z}_{\geq 0}$ (with binary marginals) that is not submodular cannot be cancelable. Since ϕ is nonsubmodular, there exists $S \subseteq T \subseteq M$ and $e \in M \setminus T$ such that $\phi(S + e) - \phi(S) < \phi(T + e) - \phi(T)$. We initialize $T' = S$ and iteratively add an element from $T \setminus S$ to T'. At the start, we have $\phi(S+e) - \phi(S) = \phi(T'+e) - \phi(T')$. We consider the first time we observe $\phi(S + e) - \phi(S) < \phi(T' + e) - \phi(T')$ after an element f is added to T'. Let U be the state of T' before f is added. Since ϕ has binary marginals, we must have $\phi(S + e) - \phi(S) = \phi(U + e) - \phi(U)$ and $\phi(S+e) - \phi(S) < \phi(U+e+f) - \phi(U+f)$. Combining the two equations and by noting that ϕ is monotone, we must have $0 \leq \phi(U + e) - \phi(U) < \phi(U + e + f) - \phi(U + f)$. Since $U + e + f$ and $U + f$ differ by only one element and ϕ has binary marginals, we must have $0 \leq \phi(U + e) - \phi(U) < \phi(U + e + f) - \phi(U + f) \leq 1$. It must be that $\phi(U + e) - \phi(U) = 0$ and $\phi(U + e + f) - \phi(U + f) = 1$. Thus, $\phi(U + e + f) > \phi(U + f)$ fails to imply $\phi(U + e) > \phi(U)$, so ϕ is not cancelable.

To show the containment is proper, we present an example of submodular set functions (with binary marginals) that is not cancelable. Let $M = \{a, b, c, d\}$ and

$$\phi(X) = \begin{cases} |X| - 1 & \text{if } \{a, b, c\} \subseteq X \\ |X| & \text{otherwise} \end{cases}.$$

It is straightforward to check (e.g., by enumerating all cases) that ϕ is submodular. It is not cancelable as $\phi(\{c,d\}) = \phi(\{b,c\}) = 2$ and $\phi(\{a,c,d\}) = 3 > 2 = \phi(\{a,b,c\})$.

Note that for set functions with non-binary marginals, the containment in Theorem 1 does not hold. A simple cancelable set function that is non-submodular is $\phi(X) = 2^{|X|}$. We list some of the useful properties of cancelable functions, whose proofs are straightforward.

Proposition 1. *Let $\phi : M \to \mathbb{Z}_{\geq 0}$ be a cancelable function.*

1. *For any $S, T \subseteq M$ and any $e \in M \setminus (S \cup T)$, if $\phi(S) = \phi(T)$, then $\phi(S + e) = \phi(T + e)$.*
2. *For any $S, T \subseteq M$ and any $U \subseteq M$ with $U \cap (S \cup T) = \emptyset$, if $\phi(S) = \phi(T)$, then $\phi(S \cup U) = \phi(T \cup U)$.*

Finally, we prove some properties for submodular functions. By Theorem 1, cancelable functions with binary marginals also satisfy these properties.

Proposition 2. *Let $\phi : M \to \mathbb{Z}_{\geq 0}$ be a submodular function with binary marginals.*

1. *For any $S, T \subseteq M$, $\phi(S) + \phi(T) \geq \phi(S \cup T) + \phi(S \cap T)$. In particular, $\phi(S) + \phi(T) \geq \phi(S \cup T)$.*
2. *If $\phi(S) = 1$ holds for some S with $|S| \geq 2$, there exists $e \in S$ s.t. $\phi(S - e) = 1$.*

Proof. 1 is a well-known alternative definition for submodularity. For 2, if $\phi(S - e) = 0$ for any $e \in S$, this means every subset of S with size $|S| - 1$ has value 0. By monotonicity of ϕ, every element of S has value 0. Then, $\phi(S) = 1$ violates the submodularity.

B   Missing Proofs in Section 4

Proof of Lemma 4: We prove by induction that $c_i(A_j) = t$ for each pair of i and j after t while-loop iterations. The base step for $t = 0$ is trivial. Suppose the claim holds for t. We will show it holds for $t + 1$. Let $(A_1, \ldots, A_n)$ be the allocation before the $(t+1)$-th iteration. By induction hypothesis, $c_i(A_j) = t$ for every pair of i and j. For each item e allocated in the $(t + 1)$-th iteration, we have $c_i(e \mid A_i) = 1$ for each agent i. By Property 1 in Proposition 1, this also implies $c_i(e \mid A_j) = 1$ for every other agent j. Thus, $c_i(A_j) = t + 1$ for every pair of i and j after the $(t + 1)$-th iteration. ∎

Proof of Proposition 4: Before entering the while-loop, exactly $|M_1|$ bundles contain one item with cost 1 from all agent's perspectives, and the remaining $n - |M_1|$ bundles are empty. The properties clearly hold. Next, suppose the properties hold before a while-loop execution, we will show that they continue to hold after. We will check all the updates, at Line 11, 14, and 16, do not violate the properties.

The update at Line 11 adds an item with marginal cost 0 to the bundle B_i, so Property 1 continues to hold. It will not invalidate Property 2 by the if-condition.

For the update at Line 14, we have $d_i(B_i \cup B_j) \leq d_i(B_i) + d_i(B_j) = 0$ by submodularity of d_i (Proposition 3), $d_i(B_i + e) \leq 1$, and $d_j(e) \leq 1$. In all cases, Property 1 continues to hold. To check Property 2, if there does not exist $j \neq i$ with $d_i(B_j) = 0$, then agent i receives $B_j + e$ and the allocation for the remaining agents is unchanged. In this case, $d_i(B_j) \geq 1$, and agent i does not envy any other agent by receiving $B_i + e$, which has cost at most 1. The EFX condition between any other agent and i still holds as i receives a superset of B_i. If $d_i(B_j) = 0$ for some j, agent i receives $B_i \cup B_j$ and agent j's bundle is updated to the singleton $\{e\}$. We have $d_i(B_i \cup B_j) \leq d_i(B_i) + d_i(B_j) = 0$, so i does not envy any other agent. On the other hand, j receives only a single item e, she will no longer envy any other agent if this item were removed. Consider any other agent k that is neither i nor j. The EFX condition between k and i continues to hold, as i now receives a superset of B_i. As for the EFX condition between k and j, we discuss two cases. If $d_k(e) = 1$, then agent k does not envy agent j as we have $d_k(B_k) \leq 1$ by Property 1. If $d_k(e) = 0$, we have $d_k(e \mid B_k) \leq d_k(e) = 0$ by submodularity. Since the if-condition at Line 10 fails (otherwise, the "else" part will not be executed), adding e to B_k will break the EFX condition between k and some other agent ℓ. By Property 1, it must be that $d_k(B_k) = 1$ and $d_k(B_\ell) = 0$. Moreover, $d_k(B_k - f) = 0$ for any $f \in B_k$ (in fact, it must be that $B_k = \{f\}$ by Property 2 in Proposition 2). In this case, the EFX condition between k and any other agent still holds.

For the update at Line 16, if we have ever reached here, it must be that $d_i(B_i) = 1$ and $d_i(B_j) = 0$. Moreover, we must have $|B_i| = 1$. Otherwise, by Property 2 of Proposition 2, the EFX property between i and j will be violated. After the update at Line 16, we have $d_i(B_i) = 0$ and $d_j(B_j) \leq 1$, and the latter holds because B_j, which is previously B_i, contains only one item. This proves Property 1. To check Property 2, i does not envy any other agents as $d_i(B_i)$ is now 0. The EFX condition between j and any other agents holds as j's bundle contains only one item now. The EFX condition between any other agent k and i or j still holds as we have only swapped the bundles of i and j.   ∎

Proof of Theorem 5: Consider n agents with $5n$ items where each agent's cost function is given by Eq. (2). It is easy to check that the only EFX allocation gives exactly 5 items to each agent. However, this allocation is Pareto-dominated by the allocation that allocates all items to a single agent.   ∎

References

1. Akrami, H., Alon, N., Chaudhury, B.R., Garg, J., Mehlhorn, K., Mehta, R.: EFX: a simpler approach and an (almost) optimal guarantee via rainbow cycle number. In: EC, p. 61. ACM (2023)

2. Amanatidis, G., et al.: Fair division of indivisible goods: Recent progress and open questions. Artif. Intell. **322**, 103965 (2023)
3. Amanatidis, G., Birmpas, G., Filos-Ratsikas, A., Hollender, A., Voudouris, A.A.: Maximum nash welfare and other stories about EFX. Theor. Comput. Sci. **863**, 69–85 (2021)
4. Amanatidis, G., Markakis, E., Ntokos, A.: Multiple birds with one stone: beating 1/2 for EFX and GMMS via envy cycle elimination. Theor. Comput. Sci. **841**, 94–109 (2020)
5. Aziz, H., Li, B., Moulin, H., Wu, X.: Algorithmic fair allocation of indivisible items: a survey and new questions. SIGecom Exch. **20**(1), 24–40 (2022)
6. Aziz, H., Lindsay, J., Ritossa, A., Suzuki, M.: Fair allocation of two types of chores. In: AAMAS, pp. 143–151. ACM (2023)
7. Babaioff, M., Ezra, T., Feige, U.: Fair and truthful mechanisms for dichotomous valuations. In: Proceedings of the AAAI Conference on Artificial Intelligence, vol. 35, pp. 5119–5126 (2021)
8. Barman, S., Narayan, V.V., Verma, P.: Fair chore division under binary supermodular costs. In: AAMAS, pp. 2863–2865. ACM (2023)
9. Barman, S., Verma, P.: Approximating nash social welfare under binary XOS and binary subadditive valuations. In: WINE. Lecture Notes in Computer Science, vol. 13112, pp. 373–390. Springer (2021)
10. Benabbou, N., Chakraborty, M., Igarashi, A., Zick, Y.: Finding fair and efficient allocations when valuations don't add up. In: SAGT. Lecture Notes in Computer Science, vol. 12283, pp. 32–46. Springer (2020)
11. Bérczi, K., et al.: Envy-free relaxations for goods, chores, and mixed items. CoRR abs/2006.04428 (2020)
12. Berger, B., Cohen, A., Feldman, M., Fiat, A.: Almost full EFX exists for four agents. In: AAAI, pp. 4826–4833. AAAI Press (2022)
13. Bhaskar, U., Sricharan, A.R., Vaish, R.: On approximate envy-freeness for indivisible chores and mixed resources. In: APPROX-RANDOM. LIPIcs, vol. 207, pp. 1:1–1:23. Schloss Dagstuhl - Leibniz-Zentrum für Informatik (2021)
14. Bogomolnaia, A., Moulin, H.: Random matching under dichotomous preferences. Econometrica **72**(1), 257–279 (2004)
15. Bu, X., Song, J., Yu, Z.: EFX allocations exist for binary valuations. In: IJTCS-FAW. Lecture Notes in Computer Science, vol. 13933, pp. 252–262. Springer (2023)
16. Caragiannis, I., Gravin, N., Huang, X.: Envy-freeness up to any item with high nash welfare: The virtue of donating items. In: EC, pp. 527–545. ACM (2019)
17. Caragiannis, I., Kurokawa, D., Moulin, H., Procaccia, A.D., Shah, N., Wang, J.: The unreasonable fairness of maximum nash welfare. ACM Trans. Economics and Comput. **7**(3), 12:1–12:32 (2019)
18. Chan, H., Chen, J., Li, B., Wu, X.: Maximin-aware allocations of indivisible goods. In: AAMAS, pp. 1871–1873. International Foundation for Autonomous Agents and Multiagent Systems (2019)
19. Chaudhury, B.R., Garg, J., Mehlhorn, K.: EFX exists for three agents. In: EC, pp. 1–19. ACM (2020)
20. Chaudhury, B.R., Garg, J., Mehlhorn, K., Mehta, R., Misra, P.: Improving EFX guarantees through rainbow cycle number. In: EC, pp. 310–311. ACM (2021)
21. Chaudhury, B.R., Kavitha, T., Mehlhorn, K., Sgouritsa, A.: A little charity guarantees almost envy-freeness. SIAM J. Comput. **50**(4), 1336–1358 (2021)
22. Ebadian, S., Peters, D., Shah, N.: How to fairly allocate easy and difficult chores. In: AAMAS, pp. 372–380. International Foundation for Autonomous Agents and Multiagent Systems (IFAAMAS) (2022)

23. Foley, D.: Resource allocation and the public sector. Yale Econ. Essays 45–98 (1967)
24. Garg, J., Murhekar, A.: Computing fair and efficient allocations with few utility values. In: SAGT. Lecture Notes in Computer Science, vol. 12885, pp. 345–359. Springer (2021)
25. Garg, J., Murhekar, A., Qin, J.: Fair and efficient allocations of chores under bivalued preferences. In: AAAI, pp. 5043–5050. AAAI Press (2022)
26. Garg, J., Murhekar, A., Qin, J.: New algorithms for the fair and efficient allocation of indivisible chores. In: IJCAI, pp. 2710–2718. ijcai.org (2023)
27. Halpern, D., Procaccia, A.D., Psomas, A., Shah, N.: Fair division with binary valuations: one rule to rule them all. In: WINE. Lecture Notes in Computer Science, vol. 12495, pp. 370–383. Springer (2020)
28. Hosseini, H., Sikdar, S., Vaish, R., Xia, L.: Fairly dividing mixtures of goods and chores under lexicographic preferences. In: AAMAS, pp. 152–160. ACM (2023)
29. Jahan, S.C., Seddighin, M., Javadi, S.M.S., Sharifi, M.: Rainbow cycle number and EFX allocations: (almost) closing the gap. In: IJCAI, pp. 2572–2580. ijcai.org (2023)
30. Kurokawa, D., Procaccia, A.D., Shah, N.: Leximin allocations in the real world. ACM Trans. Economics and Comput. 6(3-4), 11:1–11:24 (2018)
31. Kyropoulou, M., Suksompong, W., Voudouris, A.A.: Almost envy-freeness in group resource allocation. Theor. Comput. Sci. 841, 110–123 (2020)
32. Li, B., Li, Y., Wu, X.: Almost (weighted) proportional allocations for indivisible chores. In: WWW, pp. 122–131. ACM (2022)
33. Plaut, B., Roughgarden, T.: Almost envy-freeness with general valuations. SIAM J. Discret. Math. 34(2), 1039–1068 (2020)
34. Viswanathan, V., Zick, Y.: A general framework for fair allocation under matroid rank valuations. In: EC, pp. 1129–1152. ACM (2023)
35. Viswanathan, V., Zick, Y.: Yankee swap: a fast and simple fair allocation mechanism for matroid rank valuations. In: AAMAS, pp. 179–187. ACM (2023)
36. Wu, X., Zhang, C., Zhou, S.: Weighted EF1 allocations for indivisible chores. In: EC, p. 1155. ACM (2023)
37. Zhou, S., Wu, X.: Approximately EFX allocations for indivisible chores. Artif. Intell. 326, 104037 (2024)

How to Play Old Maid with Virtual Players

Kazumasa Shinagawa[1,4]([✉]) [iD], Daiki Miyahara[2,4] [iD], and Takaaki Mizuki[3,4] [iD]

[1] Ibaraki University, Hitachi, Ibaraki, Japan
kazumasa.shinagawa.np92@vc.ibaraki.ac.jp
[2] The University of Electro-Communications, Tokyo, Japan
[3] Tohoku University, Sendai, Japan
[4] National Institute of Advanced Industrial Science and Technology, Tokyo, Japan

Abstract. Old Maid is a popular card game. While typically played with three or more players, it is less enjoyable with only two people. To address this, we propose a protocol to create a virtual player, Carol, by making use of card-based cryptography when only two people, Alice and Bob, are available to play Old Maid. Specifically, we design a card-based protocol to remove any pair of cards having the same number in Carol's hand (namely, the virtual player's hand) without leaking any information about Carol's hand (more than necessary); our protocol uses additional cards aside from playing cards that are used in Old Maid. Using our protocol, without any third human player, Alice and Bob can have fun with Old Maid!

Keywords: Card-based cryptography · Cryptology · Old Maid · Card games

1 Introduction

Old Maid is a card game that is popular all over the world. It is typically played with a standard deck of 53 playing cards, including one joker. The rule is roughly as follows:

- all players are dealt an (almost) equal number of cards;
- any pair of cards having the same number is removed from a hand;
- each player takes one card from the next player's hand in turn;
- the last player to have the joker loses.

Technically, Old Maid can be played by two or more players, but preferably three or more. This is because if only two people play the game, the player holding the joker can be completely identified (by both the players), and the game tends to become monotonous and somewhat boring. On the other hand, if there are three or more players, it is not possible to immediately identify who has the joker, and the game tends to become more fluid and tactical.

© The Author(s), under exclusive license to Springer Nature Singapore Pte Ltd. 2025
B. Li et al. (Eds.): IJTCS-FAW 2024, LNCS 14752, pp. 53–65, 2025.
https://doi.org/10.1007/978-981-97-7752-5_4

1.1 What if Only Two Players are Available?

So, what should we do if there are only two people who are eager to play Old Maid? One solution would be to create a *virtual* player just like virtual players in video games, so that we have three players to play the game. In this paper, we propose a method to play Old Maid as if there were three players, even when there are only two people.

Suppose that Alice and Bob want to create a virtual player, Carol, to play Old Maid. The difficulty in creating such a virtual player Carol is how to remove any pair of cards having the same number in Carol's hand. Of course, it would be easy to achieve this if Alice and Bob were allowed to look at Carol's hand; however, to keep the game fun, they are not allowed to know any information about Carol's hand. How would they be able to remove any pair in Carol's hand (without looking at her hand)?

1.2 Contribution

In this paper, we design a cryptographic protocol that removes any pairs in Carol's hand without revealing any information about Carol's hand (more than necessary); we call it a *removal protocol*. Our removal protocol is a so-called *card-based protocol* [1,2,15], which uses a deck of physical cards to perform a cryptographic task. Our protocol takes Alice's, Bob's, and Carol's hands as input and removes any pairs in Carol's hand using some "helping cards."

More specifically, our removal protocol is a generic one: it can remove any pairs in a virtual player's hand for the case where there are one or more virtual players along with any number of real human players. The protocol uses 108 "helping cards" in addition to 54 playing cards (in a standard deck) when Old Maid is played with a single standard deck of playing cards.

After providing preliminaries in Sect. 2, we explain the rule of Old Maid precisely and give a framework for execution of the game with virtual players in Sect. 3. Next, we present our removal protocol in Sect. 4. We conclude in Sect. 5 with future directions.

1.3 Related Work

Card-based cryptography is a research area for designing cryptographic protocols such as secure computation protocols using a deck of physical cards. It is shown that any function can be securely computed using physical cards [1,16]. Although the main research topic in card-based cryptography has been the design of secure computation protocols for Boolean functions $f : \{0,1\}^n \rightarrow \{0,1\}$, various other applications have been explored in recent years. A partial list of these studies is given below.

- **Zero-Knowledge Proof for Puzzles:** Gradwohl et al. [3] proposed a zero-knowledge proof protocol for a Sudoku puzzle using a deck of cards. This protocol enables us to convince someone that we know a solution of a Sudoku puzzle without revealing any information on the solution. Subsequent studies improved on its efficiency [20,24,26] and considered other puzzles, e.g., [5,8, 19,21–23].

- **No-Fixed Point Problem:** Crépeau and Kilian [1] proposed a protocol for generating a random permutation with no fixed point. This protocol can be used to exchange gifts without receiving a gift of one's own, e.g., in a Christmas party. Recent studies improved its efficiency [9,10,17] and gave a lower bound on the number of cards required [6].
- **Werewolf Game:** Hashimoto et al. [7] proposed a secure grouping protocol. This can be used to determine a character role without a moderator in the Werewolf game. Their study has the same motivation as our study in that it applies card-based cryptography to a game. However, even with their protocol, not all parts of the Werewolf game can be played without a moderator.
- **Covert Lottery:** Shinoda et al. [25] proposed a covert lottery protocol. For a two-player board game such as Chess or Shogi, this protocol can be used to determine who makes the first move according to the players' requests without revealing them.

In this paper, we design a protocol for the card game Old Maid. Although several protocols with a standard deck of playing cards were proposed [4,11–14,18,20], they do not apply card-based cryptography to card games. Our work is of significant value in that it discovers a new application of card-based cryptography.

2   Preliminaries

In this section, we introduce the definitions of physical cards and a shuffling action, the "pile-scramble shuffle," which are necessary to describe our card-based protocols later; they follow the standard computation model of card-based protocols [15].

2.1   Cards

Our proposed protocol employs three types of cards: a standard deck of playing cards, numbered cards, and dummy cards.

A standard deck consists of 54 cards of a combination of 13 numbers and four suits along with two jokers, as follows:

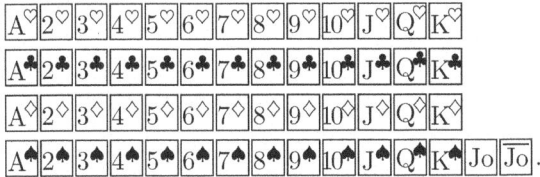

Their backs are all identical and written as ?.

The numbered cards have only numbers referred to as $\boxed{1}\boxed{2}\boxed{3}\cdots$. The cards having the same number are identical. The dummy cards have no symbol referred to as $\boxed{}$. The backs of the numbered cards and the dummy cards are all identical. We remark that the back of the standard deck and the back of the numbered cards may be different or the same. For simplicity, we assume that they are the same and written as $\boxed{?}$.

2.2 Pile-Scramble Shuffle

A shuffling action is operated on a sequence of cards, rearranging it randomly. Our proposed protocol employs the *pile-scramble shuffle* [10]. This divides a sequence into multiple piles of cards and rearranges them randomly.

Let ℓ be the number of cards and k be the number of piles (i.e., divisions) such that k is a divisor of ℓ. An ℓ-card k-pile pile-scramble shuffle is to divide a sequence of ℓ cards into k piles (each consisting of ℓ/k cards) and rearrange them completely randomly. There are $k!$ possibilities in total, but this shuffling is executed such that nobody can trace it. Note that when $k = 2$, this shuffle is called a *random bisection cut*.

For example, a six-card three-pile pile-scramble shuffle rearranges a sequence of cards as follows:

$$\begin{matrix} 1 & 2 & 3 & 4 & 5 & 6 \\ \boxed{?} & \boxed{?} & \boxed{?} & \boxed{?} & \boxed{?} & \boxed{?} \end{matrix} \rightarrow \begin{cases} \begin{matrix} 1 & 2 & 3 & 4 & 5 & 6 \\ \boxed{?} & \boxed{?} & \boxed{?} & \boxed{?} & \boxed{?} & \boxed{?} \end{matrix}, \\ \begin{matrix} 1 & 2 & 5 & 6 & 3 & 4 \\ \boxed{?} & \boxed{?} & \boxed{?} & \boxed{?} & \boxed{?} & \boxed{?} \end{matrix}, \\ \begin{matrix} 3 & 4 & 5 & 6 & 1 & 2 \\ \boxed{?} & \boxed{?} & \boxed{?} & \boxed{?} & \boxed{?} & \boxed{?} \end{matrix}, \\ \begin{matrix} 3 & 4 & 1 & 2 & 5 & 6 \\ \boxed{?} & \boxed{?} & \boxed{?} & \boxed{?} & \boxed{?} & \boxed{?} \end{matrix}, \\ \begin{matrix} 5 & 6 & 1 & 2 & 3 & 4 \\ \boxed{?} & \boxed{?} & \boxed{?} & \boxed{?} & \boxed{?} & \boxed{?} \end{matrix}, \\ \begin{matrix} 5 & 6 & 3 & 4 & 1 & 2 \\ \boxed{?} & \boxed{?} & \boxed{?} & \boxed{?} & \boxed{?} & \boxed{?} \end{matrix}. \end{cases}$$

The resulting sequence is chosen among the six sequences shown on the right side with a probability of $1/6$. This pile-scramble shuffle is denoted as follows by surrounding each pile with $|\cdot|$

$$\left| \boxed{?}\boxed{?} \middle| \boxed{?}\boxed{?} \middle| \boxed{?}\boxed{?} \right|.$$

This notation is also used when a sequence of cards is represented as a matrix. For example, the following notation shows a 20-card five-pile pile-scramble shuffle:

$$\left| \begin{matrix} \boxed{?}\boxed{?} \\ \boxed{?}\boxed{?} \end{matrix} \middle| \begin{matrix} \boxed{?}\boxed{?} \\ \boxed{?}\boxed{?} \end{matrix} \middle| \begin{matrix} \boxed{?}\boxed{?} \\ \boxed{?}\boxed{?} \end{matrix} \middle| \begin{matrix} \boxed{?}\boxed{?} \\ \boxed{?}\boxed{?} \end{matrix} \middle| \begin{matrix} \boxed{?}\boxed{?} \\ \boxed{?}\boxed{?} \end{matrix} \right|.$$

3   Old Maid

In this section, we explain the rule of Old Maid, and present a framework for playing Old Maid with virtual players.

3.1   Rule of Old Maid

First, we explain the rule of Old Maid. It is played with a standard deck of 54 playing cards. In the game, the second joker $\boxed{\overrightarrow{\text{Jo}}}$ is not used[1] and the number of required cards is 53. Suppose that there are m players $P_1, P_2, \ldots, P_m$. Hereinafter, a *pair* in a hand means a pair of cards having the same number in the hand. The game proceeds as follows:

1. **Dealing phase:** All 53 cards are randomly dealt to the players; each player receives $\lfloor 53/m \rfloor$ or $\lceil 53/m \rceil$ cards.
2. **Initial phase:** Each player removes all pairs from his/her hands. (If some player has no cards at this point, he/she wins and is out of the game).
3. **Playing phase:** From the first player P_1, each player P_i draws one card from the next player's hand in turn. If the drawn card makes a pair in the P_i's hand, P_i removes the pair. If P_i has no cards at this point, P_i wins and is out of the game.
4. The player who has the joker at the end loses.

3.2   Playing Old Maid with Virtual Players

In this subsection, we explain how to play Old Maid with virtual players, while the main protocol called a *removal protocol* will be introduced in the succeeding section.

Suppose that there are m_r real players and m_v virtual players. A typical case is $(m_r, m_v) = (2, 1)$: Two real players Alice and Bob want to play Old Maid with a virtual player Carol because two-player Old Maid is somewhat boring. Another typical case is $m_r = 1$ and $m_v \geq 1$: A real player Alice wants to play Old Maid in solitary. An extreme case is $m_r = 0$ and $m_v \geq 2$: This is a simulation of Old Maid with no real players.

Old Maid with virtual players proceeds as follows:

1. **Dealing phase:** All 53 cards are randomly dealt to the players; each player receives $\lfloor 53/(m_r + m_v) \rfloor$ or $\lceil 53/(m_r + m_v) \rceil$ cards.
2. **Initial phase:** Each real player removes all pairs from his/her hand. For each virtual player, execute a removal protocol, which removes all pairs from the virtual player's hand. (If some player has no cards at this point, he/she wins and is out of the game.)

---

[1] However, $\boxed{\overrightarrow{\text{Jo}}}$ will be used in our removal protocol presented in Sect. 4.

3. **Playing phase:** Each player P_i draws a card from the next player's hand in turn. If the drawn card makes a pair in the P_i's hand, P_i removes the pair. When P_i is a virtual player, choose a random card from the next player's hand using a shuffle and then execute a removal protocol. If P_i has no cards at this point, P_i wins and is out of the game.
4. The player who has the joker at the end loses.

4   Removal Protocol

In this section, we construct a removal protocol, which is the main protocol for playing Old Maid with virtual players. In Sect. 4.1, we present the description of our protocol. In Sects. 4.2 and 4.3, we give the efficiency evaluation and security proof of our protocol, respectively.

4.1   Protocol Description

Suppose that there are m players: P_1 is a virtual player whose hand will be examined to remove pairs by the protocol and $P_2, P_3, \ldots, P_m$ are real or virtual players. We call a numbered card $\boxed{i}$ an *i-card* hereinafter. Our removal protocol proceeds as follows.

1. Place all players' hands in a horizontal line as follows:

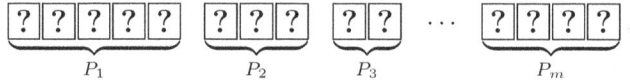

2. Place i-cards on the bottom of the P_i's cards ($1 \le i \le m$) as follows:

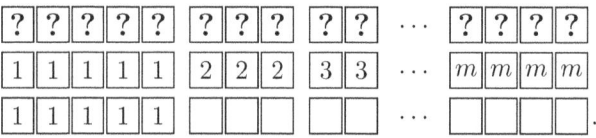

3. Place 1-cards on the bottom of the P_1's cards and dummy cards ($\boxed{}$) on the bottom of the P_i's cards ($2 \le i \le m$) as follows:

4. Place three cards $\boxed{\text{Jo}}\boxed{}\boxed{}$ next to the rightmost column as follows:

5. Turn over all face-up cards as follows:

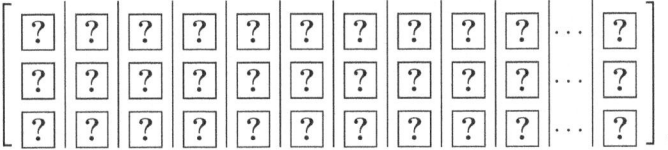

6. Repeat the following procedure for $(\alpha, \beta, \gamma, \delta) = (\spadesuit, \heartsuit, \diamondsuit, \clubsuit)$, $(\spadesuit, \diamondsuit, \heartsuit, \clubsuit)$, and $(\spadesuit, \clubsuit, \heartsuit, \diamondsuit)$.

(a) Apply a pile-scramble shuffle with each column as a pile as follows:

(b) Turn over the cards in the first row as follows:

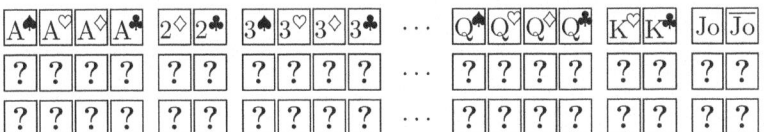

(c) Rearrange the columns so that columns with the same numbers are consecutive in the order $\alpha \rightarrow \beta \rightarrow \gamma \rightarrow \delta$. An example case for $(\alpha, \beta, \gamma, \delta) = (\spadesuit, \heartsuit, \diamondsuit, \clubsuit)$ is given as follows:

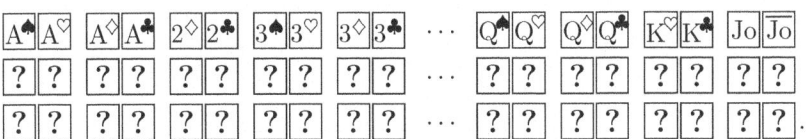

In this example, each in $\{A, 3, \ldots, Q\}$ has four cards and each in $\{2, \ldots, K\}$ has two cards. We call the former a *two-pair number* and the latter a *one-pair number*. For each two-pair number, make two piles with (α, β) and (γ, δ) as follows:

(d) Turn over all face-up cards as follows:

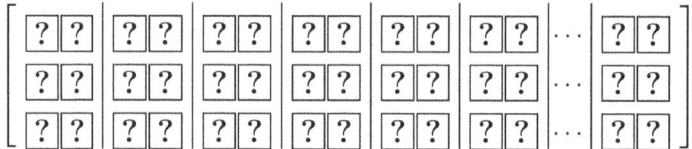

(e) Apply a pile-scramble shuffle with each pair of columns as a pile as follows:

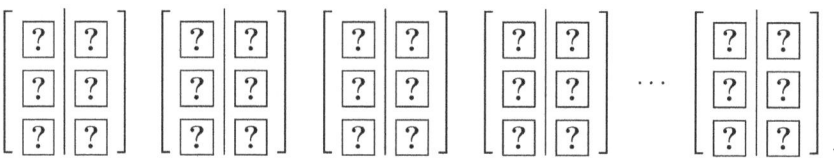

(f) Apply a random bisection cut for each pair of columns as follows:

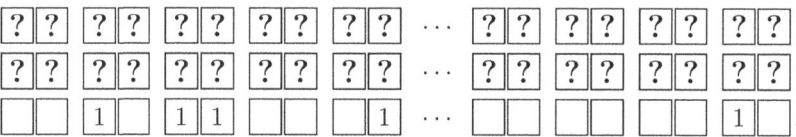

(g) Turn over the cards in the bottom row. Suppose that there are piles having 1 1 (i.e., there are pairs in P_1's hand) as follows:

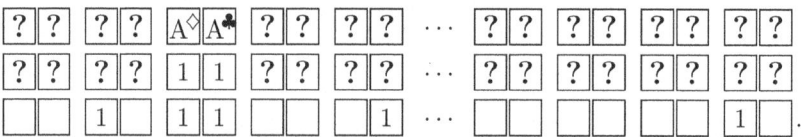

Then, open the cards in these piles as follows:

Then, the top line of these piles are removed and the other cards (i.e., the 1-cards) on these piles are returned to free cards. If there are no piles having 1 1, do nothing.

(h) If $(\alpha, \beta, \gamma, \delta) = (\spadesuit, \heartsuit, \diamondsuit, \clubsuit)$ or $(\spadesuit, \diamondsuit, \heartsuit, \clubsuit)$, turn over all face-up cards as follows:

Otherwise, return the bottom row of cards to free cards as follows:

$$\boxed{?\,?}\ \boxed{?\,?}\ \boxed{?\,?}\ \boxed{?\,?}\ \cdots\ \boxed{?\,?}\ \boxed{?\,?}\ \boxed{?\,?}\ \boxed{?\,?}$$
$$\boxed{?\,?}\ \boxed{?\,?}\ \boxed{?\,?}\ \boxed{?\,?}\ \cdots\ \boxed{?\,?}\ \boxed{?\,?}\ \boxed{?\,?}\ \boxed{?\,?}.$$

7. Apply a pile-scramble shuffle with each column as a pile as follows:

$$\left[\ \boxed{?}\ \boxed{?}\ \boxed{?}\ \boxed{?}\ \boxed{?}\ \boxed{?}\ \boxed{?}\ \boxed{?}\ \boxed{?}\ \boxed{?}\ \cdots\ \boxed{?}\ \right].$$

8. Turn over the cards in the bottom row as follows:

$$\boxed{?}\,\boxed{?}\,\boxed{?}\,\boxed{?}\,\boxed{?}\,\boxed{?}\,\boxed{?}\,\boxed{?}\,\boxed{?}\,\boxed{?}\,\boxed{?}\,\boxed{?}\,\boxed{?}\,\boxed{?}\,\boxed{?}\ \cdots\ \boxed{?}$$
$$\boxed{4}\,\boxed{6}\,\boxed{3}\,\boxed{2}\,\boxed{m}\,\boxed{2}\,\boxed{1}\,\boxed{5}\,\boxed{1}\,\boxed{}\,\boxed{4}\,\boxed{1}\,\boxed{3}\,\boxed{m}\ \cdots\ \boxed{2}.$$

Then, turn over the card above the dummy card as follows:

$$\boxed{?}\,\boxed{?}\,\boxed{?}\,\boxed{?}\,\boxed{?}\,\boxed{?}\,\boxed{?}\,\boxed{?}\,\boxed{?}\,\boxed{?}\,\boxed{\text{Jo}}\,\boxed{?}\,\boxed{?}\,\boxed{?}\,\boxed{?}\ \cdots\ \boxed{?}$$
$$\boxed{4}\,\boxed{6}\,\boxed{3}\,\boxed{2}\,\boxed{m}\,\boxed{2}\,\boxed{1}\,\boxed{5}\,\boxed{1}\,\boxed{}\,\boxed{4}\,\boxed{1}\,\boxed{3}\,\boxed{m}\ \cdots\ \boxed{2}.$$

Then, the joker and the dummy card are returned to free cards.

9. Rearrange the columns so that columns with the same numbered cards are next to each other as follows:

$$\boxed{?}\,\boxed{?}\,\boxed{?}\ \boxed{?}\,\boxed{?}\,\boxed{?}\ \boxed{?}\,\boxed{?}\ \cdots\ \boxed{?}\,\boxed{?}\,\boxed{?}\,\boxed{?}$$
$$\boxed{1}\,\boxed{1}\,\boxed{1}\ \boxed{2}\,\boxed{2}\,\boxed{2}\ \boxed{3}\,\boxed{3}\ \cdots\ \boxed{m}\,\boxed{m}\,\boxed{m}\,\boxed{m}.$$

Then, the cards above the i-cards are returned to the P_i's hand, and the cards on the bottom line are returned to free cards.

This is our removal protocol.

If the virtual player P_1 has pairs with suits $(\spadesuit, \heartsuit)$ or $(\diamondsuit, \clubsuit)$ at the beginning of the protocol, these pairs will be removed at Step 6g for the loop $(\alpha, \beta, \gamma, \delta) = (\spadesuit, \heartsuit, \diamondsuit, \clubsuit)$ because 1-cards were added to the bottom of the P_1's cards at Step 3. Similarly, if P_1 has pairs with suits $(\spadesuit, \diamondsuit)$ or $(\heartsuit, \clubsuit)$ (resp. $(\spadesuit, \clubsuit)$ or $(\heartsuit, \diamondsuit)$) at the beginning of the protocol, these pairs will be removed at Step 6g for the loop $(\alpha, \beta, \gamma, \delta) = (\spadesuit, \diamondsuit, \heartsuit, \clubsuit)$ (resp. $(\spadesuit, \clubsuit, \heartsuit, \diamondsuit)$). Therefore, all pairs in P_1's hand will be removed at the end of the protocol.

4.2 Efficiency

Let n_i $(1 \le i \le m)$ be the number of cards in P_i's hand. Let $n := \sum_{i=1}^{m} n_i$ be the total number of playing cards. In addition to the standard deck of 54 playing cards, we require $2n + 2$ helping cards, which consist of the following cards:

$$\underbrace{\boxed{1}\cdots\boxed{1}}_{2n_1}\ \underbrace{\boxed{2}\cdots\boxed{2}}_{n_2}\ \cdots\ \underbrace{\boxed{m}\cdots\boxed{m}}_{n_m}\ \underbrace{\boxed{}\cdots\boxed{}}_{n-n_1+2}.$$

Since $n \leq 53$, our removal protocol requires at most 108 helping cards.

For the number of shuffles, we require $3(\frac{n+1}{2} + 2) + 1$ shuffles, which consist of $\frac{3(n+1)}{2}$ random bisection cuts and 7 pile-scramble shuffles.

4.3 Security

Our protocol follows the computational model of card-based cryptography [15], in which the security of a card-based protocol is proven based on information theory. Thus, to prove the security of our protocol, it suffices to show that revealed cards during execution of the protocol do not leak any information beyond public information. In our proposed protocol, Steps 6b, 6g, and 8 reveal cards.

In Step 6b, the cards in the first row, i.e., all players' hands are revealed. Note that in Step 6a, a pile-scramble shuffle is applied to the first row, randomizing the order of the cards. Therefore, Step 6b reveals no additional information because all the removed cards are public information in a game of Old Maid.

In Step 6g, the cards in the bottom row are revealed. Remember that a revealed 1-card $\boxed{1}$ means that the top card in the same column comes from P_1; a dummy card $\boxed{}$ means that the top card in the same column comes from a player other than P_1. If a pair $\boxed{1}\boxed{1}$ appears, it is removed, and hence, the number of $\boxed{1}\boxed{1}$ is public information. Since the size of P_1's hand is public, the number of $\boxed{1}\boxed{1}$ immediately tells us the number of $\boxed{}\boxed{}$ as well as the number of either $\boxed{1}\boxed{}$ or $\boxed{}\boxed{1}$. Because Step 6e applies a pile-scramble shuffle with each pair of columns, the position of each pair is independent of the players' hands. Moreover, either $\boxed{1}\boxed{}$ or $\boxed{}\boxed{1}$ occurs with a probability of exactly $1/2$ because of the application of a random bisection cut in Step 6f. Therefore, revealed cards do not leak any information beyond public information.

Finally, in Step 8, all the cards in the bottom row are revealed. Remember that an i-card means that the card above it comes from P_i. Because Step 7 applies a pile-scramble shuffle, the positions of each i-card and the dummy are independent of the players' hands.

5   Conclusion

In this paper, we showed that Alice and Bob can enjoy playing Old Maid with a virtual player Carol. Technically, we designed a removal protocol, which is a card-based protocol for removing any pairs in Carol's hand without revealing any information about Carol's hand.

Although card-based cryptography has sometimes been inspired by the card game techniques, the reverse direction (i.e., applying card-based cryptography to card games) had rarely been studied. In this paper, we posed a novel problem of simulating virtual players in a card game using card-based cryptographic techniques. We name this problem a *player simulation problem* for card games. We believe that the player simulation problem for card games is a new research area of fun with cryptography and it will attract the attention of many researchers as well as many non-specialists, like physical zero-knowledge proofs.

Acknowledgements. We would like to thank Tatsuya Sasaki for his cooperation in preparing a Japanese draft version at an earlier stage of this work. This work was supported in part by JSPS KAKENHI Grant Numbers JP21K17702, JP23H00479, and JP24K02938, and JST CREST Grant Number JPMJCR22M1.

References

1. Crépeau, C., Kilian, J.: Discreet solitary games. In: Stinson, D.R. (ed.) CRYPTO 1993. LNCS, vol. 773, pp. 319–330. Springer, Heidelberg (1994). https://doi.org/10.1007/3-540-48329-2_27
2. Boer, B.: More efficient match-making and satisfiability *the five card trick*. In: Quisquater, J.-J., Vandewalle, J. (eds.) EUROCRYPT 1989. LNCS, vol. 434, pp. 208–217. Springer, Heidelberg (1990). https://doi.org/10.1007/3-540-46885-4_23
3. Gradwohl, R., Naor, M., Pinkas, B., Rothblum, G.N.: Cryptographic and physical zero-knowledge proof systems for solutions of sudoku puzzles. In: Crescenzi, P., Prencipe, G., Pucci, G. (eds.) FUN 2007. LNCS, vol. 4475, pp. 166–182. Springer, Heidelberg (2007). https://doi.org/10.1007/978-3-540-72914-3_16
4. Haga, R., Hayashi, Y., Miyahara, D., Mizuki, T.: Card-minimal protocols for three-input functions with standard playing cards. In: Batina, L., Daemen, J. (eds.) AFRICACRYPT 2022. LNCS, vol. 13503, pp. 448–468. Springer, Cham (2022). https://doi.org/10.1007/978-3-031-17433-9_19
5. Hand, S., Koch, A., Lafourcade, P., Miyahara, D., Robert, L.: Check alternating patterns: a physical zero-knowledge proof for Moon-or-Sun. In: Shikata, J., Kuzuno, H. (eds.) Advances in Information and Computer Security. LNCS, vol. 14128, pp. 255–272. Springer, Cham (2023). https://doi.org/10.1007/978-3-031-41326-1_14
6. Hashimoto, Y., Nuida, K., Shinagawa, K., Inamura, M., Hanaoka, G.: Toward finite-runtime card-based protocol for generating a hidden random permutation without fixed points. IEICE Trans. Fundam. **E101.A**(9), 1503–1511 (2018). https://doi.org/10.1587/transfun.E101.A.1503
7. Hashimoto, Y., Shinagawa, K., Nuida, K., Inamura, M., Hanaoka, G.: Secure grouping protocol using a deck of cards. IEICE Trans. Fundam. **E101.A**(9), 1512–1524 (2018). https://doi.org/10.1587/transfun.E101.A.1512
8. Hatsugai, K., Asano, K., Abe, Y.: A physical zero-knowledge proof for Sumplete, a puzzle generated by ChatGPT. In: Wu, W., Tong, G. (eds.) Computing and Combinatorics. LNCS, vol. 14422, pp. 398–410. Springer, Cham (2024). https://doi.org/10.1007/978-3-031-49190-0_29
9. Ibaraki, T., Manabe, Y.: A more efficient card-based protocol for generating a random permutation without fixed points. In: Mathematics and Computers in Sciences and in Industry (MCSI), pp. 252–257 (2016). https://doi.org/10.1109/MCSI.2016.054
10. Ishikawa, R., Chida, E., Mizuki, T.: Efficient card-based protocols for generating a hidden random permutation without fixed points. In: Calude, C.S., Dinneen, M.J. (eds.) UCNC 2015. LNCS, vol. 9252, pp. 215–226. Springer, Cham (2015). https://doi.org/10.1007/978-3-319-21819-9_16
11. Koch, A., Schrempp, M., Kirsten, M.: Card-based cryptography meets formal verification. In: Galbraith, S.D., Moriai, S. (eds.) ASIACRYPT 2019. LNCS, vol.

11921, pp. 488–517. Springer, Cham (2019). https://doi.org/10.1007/978-3-030-34578-5_18

12. Koyama, H., Miyahara, D., Mizuki, T., Sone, H.: A secure three-input AND protocol with a standard deck of minimal cards. In: Santhanam, R., Musatov, D. (eds.) CSR 2021. LNCS, vol. 12730, pp. 242–256. Springer, Cham (2021). https://doi.org/10.1007/978-3-030-79416-3_14

13. Manabe, Y., Ono, H.: Card-based cryptographic protocols with a standard deck of cards using private operations. In: Cerone, A., Ölveczky, P.C. (eds.) ICTAC 2021. LNCS, vol. 12819, pp. 256–274. Springer, Cham (2021). https://doi.org/10.1007/978-3-030-85315-0_15

14. Mizuki, T.: Efficient and secure multiparty computations using a standard deck of playing cards. In: Foresti, S., Persiano, G. (eds.) CANS 2016. LNCS, vol. 10052, pp. 484–499. Springer, Cham (2016). https://doi.org/10.1007/978-3-319-48965-0_29

15. Mizuki, T., Shizuya, H.: A formalization of card-based cryptographic protocols via abstract machine. Int. J. Inf. Secur. **13**(1), 15–23 (2014). https://doi.org/10.1007/s10207-013-0219-4

16. Mizuki, T., Sone, H.: Six-card secure AND and four-card secure XOR. In: Deng, X., Hopcroft, J.E., Xue, J. (eds.) FAW 2009. LNCS, vol. 5598, pp. 358–369. Springer, Heidelberg (2009). https://doi.org/10.1007/978-3-642-02270-8_36

17. Murata, S., Miyahara, D., Mizuki, T., Sone, H.: Efficient generation of a card-based uniformly distributed random derangement. In: Uehara, R., Hong, S.-H., Nandy, S.C. (eds.) WALCOM 2021. LNCS, vol. 12635, pp. 78–89. Springer, Cham (2021). https://doi.org/10.1007/978-3-030-68211-8_7

18. Niemi, V., Renvall, A.: Solitaire zero-knowledge. Fundam. Inf. **38**(1,2), 181–188 (1999). https://doi.org/10.3233/FI-1999-381214

19. Robert, L., Miyahara, D., Lafourcade, P., Mizuki, T.: Physical ZKP protocols for Nurimisaki and Kurodoko. Theor. Comput. Sci. **972**, 114071 (2023). https://doi.org/10.1016/j.tcs.2023.114071

20. Ruangwises, S.: Two standard decks of playing cards are sufficient for a ZKP for Sudoku. New Gener. Comput. **40**, 49–65 (2022). https://doi.org/10.1007/s00354-021-00146-y

21. Ruangwises, S.: Physical zero-knowledge proof for ball sort puzzle. In: Della Vedova, G., Dundua, B., Lempp, S., Manea, F. (eds.) CiE 2023. LNCS, vol. 13967, pp. 246–257. Springer, Cham (2023). https://doi.org/10.1007/978-3-031-36978-0_20

22. Ruangwises, S.: Physical zero-knowledge proofs for Five Cells. In: Aly, A., Tibouchi, M. (eds.) LATINCRYPT 2023. LNCS, vol. 14168, pp. 315–330. Springer, Cham (2023). https://doi.org/10.1007/978-3-031-44469-2_16

23. Ruangwises, S.: Physically verifying the first nonzero term in a sequence: physical ZKPs for ABC end view and Goishi Hiroi. In: Li, M., Sun, X., Wu, X. (eds.) IJTCS-FAW 2023. LNCS, vol. 13933, pp. 171–183. Springer, Cham (2023). https://doi.org/10.1007/978-3-031-39344-0_13

24. Sasaki, T., Miyahara, D., Mizuki, T., Sone, H.: Efficient card-based zero-knowledge proof for Sudoku. Theor. Comput. Sci. **839**, 135–142 (2020). https://doi.org/10.1016/j.tcs.2020.05.036

25. Shinoda, Y., Miyahara, D., Shinagawa, K., Mizuki, T., Sone, H.: Card-based covert lottery. In: Maimut, D., Oprina, A.-G., Sauveron, D. (eds.) SecITC 2020. LNCS, vol. 12596, pp. 257–270. Springer, Cham (2021). https://doi.org/10.1007/978-3-030-69255-1_17

26. Tanaka, K., Mizuki, T.: Two UNO decks efficiently perform zero-knowledge proof for Sudoku. In: Fernau, H., Jansen, K. (eds.) FCT 2023. LNCS, vol. 14292, pp. 406–420. Springer, Cham (2023). https://doi.org/10.1007/978-3-031-43587-4_29

Algorithms for Optimally Shifting Intervals Under Intersection Graph Models

Nicolás Honorato-Droguett[1]([⊠]), Kazuhiro Kurita[1][iD], Tesshu Hanaka[2][iD], and Hirotaka Ono[1][iD]

[1] Nagoya University, Nagoya, Japan
honorato.droguett.nicolas.n7@s.mail.nagoya-u.ac.jp,
kurita@i.nagoya-u.ac.jp, ono@nagoya-u.ac.jp
[2] Kyushu University, Fukuoka, Japan
hanaka@inf.kyushu-u.ac.jp

Abstract. We propose a new model for graph editing problems on intersection graphs. In well-studied graph editing problems, adding and deleting vertices and edges are used as graph editing operations. As a graph editing operation on intersection graphs, we propose moving objects corresponding to vertices. In this paper, we focus on interval graphs as an intersection graph. We give a linear-time algorithm to find the total moving distance for transforming an interval graph into a complete graph. The concept of this algorithm can be applied for (i) transforming a unit square graph into a complete graph over L_1 distance and (ii) attaining the existence of a k-clique on unit interval graphs. In addition, we provide LP-formulations to achieve several properties in the associated graph of unit intervals.

Keywords: Intersection graphs · Optimisation · Graph edit distance

1 Introduction

The *graph editing problem* is a problem of editing a graph G so that G satisfies a graph property Π. Common graph editing operations in well-studied graph editing problems are vertex and edge deletion/insertion. The minimum number of such operations required for G to satisfy Π is called *graph edit distance (GED)* [14]. When Π contains only the target graph, our problem is equivalent to the problem addressed in [14]. GED is a metric used to measure the dissimilarity between two given graphs, defined by Sanfeliu and Fu [15]. The problem of determining GED can be defined as determining the edit distance between two graphs G_1 and G_2, defined as the minimum number of edit operations to

This work is partially supported by JSPS KAKENHI Grant Numbers JP20H05967, JP21K19765, JP21K17707, JP21K17812, JP22H00513, JP22H03549, and JP23H04388, JST CREST Grant Number JPMJCR18K3, and JST ACT-X Grant Number JPM-JAX2105.

B. Li et al. (Eds.): IJTCS-FAW 2024, LNCS 14752, pp. 66–78, 2025.
https://doi.org/10.1007/978-981-97-7752-5_5

transform G_1 into G_2. The edit operations of this problem are the deletion and insertion of an isolated vertex and an edge. The problem of determining the GED is known to be NP-hard [17]. However, solutions to this problem continue to be designed due to the growing interest in GED as a pairwise graph similarity measure in a wide range of fields such as pattern recognition, image analysis, and molecular analysis [9,10,16]. For instance, in a face recognition scenario, if a facial pattern can be modelled as a graph, and an image is also represented as a graph, then GED can be employed on these graphs to quantify the similarity of the image and the facial model [16].

In this paper, we propose a new graph edit operation for the graph edit problem on intersection graphs. Given a collection of geometric objects, an intersection graph $G = (V, E)$ is a graph such that V is a collection of geometric objects represented by vertices and E has an edge if two geometric objects intersect. In this graph model, one natural graph edit operation is moving a geometric object. We consider this moving operation as a graph editing operation and address the problem of measuring the similarity between a given graph and a specific graph property. The similarity is measured by calculating the total moving distance, which is the sum of the distance values applied to each object in the collection. We formulate the problem as follows.

Problem 1. Given a collection of objects and a graph property Π, minimise the total moving distance of the objects so that the resulting intersection graph satisfies Π.

In the above context, we primarily focus on (unit) intervals. This intersection graph is called *(unit) interval graph*. We identify an interval graph with a collection of intervals. We formally define these graphs in Sect. 2 and summarise our results in Table 1.

Table 1. Summary of our results. In this table, IG, UIG, and USG (L_1) are abbreviations of interval graphs, unit interval graphs, and unit square graphs for L_1 distance, respectively.

Object	Property of target graph	Running time
IG	complete	$O(n)$
UIG	edgeless	Poly. (LP)
	acyclic	Poly. (LP)
	has a k-clique	$O(n \log n)$
	does not have a k-clique	Poly. (LP)
	k-connected	Poly. (LP)
USG (L_1)	complete	$O(n)$

Problem 1 is similar to the System of q-Distant Representatives (Sq-DR), introduced by Fiala et al. [7] as an extension of the system of distinct representatives [11]. Given a universe X equipped with a metric denoted as dist, a parameter $q > 0$ and a family $\mathcal{M} = \{M_i \subseteq X, i \in I\}$ where I is the index set, a mapping $f : I \to X$ is called a *System of q-Distant Representatives* if $f(i)$ is contained in M_i for each $i \in I$ and $\text{dist}(f(i), f(j)) \geq q$ for every $i, j \in I$, $i \neq j$. Deciding if a family $\mathcal{M}$ has a System of q-Distant Representatives is NP-complete when $X = \mathbb{R}^2$, each M_i is a unit disk, dist is the Euclidean metric L_2, and $q = 1$ [7]. However, it becomes tractable if $X = \mathbb{R}$ and the subsets are restricted to closed intervals since it reduces the scheduling problem. As for the similarity between Sq-DR and Problem 1, Sq-DR includes the following problem. Let $\{C_1, \ldots, C_n\}$ be a collection of disks with the radius $q/2$. By moving each C_i by a distance of at most d over Euclidean distance, all disks are moved so they do not overlap. When $X = \mathbb{R}^2$, each C_i corresponds to M_i and moving each C_i corresponds to a mapping f, Sq-DR is equivalent to this problem. In other words, the problems that minimise the maximum moving distance are contained in the System of q-Distant Representatives as special subclasses. On the other hand, our problem minimises the sum of moving distance.

As another type of problem that minimises the maximum moving distance, Fomin et al. [8] addressed the problem of *disk dispersal*, where a family $\mathcal{S}$ of n disks, an integer $k \geq 0$ and a real $d \geq 0$ are given, and the goal is to decide whether it is possible to obtain a disjoint arrangement by moving at most k disks by at most d distance each. According to the results of Fiala et al. [7], this problem is NP-hard when $d = 2$ and $k = n$. On the other hand, it was shown by Fomin et al. [8] that it is FPT when parameterised by $k + d$ and W[1]-hard when parameterised by k when the movement of disks is restricted to rectilinear directions. While some of the cases presented in these works remain open, the system of distant representatives is continuously addressed in the literature [1,5].

This paper is organised as follows. Section 2 formally describes the definitions needed to address the above ideas. In Sect. 3, we show that Problem 1 is solvable in (i) $O(n)$ time for satisfying the property Π_{comp} given an interval graph over Euclidean distance and (ii) $O(n)$ time for unit square graphs for satisfying Π_{comp} over L_1 distance. We use the standard word-RAM model. In Sect. 4, we present an $O(n \log n)$-time algorithm for satisfying $\Pi_{k\text{-clique}}$ given a unit interval graph over Euclidean distance. In Sect. 5, we give LP-formulations for satisfying several fundamental graph properties. Lastly, we present our concluding remarks in Sect. 6.

Due to space limitations, we omit all proofs.

2   Preliminaries

In this section, we give general definitions that are used throughout the paper. We refer to the basic terminology of graphs, linear programming, and convex functions described in the textbooks [2–4].

An *interval* I is a closed line segment of length $\text{len}(I) \in \mathbb{R}^+$ on the real line. An interval such that $\text{len}(I) = 1$ is called a *unit interval*. The *left endpoint* $\ell(I)$ of an interval I is the point that satisfies $\ell(I) \leq y$ for any $y \in I$. Similarly, the *right endpoint* $r(I)$ of I is the point that satisfies $y \leq r(I)$ for any $y \in I$. The *centre* of I is the point $(r(I) - \ell(I))/2$ and we denote it as $c(I)$.

The *left interval set* of a collection of intervals $\mathcal{I}$ is the subcollection of intervals to the "left" of a given point x. That is, $L(\mathcal{I}, x) = \{I \in \mathcal{I} : r(I) < x\}$. Similarly, the *right interval set* is defined as $R(\mathcal{I}, x) = \{I \in \mathcal{I} : \ell(I) > x\}$. Moreover, the *leftmost endpoint* $\ell(\mathcal{I})$ of a collection of intervals $\mathcal{I}$ is defined as $\min_{I \in \mathcal{I}} \ell(I)$. Similarly, the *rightmost endpoint* $r(\mathcal{I})$ of $\mathcal{I}$ is defined as $\max_{I \in \mathcal{I}} r(I)$. The *set of endpoints* $\mathcal{E}(\mathcal{I})$ of $\mathcal{I}$ is defined as $\mathcal{E}(\mathcal{I}) = \bigcup_{I \in \mathcal{I}} (\ell(I) \cup r(I))$. The subcollection $\mathcal{I}(i, k)$ is the subcollection $\{I_i, I_{i+1}, \ldots, I_{i+k-1}\}$. The endpoint $x_k^{\mathcal{I}}$ is the k-th endpoint of $\mathcal{E}(\mathcal{I})$. For an arbitrary point x, we define two subsets of intervals $\mathcal{E}_\ell(\mathcal{I}, x) = \{I \in \mathcal{I} : \ell(I) = x\}$ and $\mathcal{E}_r(\mathcal{I}, x) = \{I \in \mathcal{I} : r(I) = x\}$ to represent the subcollection of intervals of $\mathcal{I}$ having x as the left and right endpoint, respectively.

A *unit square* S is a regular quadrilateral where each side has length 1, and its centre is positioned at $(s_x, s_y) \in \mathbb{R}^2$.

An *edgeless graph* is a graph $G = (V, E)$ such that $E = \emptyset$. A *complete graph* is a graph $G = (V, E)$ such that $E = \binom{V}{2}$. A *k-clique* of a graph $G = (V, E)$ is a subset $W \subseteq V$ such that $|W| = k$ and for all $u, v \in W$, $u \neq v$, $(u, v) \in E$, for $k \leq n$. If such W exists in V, we say that G *contains a k-clique*. An *interval graph* is an intersection graph $G = (V, E)$ where the vertex set $V = \{v_1, \ldots, v_n\}$ corresponds to a collection of intervals $\mathcal{I} = \{I_1, \ldots, I_n\}$. There is an edge $(v_i, v_j) \in E$ if and only if $I_i \cap I_j \neq \emptyset$, for any $1 \leq i, j \leq n$, $i \neq j$. An interval graph where for all $I \in \mathcal{I}$, $\text{len}(I) = 1$, is called *unit interval graph*. A *unit square graph* is an intersection graph $G = (V, E)$ where the vertex set $V = \{v_1, \ldots, v_n\}$ corresponds to an unit square collection $\mathcal{S} = \{S_1, \ldots, S_n\}$. There is an edge $(v_i, v_j) \in E$ if and only if $S_i \cap S_j \neq \emptyset$, for any $1 \leq i, j \leq n$, $i \neq j$.

An (infinite) set of graphs Π is a *graph property* (or simply a *property*), and we say that G *satisfies* Π if $G \in \Pi$. In this paper, we deal with the following properties. (i) $\Pi_{\text{comp}} = \{G : G \text{ is a complete graph.}\}$, (ii) $\Pi_{\text{edgeless}} = \{G : G \text{ is an edgeless graph.}\}$, (iii) $\Pi_{\text{acyc}} = \{G : G \text{ is an acyclic graph.}\}$, (iv) $\Pi_{k\text{-clique}} = \{G : G \text{ contains a k-clique.}\}$, (v) $\overline{\Pi}_{k\text{-clique}} = \{G : G \notin \Pi_{k\text{-clique}}\}$, and (vi) $\Pi_{k\text{-conn}} = \{G : G \text{ is a k-connected graph.}\}$.

A *linear function* f for a given set of real numbers $a_1, a_2, \ldots, a_n$ and a set of variables $x_1, x_2, \ldots, x_n$ as $f(x_1, \ldots, x_n) = \sum_{j=1}^{n} a_j x_j$. Similarly, a *linear inequality* are equations of the form $f(x_1, \ldots, x_n) \leq b$ and $f(x_1, \ldots, x_n) \geq b$. We use linear inequalities to denote *linear constraints*. A *linear-programming problem* is the problem of either minimising or maximising a linear function subject to a finite set of linear constraints. We refer in this paper to a *linear program* as an instance of a linear programming problem described as:

$$\begin{aligned} \text{minimise} \quad & f(x_1, x_2, \ldots, x_n) \\ \text{subject to} \quad & \mathcal{C} \end{aligned} \tag{1}$$

where $\mathcal{C}$ is a finite collection of linear constraints. We avoid expressly showing linear programs in their standard form to maintain legibility in the present work. The linear function $f(x_1, x_2, \ldots, x_n)$ in (1) is called the *objective function*. An

n-vector $\boldsymbol{x}$ representing an input of f is called a *feasible solution* if it satisfies all constraints contained in $\mathcal{C}$. A feasible solution $\boldsymbol{x}$ has an *objective value* $f(\boldsymbol{x})$. If the objective value $f(\boldsymbol{x})$ is minimum (maximum) among all feasible solutions, then $\boldsymbol{x}$ is an *optimal solution* of the linear program.

A set C is *convex* if the line segment between two arbitrary points in C lies entirely in C. Such a set is called a *convex set*. A function $f : \mathbb{R}^n \to \mathbb{R}$ is *convex* if its domain $\mathrm{dom}f$ is a convex set and if for all $x, y \in \mathrm{dom}f$ and θ such that $0 \leq \theta \leq 1$, the inequality $f(\theta x + (1-\theta)y) \leq \theta f(x) + (1-\theta)f(y)$ holds. A function satisfying the above is called a *convex function*. The *graph* of a function f is the set $\{(x, f(x)) \,|\, x \in \mathrm{dom}f\}$. We mainly refer to the geometric interpretation of the above inequality, which says that if f is convex, then an arbitrary line segment from point $(x, f(x))$ to point $(y, f(y))$ lies above the graph of f. The *epigraph* of a function $f : \mathbb{R}^n \to \mathbb{R}$ is the set $\mathrm{epi}f = \{(x, t) \,|\, x \in \mathrm{dom}f, \ f(x) \leq t\}$. A function is convex if and only if its epigraph is a convex set.

3 Linear-Time Algorithms for Satisfying Π_{comp}

This section presents two linear-time algorithms for satisfying Π_{comp} on interval graphs and unit square graphs over Euclidean and L_1 distances, respectively. We remark that our algorithm runs even if the given collections of intervals and unit squares are not sorted. Algorithms in this section give the movement that has to be applied to each interval for satisfying Π_{comp} while minimising the total moving distance. First, we give a linear-time algorithm for interval graphs.

3.1 Interval Graphs

Observe that if an interval graph is complete, then all intervals share at least one point. We call such a point as a *gathering point*. Furthermore, if the value of a gathering point x is known and the intervals are scattered over the line, then intervals in $L(\mathcal{I}, x)$ have to be moved to the right, and intervals $R(\mathcal{I}, x)$ have to be moved to the left. We try to find an optimal x, which is called an *optimal gathering point*. We show that $\mathcal{E}(\mathcal{I})$ contains at least one optimal gathering point. We define a function to calculate the moving distance for an arbitrary interval. In particular, the moving distance function is defined as:

Definition 1. *The* moving distance *of an interval* $I \in \mathcal{I}$ *to a point* x *is a function* $d_I : \mathbb{R} \to \mathbb{R}$ *defined as:*

$$
d_I(x) = \begin{cases} m_I(x - c(I) - \frac{\mathrm{len}(I)}{2}), & x - \frac{\mathrm{len}(I)}{2} - c(I) > 0 \\ m_I(c(I) - x - \frac{\mathrm{len}(I)}{2}), & x + \frac{\mathrm{len}(I)}{2} - c(I) < 0 \\ 0, & otherwise, \end{cases} \tag{2}
$$

where $m_I \in \mathbb{R}^+$ *is an arbitrary constant describing the slope on both sides. The* total moving distance *of a collection of intervals is then defined as:* $D(\mathcal{I}, x) = \sum_{I \in \mathcal{I}} d_I(x)$.

Figure 1 illustrates the curve of an arbitrary moving distance function. In simple words, if interval I is moved from the left to x, its moving distance is $m_I(x - c(I) - \text{len}(I)/2)$; if it is moved to x from the right, its moving distance is the value given by $m_I(c(I) - x - \text{len}(I)/2)$. In all other cases, the interval intersects x, and hence its moving distance is 0. For a given interval graph, we say that the *moving distance function is uniform* when $m_I = m_J$ for any $I, J \in \mathcal{I}$.

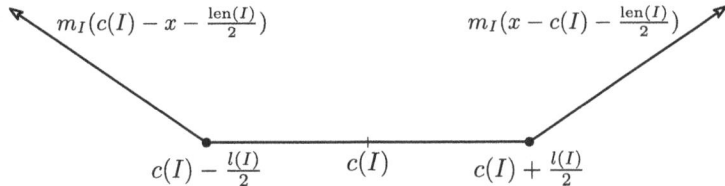

Fig. 1. Curve of the moving distance function of an arbitrary interval I with centre $c(I)$.

We state the following property of the above function.

Lemma 1. *The moving distance $d_I(x)$ is a convex function.*

Corollary 1. *The total moving distance function $D(\mathcal{I}, x)$ is convex and piecewise linear. In other words, the slope of linear functions defining $D(\mathcal{I}, x)$ is non-decreasing.*

In what follows, we define a total moving distance function $D(\mathcal{I}, x)$ as a sequence of linear functions $D_1(\mathcal{I}, x), \ldots, D_{2|\mathcal{I}|+1}(\mathcal{I}, x)$. For $1 \leq i < j \leq 2|\mathcal{I}|+1$, the slope of $D_i(\mathcal{I}, x)$ is smaller than the slope of $D_j(\mathcal{I}, x)$. Since the total moving distance function is convex, optimal gathering points are located within a range $e_l \leq x \leq e_r$, where $e_l, e_r \in \mathbb{R}$ and e_l, e_r are also optimal gathering points. In the subsequent paragraphs, e_l is the *leftmost optimal gathering point* and e_r is the *right most optimal gathering point*. More formally, for any optimal gathering point x, e_l satisfies $e_l \leq x$ and e_r satisfies $x \leq e_r$. The above corollary is used in the algorithm later. Next, we tighten the range of where an optimal gathering point can be found.

Lemma 2. *Let $D(\mathcal{I}, x)$ be a total moving distance function defined by linear functions $D_1(\mathcal{I}, x), \ldots, D_{2|\mathcal{I}|+1}(\mathcal{I}, x)$, α be an integer that satisfies the slope of $D_\alpha(\mathcal{I}, x)$ is non-positive and $D_{\alpha+1}(\mathcal{I}, x)$ is non-negative, and y be a real number in the range of $D_\alpha(\mathcal{I}, x)$. Then, y is an optimal gathering point.*

The above lemma implies the following corollary.

Corollary 2. *There is an optimal gathering point x such that x is an endpoint of an interval in $\mathcal{I}$.*

Algorithm 1: A linear-time algorithm for $\Pi = \Pi_{\text{comp}}$.

---

1 **Procedure** FindingOptimalGatheringPoint($\mathcal{I}$)
2     $\ell \leftarrow \ell(\mathcal{I})$ and $r \leftarrow r(\mathcal{I})$.
3   x is the median of $\mathcal{E}(\mathcal{I})$.
4   **while** $D(\mathcal{I}, x) > D(\mathcal{I}, r(L(\mathcal{I}, x))) \lor D(\mathcal{I}, x) > D(\mathcal{I}, \ell(R(\mathcal{I}, x)))$ **do**
5   **if** $D(\mathcal{I}, r(L(\mathcal{I}, x))) < D(\mathcal{I}, x)$ **then** $r \leftarrow x$
6   **if** $D(\mathcal{I}, \ell(R(\mathcal{I}, x))) < D(\mathcal{I}, x)$ **then** $\ell \leftarrow x$
7   x is the median of $\{e \in \mathcal{E}(\mathcal{I}) : \ell \leq e \leq r\}$.
8   **return** x

We are now ready to describe a linear-time algorithm for getting an optimal gathering point x. Our algorithm is based on a binary search approach. From the convexity of the total moving distance function and Corollary 2, x has to be between the leftmost and rightmost endpoint of $\mathcal{I}$, namely $\ell = \ell(\mathcal{I})$ and $r = r(\mathcal{I})$, respectively. We initially define the point x as the median point of $\mathcal{E}(\mathcal{I})$. Next, we inspect $\mathcal{I}$ to obtain $r(L(\mathcal{I}, x))$ and $\ell(R(\mathcal{I}, x))$, respectively. We then calculate $D(\mathcal{I}, x)$, $D(\mathcal{I}, r(L(\mathcal{I}, x))$ and $D(\mathcal{I}, \ell(R(\mathcal{I}, x)))$ and compare their values. This comparison leads to the procedure described in Algorithm 1.

We now show the correctness of this algorithm. We claim that an optimal gathering point is found by repeating the above procedure. To show the correctness of Algorithm 1, we give a characterization of optimal gathering points.

By Corollary 2 and Lemma 2, it is clear that by checking the endpoints of the intervals and its next and previous endpoints, an optimal gathering point can be found. We show that Algorithm 1 finds such a point.

Lemma 3. *Let x be a point found by Algorithm 1, ℓ be the endpoint $r(L(\mathcal{I}, x))$ and r be the endpoint $\ell(R(\mathcal{I}, x))$. Then, $D(\mathcal{I}, x) \leq D(\mathcal{I}, r)$ and $D(\mathcal{I}, x) \leq D(\mathcal{I}, \ell)$ hold, that is, x is an optimal gathering point.*

Lastly, we analyse the time complexity of our algorithm. Algorithm 1 calculates the total moving distance for several points. Straightforwardly, it can be done in $O(n)$ time. To speed up this calculation, we decompose the cost function $D(\mathcal{I}, x)$ into three parts $D(L, x)$, $D(R, x)$, and $D(\mathcal{I}, x)$, where L and R are a collection of left/right part intervals. Using this decomposition technique, we speed up the calculation of each step.

Theorem 1. *Given a collection of intervals, an optimal gathering point can be found in $O(n)$ time.*

All Slopes Are the Same. We turn our attention to a specific yet sufficiently general case of the above problem setting. In the definition of the moving distance, we allow each interval I to have a distinct m_I. We consider a case such that, for any $I, J \in \mathcal{I}$, $m_I = m_J$. In this case, it is not necessary to look for optimal gathering points as they correspond to the n- and $n + 1$-th endpoints, where n is the number of intervals.

Theorem 2. *If the moving distance function of each interval in $\mathcal{I}$ is uniform, then the n- and n + 1-th endpoints are optimal gathering points.*

3.2   Unit Square Graphs

We show that Problem 1 applied to unit square graphs is linear-time solvable for $\Pi_{\texttt{comp}}$. Recall that over L_1 distance, the distance between two points $p_1 = (x_1, y_1)$, $p_2 = (x_2, y_2)$ in the plane is $|x_2 - x_1| + |y_2 - y_1|$. For each unit square S, we define the *moving distance function* d_S as follows:

Definition 2. *The* moving distance *of an unit square $S \in \mathcal{S}$ with centre (s_x, s_y) to a point $p = (p_x, p_y)$ is a function $d_S : \mathbb{R}^2 \to \mathbb{R}$ defined as:*

$$d_S(p) = m_S(\max(|s_x - p_x| - 1/2, 0) + \max(|s_y - p_y| - 1/2, 0))$$

where $m_S \in \mathbb{R}^+$ is an arbitrary constant describing the slope. The total moving distance *of a collection of squares is then defined as $D(\mathcal{S}, p) = \sum_{S \in \mathcal{S}} d_S(p)$.*

For a given collection of unit squares $\mathcal{S}$, we give a linear-time algorithm that finds a point p that minimises the total moving distance function. Similarly as for intervals, the point p is an *optimal gathering point* if p minimises the total moving distance function.

Theorem 3. *Given a collection of unit squares $\mathcal{S}$, an optimal gathering point can be found in $O(n)$ time for satisfying Π_{comp} over L_1 distance.*

4   Satisfying $\Pi_{k\text{-clique}}$ on Unit Interval Graphs in $O(n \log n)$ Time

In this section, we address the complexity of the case where a k-clique is desired on a given graph. It turns out that this case can be efficiently solvable if the moving distance function is uniform. In particular, given a collection of unit intervals $\mathcal{I}$ where the moving distance function is uniform, we introduce an $O(n \log n)$-time algorithm to solve Problem 1 for satisfying $\Pi_{k\text{-clique}}$. We first define the following lemma:

Lemma 4. *Given a collection of unit intervals $\mathcal{I}$ such that it is ordered by the centre of the intervals, an optimal solution for satisfying $\Pi_{k\text{-clique}}$ consists of moving intervals $I_i, \ldots, I_{i+k-1}$ to a gathering point x, where $1 \le i \le n - k + 1$.*

The previous lemma indicates that examining sequences of k consecutive intervals of $\mathcal{I}$ is sufficient to obtain a k-clique using the minimum moving distance. Furthermore, we deduce from Theorem 2 that the k- and $k + 1$-th endpoints are optimal gathering points. Thus we only need to check these endpoints on each sequence to obtain an optimal solution.

The outline of our algorithm is as follows. Let x_1 and x_2 be the optimal gathering points of $\mathcal{I}(1, k)$ and $\mathcal{I}(2, k)$. From Theorem 2, x_1 and x_2 are medians

---

Algorithm 2: An $O(n \log n)$-time algorithm for $\Pi = \Pi_{k\text{-clique}}$

---

1 **Procedure** FindingOptimalClique($\mathcal{I}$,k)
2   Sort the interval set $\mathcal{I}$ by the centre of the intervals.
3     $D_{\text{act}} \leftarrow D(\mathcal{I}(1,k), x_k^{\mathcal{I}(1,k)})$
4   Calculates $D(\mathcal{I}(1,k), x_{k+1}^{\mathcal{I}(1,k)})$ using D_{act}
5     $x_{\text{act}} = x_{k+1}^{\mathcal{I}(1,k)}$
6     $D_{\text{min}} \leftarrow D_{\text{act}}$, $\mathcal{I}_{\text{min}} \leftarrow \mathcal{I}(1,k)$, $x_{\text{min}} \leftarrow x_{\text{act}}$
7   **while** $2 \leq i \leq n-k+1$ **do**
8        $x_{\text{act}} = x_k^{\mathcal{I}(i,k)}$
9        $D_{\text{act}} \leftarrow D(\mathcal{I}(i,k), x_k^{\mathcal{I}(i,k)})$ using $D(\mathcal{I}(i-1,k), x_{k+1}^{\mathcal{I}(i-1,k)})$
10   **if** $D_{\text{act}} < D_{\text{min}}$ **then** $D_{\text{min}} \leftarrow D_{\text{act}}$, $x_{\text{min}} \leftarrow x_{\text{act}}$, $\mathcal{I}_{\text{min}} \leftarrow \mathcal{I}(\lceil i/2 \rceil)$
11   Calculates $D(\mathcal{I}(i,k), x_{k+1}^{\mathcal{I}(i,k)})$ using D_{act}
12        $i \leftarrow i+1$
13   **return** $(x_{\text{min}}, \mathcal{I}_{\text{min}})$

---

of $\mathcal{I}(1,k)$ and $\mathcal{I}(2,k)$. First, we find x_1 and calculate $D(\mathcal{I}(1,k), x_1)$. Since the difference between $\mathcal{I}(1,k)$ and $\mathcal{I}(2,k)$ is two intervals, we can find x_2 and calculate $D(\mathcal{I}(2,k), x_2)$ efficiently. For this purpose, we give formulas for updating the total moving distance.

Lemma 5. *The total moving distance $D(\mathcal{I}, x_{n+1}^{\mathcal{I}})$ equals $D(\mathcal{I}, x_n^{\mathcal{I}}) + (x_{n+1}^{\mathcal{I}} - x_n^{\mathcal{I}})(|L(\mathcal{I}, x_{n+1}^{\mathcal{I}})| - |R(\mathcal{I}, x_n^{\mathcal{I}})|)$.*

Thus we can calculate the total moving distance efficiently when a gathering point is next to a median. As another update formula, we consider the update formula when one interval is added and one interval is deleted.

Lemma 6. *Let $\mathcal{I}$ be a collection of intervals $\{I_1, \ldots, I_n\}$ sorted by the centre, I_{n+1} be an interval such that $c(I_n) \leq c(I_{n+1})$, and $\mathcal{J}$ be $(\mathcal{I} \setminus \{I_1\}) \cup \{I_{n+1}\}$. If $x_n^{\mathcal{J}} \geq x_{n+1}^{\mathcal{I}}$ holds, then the total moving distance $D(\mathcal{J}, x_n^{\mathcal{J}})$ equals $D(\mathcal{I}, x_{n+1}^{\mathcal{I}}) + |x_{n+1}^{\mathcal{I}} - x_n^{\mathcal{J}}|(|L(\mathcal{J}, x_n^{\mathcal{J}})| - |R(\mathcal{J}, x_{n+1}^{\mathcal{I}})|) + d_{I_{n+1}}(x_{n+1}^{\mathcal{I}}) - d_{I_1}(x_{n+1}^{\mathcal{I}})$.*

Lemma 7. *Let $\mathcal{I} = \{I_1, \ldots, I_n\}$ be an ordered collection of unit intervals by the centre, I_{n+1} be an unit interval such that $c(I_n) \leq c(I_{n+1})$, $\mathcal{J}$ be the collection of unit intervals $(\mathcal{I} \setminus \{I_1\}) \cup I_{n+1}$. If $x_n^{\mathcal{J}} < x_{n+1}^{\mathcal{I}}$ holds, then the total moving distance $D(\mathcal{J}, x_n^{\mathcal{J}})$ equals $D(\mathcal{I}, x_{n+1}^{\mathcal{I}}) + \delta(|R(\mathcal{J}, x_n^{\mathcal{J}})| - |L(\mathcal{J}, x_{n+1}^{\mathcal{I}})|) + d_{I_{n+1}}(x_{n+1}^{\mathcal{I}}) - d_{I_1}(x_{n+1}^{\mathcal{I}})$.*

Lemmas 6 and 7 provide a method to update the total moving distance. To update the total moving distance efficiently based on Lemmas 5, 6 and 7, we need to calculate point $x_j^{\mathcal{I}(i,k)}$ for any j and the number of left/right intervals of a given point x. It can be done using a binary search tree [3] and we can show the following theorem for Algorithm 2.

Theorem 4. *Given a collection of unit intervals $\mathcal{I}$, if the moving distance function is uniform, then an optimal solution can be found in $O(n \log n)$ time.*

5   LP-Formulations for Other Fundamental Graph Properties on Unit Interval Graphs

This section addresses the remaining properties, $\Pi_{\texttt{edgeless}}$, $\Pi_{\texttt{acyc}}$, $\overline{\Pi}_{k\texttt{-clique}}$, and $\Pi_{k\texttt{-conn}}$ on unit interval graphs. We show that the problems for satisfying these properties can be formulated using linear programming. It implies that these problems can be solved by using polynomial-time algorithms for linear programming.

In the following subsections, we use the absolute value in the objective function to express the formulations. This formulation differs from the formulation in (1), but we assert that a formulation with an absolute value in the objective function can easily be transformed into a linear program. For each variable x_i of the objective function in an absolute value, we add two new variables x_i' and x_i'' and replace x_i by $x_i' - x_i''$ and $|x_i|$ by $x_i' + x_i''$, for $1 \leq i \leq n$. We also add two constraints $x_i' \geq 0$ and $x_i'' \geq 0$ to $\mathcal{C}$. The optimal solutions after transforming the program are the same if at least one of the values x_i' and x_i'' is zero. That is, $x_i = x_i'$ if $x_i \geq 0$ and $x_i = -x_i''$ if $x_i \leq 0$. For instance, if $x_i', x_i'' > 0$ and we take $\delta = \min\{x_i', x_i''\}$, then $x_i' - x_i'' = (x_i' - \delta) - (x_i'' - \delta)$ but $x_i' + x_i''$ is reduced. Consequently, a solution where $x_i', x_i'' > 0$ contradicts the minimisation of the linear function.

We also add an infinitesimal value $\varepsilon > 0$ to formulations that contain strict inequalities. We treat ε as a symbolic infinitesimal parameter rather than calculating an explicit value. Linear programs only admit weak constraints ($\leq, \geq =$), as strict constraints ($<, >, \neq$) lead to formulations with no optimal solution. However, if we add the infinitesimal ε to strict constraints, then we obtain a set of linear weak constraints equivalent to the original set of constraints (see [6,13] for details).

5.1   The Case $\Pi = \Pi_{\texttt{edgeless}}$

We show that Problem 1 can be formulated as a linear problem for satisfying $\Pi_{\texttt{edgeless}}$. Let $\mathcal{I} = \{I_1, \ldots, I_n\}$ be a collection of unit intervals such that $c(I_i) \leq c(I_{i+1})$ for all $1 \leq i \leq n - 1$. We move each interval such that every pair of intervals has no intersections. Moving intervals may change the order of $\mathcal{I}$ for satisfying Π, with total moving distance D. However, the following lemma ensures that intervals in $\mathcal{I}$ can be moved by at most D while preserving the order, for satisfying any property Π.

Lemma 8. *Let $D = (d_1, \ldots, d_n)$ be a vector describing the distances applied to each interval and $\mathcal{I}^D = \{I_1^D, \ldots, I_n^D\}$ be the collection of intervals such that each centre is $c(I_i) + d_i$. If $\mathcal{I}$ has a pair of intervals I and J satisfying $c(I) \leq c(J)$ and $c(I) + d_i > c(J) + d_j$, then there is the same collection of intervals such that the total moving distance is at most $\sum_{d_i \in D} |d_i|$.*

Lemma 8 implies that $\mathcal{I}$ has no pair of intervals that intersect when the following inequality holds: $(c(I_{i+1}) + d_{i+1}) - (c(I_i) + d_i) > 1$ for $1 \leq i \leq n - 1$. It follows that the following linear program with $O(n)$ constraints can be defined:

$$\text{minimise} \sum_{i=1}^{n} |x_i|$$
$$\text{subject to } (x_{i+1} + c(I_{i+1})) - (x_i + c(I_i)) - \varepsilon \geq 1$$

Theorem 5. *An optimal solution for satisfying $\Pi_{edgeless}$ can be found in polynomial time by linear programming.*

5.2 The Case $\Pi = \Pi_{acyc}$

We assume $\mathcal{I} = \{I_1, \ldots, I_n\}$ is sorted by the centres of each I_i. Moreover, we deduce from Lemma 8 that the order of $\mathcal{I}$ does not change. A graph is *chordal* if it has no induced cycle with four or more vertices. Since an interval graph is chordal [12], we can take any 3-tuple from $\mathcal{I}$ of the form (I_{i-1}, I_i, I_{i+1}). This tuple does not make a cycle if at least one pair of intervals does not intersect. Thus the condition $c(I_{i+2}) - c(I_i) > 1$ for $1 \leq i \leq n - 2$, assures an acyclic unit interval graph. The following linear program with $O(n)$ constraints is defined:

$$\text{minimise} \sum_{i=1}^{n} |x_i|$$
$$\text{subject to } (x_{i+2} + c(I_{i+2})) - (x_i + c(I_i)) - \varepsilon \geq 1$$

Theorem 6. *An optimal solution for satisfying Π_{acyc} can be found in polynomial time by linear programming.*

5.3 The Case $\Pi = \overline{\Pi}_{k\text{-clique}}$

We generalise the idea presented for the acyclic case and show a linear program to obtain the minimum moving distance for satisfying $\overline{\Pi}_{k\text{-clique}}$.

Lemma 9. *Given a unit interval graph $(G, \mathcal{I})$, a k-clique exists in G if and only if $c(I_{i+k-1}) - c(I_i) \leq 1$ for $1 \leq i \leq n - k + 1$.*

By Lemma 9, a k-clique exists if the distance between $c(I_i)$ and $c(I_{i+k})$ is at most 1, for $1 \leq i \leq n - k$. Conversely, a k-clique does not exist between I_i and I_{i+k} if I_i and I_{i+k} do not intersect. Thus we check whether the following constraint is true on every such a pair over $\mathcal{I}$: $c(I_{i+k-1}) - c(I_i) > 1$, $1 \leq i \leq n - k + 1$. It leads to the following $O(n)$-size linear program.

$$\text{minimise} \sum_{i=1}^{n} |x_i|$$
$$\text{subject to } (c_{i+k-1} + x_{i+k-1}) - (c_i + x_i) - \varepsilon \geq 1$$

Theorem 7. *An optimal solution for satisfying $\overline{\Pi}_{k\text{-clique}}$ can be found in polynomial time by linear programming.*

5.4   The Case $\Pi = \Pi_{k\text{-conn}}$

In this section, we show a linear program to solve Problem 1 for satisfying $\Pi_{k\text{-conn}}$. We first show the following lemma.

Lemma 10. *A unit interval graph G of $\mathcal{I}$ is k-connected if and only if all pairs of the form I_i, I_{i+k+1} share at least one point, for $1 \leq i \leq n - (k+1)$.*

Using the above lemma, satisfying $\Pi_{k\text{-conn}}$ can be achieved in polynomial time by the following linear program:

$$\text{minimise} \ \sum_{i=1}^{n} |x_i|$$
$$\text{subject to } (c_{i+k+1} + x_{i+k+1}) - (c_i + x_i) \leq 1,$$

Theorem 8. *An optimal solution for satisfying $\Pi_{k\text{-conn}}$ can be found in polynomial time by linear programming.*

6   Concluding Remarks

We have defined an optimisation problem described in Problem 1 and presented several results, as summarised in Table 1. In our present work, the main result is the definition of a new edit model to characterise different scenarios of the presented problem. Several algorithmic results are presented for solving cases of fundamental graph properties.

First, we showed a linear-time algorithm to achieve graph completeness on interval graphs using the minimum moving distance, over Euclidean distance. By applying this algorithm, we demonstrated that graph completeness can be achieved in linear time on unit square graphs over L_1 distance. We also depicted the contrast for cases where the moving distance is uniform, showing that in those cases the optimal gathering points are always the n-th and $n+1$-th endpoints. Using this property, we presented a $O(n \log n)$-time algorithm for achieving a k-clique on unit interval graphs using the minimum moving distance, over Euclidean distance. In addition, we describe $O(n)$-size linear programs to achieve distinct graph properties on unit interval graphs in polynomial time.

References

1. Biedl, T., Lubiw, A., Naredla, A.M., Ralbovsky, P.D., Stroud, G.: Distant representatives for rectangles in the plane. In: 29th Annual European Symposium on Algorithms (ESA 2021). Schloss Dagstuhl - Leibniz-Zentrum für Informatik (2021). https://doi.org/10.4230/LIPICS.ESA.2021.17
2. Boyd, S., Vandenberghe, L.: Convex Optimization. Cambridge University Press, Cambridge (2004)
3. Cormen, T.H., Leiserson, C.E., Rivest, R.L., Stein, C.: Introduction to Algorithms, 3rd edn. The MIT Press, Cambridge (2009)

4. Diestel, R.: Graph Theory. Graduate Texts in Mathematics, vol. 173, 4th edn. Springer-Verlag, Berlin Heidelberg (2010)

5. Dumitrescu, A., Jiang, M.: Systems of distant representatives in Euclidean space. J. Combin. Theory Ser. A **134**, 36–50 (2015). https://doi.org/10.1016/j.jcta.2015.03.006

6. Dutertre, B., de Moura, L.: A fast linear-arithmetic solver for DPLL(T). In: Ball, T., Jones, R.B. (eds.) CAV 2006. LNCS, vol. 4144, pp. 81–94. Springer, Heidelberg (2006). https://doi.org/10.1007/11817963_11

7. Fiala, J., Kratochvíl, J., Proskurowski, A.: Systems of distant representatives. Discret. Appl. Math. **145**(2), 306–316 (2005). https://doi.org/10.1016/j.dam.2004.02.018

8. Fomin, F.V., Golovach, P.A., Inamdar, T., Saurabh, S., Zehavi, M.: Kernelization for spreading points (2023). https://doi.org/10.48550/ARXIV.2308.07099

9. Gao, X., Xiao, B., Tao, D., Li, X.: A survey of graph edit distance. Pattern Anal. Appl. **13**(1) (2009). https://doi.org/10.1007/s10044-008-0141-y

10. Garcia-Hernandez, C., Fernández, A., Serratosa, F.: Ligand-based virtual screening using graph edit distance as molecular similarity measure. J. Chem. Inf. Model. **59**(4), 1410–1421 (2019). https://doi.org/10.1021/acs.jcim.8b00820

11. Hall, P.: On representatives of subsets. J. London Math. Soc. **s1-10**(1), 26–30 (1935). https://doi.org/10.1112/jlms/s1-10.37.26

12. Jungnickel, D.: Graphs, Networks and Algorithms. Springer, Heidelberg (2013). https://doi.org/10.1007/978-3-642-32278-5

13. Nalbach, J., Ábrahám, E., Kremer, G.: Extending the fundamental theorem of linear programming for strict inequalities. In: Proceedings of the 2021 on International Symposium on Symbolic and Algebraic Computation, ISSAC 2021. ACM (2021). https://doi.org/10.1145/3452143.3465538

14. Riesen, K.: Structural Pattern Recognition with Graph Edit Distance. Springer, Cham (2015). https://doi.org/10.1007/978-3-319-27252-8

15. Sanfeliu, A., Fu, K.S.: A distance measure between attributed relational graphs for pattern recognition. IEEE Trans. Syst. Man. Cybern. **SMC-13**(3) (1983). https://doi.org/10.1109/tsmc.1983.6313167

16. Xiao, B., Gao, X., Tao, D., Li, X.: Biview face recognition in the shape–texture domain. Pattern Recogn. **46**(7), 1906–1919 (2013). https://doi.org/10.1016/j.patcog.2012.12.009

17. Zeng, Z., Tung, A.K.H., Wang, J., Feng, J., Zhou, L.: Comparing stars: on approximating graph edit distance. Proc. VLDB Endow. **2**(1), 25–36 (2009). https://doi.org/10.14778/1687627.1687631

On the Fine-Grained Complexity
of Approximating Max k-Coverage

Haoqi Wang[✉]

Nanjing University, Nanjing, China
whq1446849503@gmail.com

Abstract. The max k-coverage problem asks us to select k subsets from a collection S of subsets of a universe U, such that the union of these k subsets covers as many elements of U as possible. We prove that for every $\lambda \in (0,1)$ and $k \in \mathbb{N}$, there exists an algorithm which, given a max k-coverage instance with input size N, finds k sets that cover at least $(1 - 1/e + \lambda/e) \cdot opt$ elements for max k-coverage in $O(N^{\lambda k + O(1)})$-time, where opt is the number of elements covered by the optimal solution. On the other hand, we show that, assuming the Gap Exponential Time Hypothesis, while λ is a constant, the fastest $(1 - 1/e + \lambda)$-approximation algorithm for max k-coverage running in $O(N^{\alpha k + o(1)})$ time satisfies $\alpha \geq \Omega(\lambda^4)$. While non constant $\lambda = \lambda(k)$ is a small function of k, the fastest $(1 - 1/e + \lambda)$-approximation for max k-coverage running in time $O(N^{\alpha k + o(1)})$ satisfies $\alpha \geq \Omega(\lambda^{\frac{1}{\lambda}})$.

Keywords: max k-coverage · complexity theory · approximation algorithm

1 Introduction

In the max k-coverage problem, we are given a set of elements and a collection of subsets of these elements, and the goal of max k-coverage problem is to find k subsets that maximize the size of their union. This problem is NP-hard. The fastest exact algorithm for the max k-coverage problem right now is in $O(N^{(k+o(1))})$ time, where N denotes the size of input [4]. Under the Strong Exponential Time Hypothesis, there is no exact algorithm for max k-coverage that runs faster than $O(N^{k-\epsilon})$ time [12].

One way to handle such a hard problem is to use approximation algorithms. It is well known that the simple greedy algorithm can achieve a $(1 - 1/e)$-approximation ratio [7]. On the other hand, the PCP Theorem implies that there exists ϵ that max k-coverage is NP-hard to approximate to within a factor $(1 - \epsilon)$ [1].

Feige proved the NP-hardness of approximation within a factor of $1 - 1/e + \epsilon$ for any constant $\epsilon > 0$ [5]. Pasin improved the result and showed that there is no $T(k) \cdot N^{o(k)}$ time approximation algorithm in a factor of $1 - 1/e + \epsilon$ under Gap-ETH [10], where $T(k)$ can be any function of k. Gap-ETH states that for

B. Li et al. (Eds.): IJTCS-FAW 2024, LNCS 14752, pp. 79–93, 2025.
https://doi.org/10.1007/978-981-97-7752-5_6

any input 3-SAT instance, there exists constant c, ϵ that no algorithm runs in $O(2^{cn})$ time can distinguish the following two cases: all the clauses in 3-SAT is satisfied, or no more than ϵ fraction of the clauses is satisfied, where n stands for the numbers of variables appear in the 3-SAT instance.

It is natural to ask whether there exists an approximation algorithm for max k-coverage in $O(N^{k-o(1)})$. In [2], the authors asked the following open problem:

Question 1. Can we determine the optimal exponent α of the fastest λ-approximation for max k-coverage running in time $O(N^{\alpha k \pm o(1)})$ for $k \geq 2$ and $1 \leq \lambda$, assuming plausible fine-grained hardness assumptions?

In order to answer Question 1 for $\lambda > 1 - 1/e$ cases, we need to examine the hardness result as well as the algorithms for max k-coverage in $O(N^{\alpha k})$ time, where $\alpha \in (0, 1)$.

Given a 3-SAT instance, using the reduction in [10] we can run the algorithm on the resulting max k-coverage instance. The result shows that, if there exists a $N^{o(k)}$ time algorithm can approximate max k-coverage within $(1 - 1/e)$ then it will violate Gap-ETH. By choosing a smaller gap and a larger size for the max k-coverage instance, a stronger running time lower bounds can be achieved. That means, we can get a hardness result that, for $\lambda > 1 - 1/e$, there exist $\alpha = \alpha(\lambda)$ that, max k-coverage has no λ-approximation algorithm in $O(N^{\alpha k})$ running time.

In this paper, by taking a closer look at the previous reduction [10], the following theorem follows.

Theorem 1. *Assuming Gap-ETH, let* $\alpha = \alpha(\lambda)$ *be the parameter of the fastest* $(1 - 1/e + \lambda)$*-approximation for max k-coverage running in time* $O(N^{\alpha k + o(1)})$*. When* λ *is a constant that does not depend on* k*, we have*

$$\alpha(\lambda) \geq \Omega(\lambda^4). \tag{1}$$

And when λ is smaller, the following theorem follows.

Theorem 2. *Assuming Gap-ETH, let* $\alpha = \alpha(\lambda)$ *be the parameter of the fastest* $(1 - 1/e + \lambda)$*-approximation for max k-coverage running in time* $O(N^{\alpha k + o(1)})$*, while* $\lambda = \lambda(k)$*. We have*

$$\alpha(\lambda) \geq \Omega(\lambda^{\frac{1}{k}}). \tag{2}$$

Furthermore, We present an algorithm that combines a greedy heuristic with a brute-force search to solve the problem.

Theorem 3. *Let* $\alpha = \alpha(\lambda)$ *be the parameter of the fastest* $(1 - 1/e + \lambda)$*-approximation for max k-coverage running in time* $O(N^{\alpha k + O(1)})$*. We have*

$$\alpha(\lambda) \leq e\lambda. \tag{3}$$

That is, there exists an $O(N^{e\lambda k + O(1)})$ *time* $1 - 1/e + \lambda$*-approximation algorithm for max k-coverage.*

1.1 Related Work

Max k-coverage is one of the representative problem called submodular maximization. The problem of maximizing a submodular function is of central importance, and max k-coverage is a monotone submodular maximization problem. There is a simple greedy algorithm that has a $(1 - 1/e)$-approximation for max k-coverage. More generally, max k-coverage can be viewed as maximization of a monotone submodular function under a cardinality constraint [11], and a $(1 - 1/e)$-approximation can be achieved for monotone submodular maximization under a knapsack constraint [13].

It is hard to approximate max k-coverage problem. The parameterized inapproximability established that max k-coverage parameterized by k belongs to W[P] [6]. Moreover, Pasin shows that max k-coverage does not have $(1 - 1/e + \epsilon)$-approximation algorithm in $T(k) \cdot N^{o(k)}$ time assuming Gap-ETH for any constant $\epsilon > 0$. The proof of the formal result relies on the STAV agreement testing theorem, which states that if a collection of local functions passes some local agreement tests with high probability, then there exists a global function that agrees with most of them [3].

The study in [2] shows that, the optimization problems can be sorted by their best approximation results that they can achieve, and for some optimization problem, there exists a best approximate value significantly faster than exhaustive search. For max k-coverage, it satisfies that for some constant ϵ, there exists $\delta \in [0, 1]$ that max k-coverage cannot be ϵ-approximate in $O(N^{\delta k})$ time.

A related problem of max k-coverage is the k-set cover problem. With the same instance of max k-coverage, the goal of set cover problem is to find the smallest set that covers the whole universe, while k-set cover requires a subcollection with size k that covers the whole universe. The k-set cover problem parameterized by k is $W[2]$ hard [9]. Moreover, there exists relation between the algorithm of max k-coverage and k-set cover. The approximation guarantee with an α-approximate oracle would be $\alpha(\ln n + 1)$ for k-set cover, and $(1 - 1/e^{\alpha})$ for max k-coverage [8].

2 Preliminaries

A max k-coverage problem can be defined as follows: given a universe U, a collection S of subsets of U, and a positive integer k. The goal of the max k-coverage problem is to find k subsets $S_1, \ldots, S_k \in S$ that maximizes $|S_1 \cup \ldots \cup S_k|$.

Pasin's reduction in [10] starts from 3-SAT. A 3-SAT problem can be defined as follows: given a 3-CNF formula Φ consists of a set of variables V and a set of clauses C, and each clause has 3 variables from V. The 3-SAT problem is to determine if there exists a satisfied assignment for 3-CNF formula.

Our results are based on the Gap Exponential Time Hypothesis. The Gap Exponential Time Hypothesis can be stated in terms of 3-SAT as follows.

Definition 1 (Gap Exponential-Time Hypothesis (Gap-ETH)). *For some constant* $c, \epsilon > 0$, *no algorithm can, given a 3-SAT formula* Φ *on* n *variables and* $m = O(n)$ *clauses, distinguishes between the following cases correctly with probability* $\geq 2/3$ *in* $O(2^{cn})$ *time:*

- *$val(\Phi) = 1$, and*
- *$val(\Phi) < 1 - \epsilon$,*

where $val(\Phi)$ *is the fraction of the clauses in* Φ *that is satisfied.*

The first step of the reduction is from 3-SAT to label cover. A label cover problem can be defined as follows:

Definition 2 (Label Cover). *a label cover instance* $\mathbf{L} = \{U, V, E, \Sigma_u, \Sigma_v, \pi\}$ *consists of a graph and constraints on the graph, which can be defined as follows:*

- *Bipartite graph* (U, V, E), U *refers to the left nodes and* V *refers to the right nodes,* E *refers to the edges of the graph.*
- *For each* $u \in U$, *there is an alphabet* Σ_u *for* U *and for each* $v \in V$, *alphabet* Σ_v *for* V.
- *For each* $e = (u, v)$, *constraint* $\pi_e : \Sigma_u \rightarrow \Sigma_v$.

The goal of the label cover is to find a label $\sigma = \{u \in U | \sigma_u\} \cup \{v \in V | \sigma_v\}$, where $\sigma_u \in \Sigma_u, \sigma_v \in \Sigma_v$. The label satisfies for each $e = (u, v) \in L, \pi_e(\sigma_u) = \sigma_v$. Define the value of label cover instances as follows: the fraction of the constraints in L that is satisfied i.e.$val(L) = |\{e \in E | \pi_e(\sigma_u) = \sigma_v\}|/|E|$. Then the ψ-gap label cover can be defined as follows: given a label cover instance L, the goal is to distinguish between $val(L) = 1$ and $val(L) < \psi$.

For the convenience of the later proof, we will use the notion of weak agreement value instead of standard value for the soundness of label cover. Define weak value as follows:

Definition 3 (Weak value). *The weak value of a left labeling* $\sigma = (\sigma_u)$ *is defined as below:*

$$ weak\text{-}val(\sigma) = \frac{|\{v \in V | \exists u_1 \neq u_2 \in N(v), \pi_{(u_1,v)}(\sigma_{(u_1)}) = \pi_{(u_2,v)}(\sigma_{(u_2)})\}|}{|V|} \quad (4) $$

Furthermore, to achieve the reduction from label cover to max k-coverage in this paper, our resulting label cover instance will have the following properties:

- For the set of clauses M in 3-SAT instance, the left vertices in label cover instance pick random collection of clauses $\mathbf{T} = \{T_1, T_2, \dots, T_k\}$ where each clause in M is included in T_i independently with possibility $p = C/k$, where C is a large constant.
- The right vertices in label cover instance is $\binom{\mathbf{T}}{t}$, which contains all t-size subcollection of $\mathbf{T}$.
- Alphabet Σ_{T_i} for each vertex is the set of assignments to $var(T_i)$ that satisfies T_i, where $var(T)$ denotes the set of variables appearing in T.

Suppose that we start the reduction from 3-SAT problem with m clauses, then the size of each collection T_i is $O(pm)$ with high probability. The size of the left alphabet sets are $2^{O(pm)}$, which is also the size of the label cover instance.

The proof for the soundness of label cover based on the agreement testing theorem. Define disagreement of the two sets below:

Definition 4 (Disagreement). *For two functions $f_1 : S_1 \rightarrow \{0,1\}$ and $f_2 : S_2 \rightarrow \{0,1\}$, we define disagreement of f_1, f_2 as $disagr(f_1, f_2) = |\{u \in S_1 \cup S_2 \mid f_1(u) \neq f_2(u)\}|$.*

And we can define t-wise agreement with the same structure:

Definition 5 (t-wise Agreement). *let $\mathbf{S}$ be the collection of the sets in the layered agreement system. Consider a collection of functions $F = f_{S S \in \mathbf{S}}$ where f_S is a function $f_S : S \rightarrow \{0,1\}$. The t-wise weak agreement probability of F is defined as*

$$t\text{-}agr(F) = \Pr_{S_1,\ldots,S_t \subseteq S}[\exists i \neq j \in [t] \ , \ f_{S_i}|_{S_1 \cup \ldots \cup S_t} = f_{S_j}|_{S_1 \cup \ldots \cup S_t}], \qquad (5)$$

where $f_S|_A$ denotes $\{S \cap A | S \in f_S\}$.

To make the agreement testing theorem works, the set of the system needs to satisfy the following restrictions. It is proved that even the subsets of variables appearing in random subsets of clauses satisfy these properties with strong parameters.

Definition 6 (Uniformity). *A set system (U, S) is (γ, μ)-uniform if, for at least $(1 - \mu)$ fraction of elements $u \in U$, u appears in at least γ fraction of the subsets in S.i.e.$|\{u \in U \mid \Pr_S[u \in S] > \gamma\}| > (1 - \mu)|U|$.*

Definition 7 (Strong Intersection Disperser). *A set system (U, S) is an (k, t, η)-strong intersection disperser if, for any k distinct subcollections $S_1, ..., S_k \subseteq \mathbf{S}$ each of size at most t, we have*

$$|U \setminus (\bigcup_{j \in [k]} (\bigcap_{s \in S_i} s))| \leq \eta|U|.$$

We have U be any universe of size m, and let $0 < p, \alpha < 1$ and $l \in N$ be any constants. Let $\mathbf{S} = \{S_1, \ldots, S_k\}$ be a set system generated by including each element $u \in U$ in each set H_i with probability p. Then, if m is large enough then $\mathbf{S}$ is a $(\alpha k^l, l, 2l \cdot e^{-\alpha pk/l})$-strong intersection disperser with high probability.

The agreement theorem shows that given a $\mathbf{S}$ that most pairs in $\mathbf{S}$ has small disagreement, we can find a global function g that approximately agrees with most of the local functions in the subcollection.

Lemma 1 ([10], Lemma 13). *Let F be a collection of functions $\{f_S\}_{S \subseteq \mathbf{S}}$ that satisfies $\Pr_{S_1, S_2 \in \mathbf{S}}[disagr(f_{S_1}, f_{S_2}) < \eta n] \geq 1 - \kappa$ and $|S_1 \cap S_2| \leq \rho n$, then there exists a global function $g : [n] \rightarrow \{0,1\}$ such that*

$$E(disagr(g, f(S))) \leq n \cdot \sqrt{\rho \kappa} + \eta.$$

Furthermore, we can state the agreement theorem as follows.

Theorem 4 ([10], Theorem 11). *Let* **S** *be a collection of* k *random subsets of* $[n]$ *where each element is included with probability* p. *If* **S** *satisfies* $\{S_{S \in \mathbf{S}}\}/\{S_i, S_j\}|_{S_i \cap S_j}$ *is an* $(\frac{\alpha}{(10t)^{2t}} \cdot k^{2t-3}, 2t-3, \eta)$-*strong intersection disperser; and* $|S_1 \cap S_2| \leq \rho n$ *Then, the following holds with probability: for any collection* $F = \{F_{\mathbf{S}}\}$ *of functions* $f_S : S \to \{0,1\}$ *such that* $t\text{-}agr(F) \geq \delta$, *there exists a subcollection* $\mathbf{S}' \subseteq \mathbf{S}$ *with size* $\frac{\delta k}{8t^2}$ *and global function* $g : [n] \to \{0,1\}$ *such that*

$$E_{S \in \mathbf{S}'}(disagr(g, f(S))) \leq n \cdot \sqrt{\rho} \cdot \sqrt{\frac{2048t^8 a}{\delta^4}} + \eta.$$

3   Greedy After Brute Force

In this section, we present an algorithm for max k-coverage that proves Theorem 3. Our algorithm works as follows. We first enumerate all possible d-tuples of sets from S, where $d < k$, and select them as part of the cover. Then we apply the greedy algorithm to choose the remaining $k - d$ sets from S that maximize the number of covered elements. That is, first for all d-tuples $C_a \subseteq S$ with $|C_a| = d$, let $S' = S \setminus C_a$ and $U' = U \setminus N(C_a)$, where $N(C_a)$ is the set of elements covered by C_a and use greedy algorithm to find $C_b \subseteq S'$ with $|C_b| = k - d$ to maximize $|N(C_b)|$. Finally, we pick $C = C_a \cup C_b$ that gives the largest cover as the solution.

Algorithm 1. Algorithm for gap max k-coverage

Input: $k \in \mathbb{N}^+$, bipartite graph $G = (S, U)$
Output: true, if there exists k vertices in S covers U; false, if any k vertices in S covers at most δ fraction of U.

1: $C \leftarrow \varnothing$;
2: **for all** $C_a \subseteq S$ with $|C_a| = d$ **do**
3:     $S' \leftarrow S/C_a$; $U' \leftarrow U/N(C_a)$; $C_b \leftarrow \varnothing$;
4:     **for** $i \in [k - d]$ **do**
5:         Find a vertex $s \in S' \setminus C_b$ with maximum $|N(C_b \cup \{s\})| - |N(C_b)|$;
6:         $C_b \leftarrow C_b \cup \{s\}$;
7:     **end for**
8:     **if** $|N(C_a \cup C_b)| > |N(C)|$ **then**
9:         $C = C_a \cup C_b$;
10:    **end if**
11: **end for**
12: **if** $N(C) > \delta|U|$ **then**
13:    **return** true;
14: **else**
15:    **return** false;
16: **end if**

The algorithm runs in $O(n^{d+O(1)})$ time. In order to analyze the approximation ratio, we have the following theorem:

Theorem 5. *Let $C_a \subseteq S$ and $|C_a| = d$ be the set found in the algorithm that maximize $|N(C_a)|$. We have $|N(C_a)| \geq \frac{d}{k}Opt(G,k)$, where $Opt(G,k)$ denotes the optimal size of cover for the max k-coverage problem with bipartite graph G and parameter k.*

Proof. Define $C \subseteq S$ and $|C| = k$ that satisfies $|N(C)| = Opt(G,k)$, then through pigeonhole principle, there exists $s_1 \in C$ that $|N(C \setminus s_1)|$ covers at least $(k-1)/k$ fraction of $Opt(G,k)$, and $s_2 \in C \setminus s_1$ that covers at least $(k-2)/(k-1)$ fraction of $N(C \setminus s_1)$, that is to say, $|N(C \setminus \{s_1 \cup s_2\})|$ covers at least $(k-2)/k$ fraction of $Opt(G,k)$. Define s_i in turn and let $C \setminus \{s_1, s_2, \ldots, s_{k-d}\} = |C_a|$ then we have $|N(C_a)| \geq \frac{d}{k}Opt(G,k)$.

Furthermore, the greedy algorithm can find a cover that covered at least $(1 - \frac{1}{e})$ fraction of the remaining uncovered vertices. So the approximation ratio of the algorithm can be written as $\frac{d}{k} + \frac{k-d}{k}(1 - \frac{1}{e}) = 1 - \frac{k-d}{ek}$. Let $\lambda = \frac{d}{ek}$, the approximation ratio can be written as $\delta = 1 - \frac{1}{e} + \lambda$. This result leads to the proof of the Theorem 3.

Proof (Proof of Theorem 3). We have there exists an $1 - \frac{1}{e} + \lambda$-approximation algorithm for max k-coverage running in $O(n^{\lambda ek + O(1)})$ time, which means the fastest λ-approximation for max k-coverage runs in at least $O(n^{\lambda ek + O(1)})$ time. This implies $\alpha(\lambda) \leq e\lambda$.

4   Lower Bounds for Max k-Coverage

The main goal of this section is to give the proof of the Theorem 1 and 2. We present a hardness result for the max k-coverage problem with an approximation ratio of $(1 - (1 - 1/t)^t + \lambda)$, where λ and $t \ll k$ are constants that affects the running time of the algorithm.

Theorem 6. *Assuming Gap-ETH, for any constant $t \in \mathbb{N}$ and $t \geq 2$, there exists an constant $R(t)$ that there is no $(1 - (1 - 1/t)^t + \lambda)$-approximation algorithm for max k-coverage in $T(k) \cdot N^{\frac{\lambda^4 k}{R(t)}}$ running time, where N denotes the size of the input and $T(k)$ can be any function of k.*

4.1 $N^{o(\lambda^4 k)}$ Lower Bound for Max k-Coverage

The proof starts from Pasin's reduction in [10]. Pasin's argument shows that for any constant $\delta \in (0,1)$, no algorithm in $T(k) \cdot N^{o(k)}$ time can distinguish label cover instance L within $val(L) = 1$ and $weak\text{-}val(L) \leq \delta$, since the size of the label cover instance N is only $2^{O(m/k)}$, and assuming Gap-ETH, no algorithm for 3-SAT runs in $2^{o(n)}$ time.

We show that the no algorithm for label cover runs in $N^{R \cdot \delta^4 k}$ time can distinguish label cover instance L within $val(L) = 1$ and $weak\text{-}val(L) \leq \delta$, where R is a small constant. To obtain the hardness result, we need to take a closer look to the size of label cover instance N. The size of label cover instance depends on p, the possibility that each clause in 3-SAT instance is included in the left vertex of label cover. And we have the following theorem about p holds.

Theorem 7. *For any constant $\epsilon \in (0,1)$ and $\Delta, t \in \mathbb{N}$ that $\Delta, t > 2$, there exists a constant $C = C(\epsilon, \Delta, t)$ such that, given a 3-SAT instance Φ with $val(\Phi) = 1-\epsilon$ and each variable in Φ appears in at most Δ clauses, and L be a label cover instance such that the constraint graph (U, V) is bi-regular with right degree t, and the left vertices is a collection of k random subsets of clauses where each clause is included in each subset with probability p. If p satisfies $p \geq \frac{C}{\delta^4 k}$, then for any sufficiently large k, label cover instance satisfies weak-$val(L) < \delta$.*

Proof. The proof starts from Theorem 14 in [10]. For any $0 < \eta, \rho, \alpha, \gamma, \mu, \delta < 1$ and $t, k, n, d \in N$ such that $t > 2$ and $k > 10t/\alpha$, if the left vertices set of label cover $\mathbf{T} = \{T_1, ..., T_k\}$ satisfies:

- $(\{var(T)|T \in \mathbf{T}\} \setminus \{var(T_i), var(T_j)\})|_{var(T_i) \cap var(T_j)}$ is an $(\frac{\alpha}{(10t)^{2t}} \cdot k^{2t-3}, 2t - 3, \eta)$-strong intersection disperser, where $var(T)$ denotes the set of variables appearing in T, and $\mathbf{S}|_A$ denotes $\{S \cap A | S \in \mathbf{S}\}$ for a collection of subsets $\mathbf{S}$;
- $|var(T_i) \cap var(T_j)| \geq \rho n (i \neq j)$;
- Any subcollection of $\mathbf{T}$ with size at least $\frac{\delta k}{8t^2}$ is (γ, μ)-uniform;

then $val(\Phi) < 1 - \mu - (3\Delta\theta/\gamma)$ implies weak-$val(L) < \delta$, where θ is the parameter that satisfies for every labeling σ, there exist function g that $t\text{-}agr(F) \geq \delta$ implies $E_{T \in \mathbf{T}}(disagr(g, \sigma(T))) \leq \theta$. From Theorem 4, we can let $\theta = \sqrt{\rho(\frac{2048t^8\alpha}{\delta^4} + \eta)}$. That means $val(\Phi) < 1 - \mu - (3\Delta/\gamma)\sqrt{\rho(\frac{2048t^8\alpha}{\delta^4} + \eta)}$ implies weak-$val(L) < \delta$.

From Proposition 17, 18 and Lemma 19 in [10], with high probability, the following parameters satisfies the following properties:

- $(\{var(T)|T \in \mathbf{T}\} \setminus \{var(T_i), var(T_j)\})|_{var(T_i) \cap var(T_j)}$ is an $(\epsilon(k - 2)^l, l, 6\Delta l \cdot e^{-\epsilon p(k-2)/l})$-strong intersection disperser.
- $|var(T_i) \cap var(T_j)| \geq 18p^2\Delta^2 n$.
- Any subcollection of $\mathbf{T}$ of size at least $8\ln(2/\mu)/p$ is $(p/2, \mu)$-uniform.

To investigate the correlation between running time and soundness, we need to fix the parameters that appear in the theorem.

Assume that $(\{var(T)|T \in \mathbf{T}\} \setminus \{var(T_i), var(T_j)\})|_{var(T_i) \cap var(T_j)}$ is an $(\epsilon(k - 2)^l, l, 6\Delta l \cdot e^{-\epsilon p(k-2)/l})$-strong intersection disperser, we can set $l = 2t - 3$ and $\eta = 6\Delta(2t - 3) \cdot e^{-\kappa p(k-2)/(2t-3)}$, then $(\{var(T)|T \in \mathbf{T}\} \setminus \{var(T_i), var(T_j)\})|_{var(T_i) \cap var(T_j)}$ is an $(\frac{\alpha}{(10t)^{2t}} \cdot k^{2t-3}, 2t - 3, \eta)$-strong intersection disperser.

Also, let $\rho = 18p^2\Delta^2 n$, we have $|var(T_i) \cap var(T_j)| \geq \rho n (i \neq j)$ holds.

Let $\gamma = p/2$, this implies that for any sufficiently large k, any subcollection of $\mathbf{T}$ with size at least $\frac{\delta k}{8t^2}$ is (γ, μ)-uniform.

In order to facilitate calculation, define $\kappa = \frac{\ln c(2t-3)}{p(k-2)}$, so we have:

- $l = 2t - 3$,
- $\eta = 6\Delta(2t - 3) \cdot e^{-\kappa p(k-2)/(2t-3)}$,
- $\gamma = p/2$,
- $\rho = 18p^2\Delta^2 n$,

$- \kappa = \frac{\ln c(2t-3)}{p(k-2)}.$

For $val(\varPhi) < 1 - \mu - (3\Delta/\gamma)\sqrt{\rho(\frac{2048t^8\alpha}{\delta^4} + \eta)}$, we can rearrange the terms on the right-hand side as follows:

$$1 - \mu - (3\Delta/\gamma)\sqrt{\rho(\frac{2048t^8\alpha}{\delta^4} + \eta)}$$

$$= 1 - \mu - (36\Delta^2)\sqrt{(\frac{1024t^8(10t)^{2t}(k-2)^{2t-3}\kappa}{\delta^4 \cdot k^{2t-3}} + 3\Delta(2t-3) \cdot e^{-\kappa p(k-2)/(2t-3)})}$$

$$= 1 - \mu - (36\Delta^2)\sqrt{(\frac{1024t^8(10t)^{2t}(k-2)^{2t-4}(2t-3)\ln c}{p \cdot \delta^4 \cdot k^{2t-3}} + 3\Delta(2t-3)/c)}.$$

Let $p \geq \frac{(36\Delta^2)^2 2048t^8(10t)^{2t}(2t-3)\ln(c)}{\delta^4\epsilon^2 k}, c = 6\Delta(2t-3)\epsilon^2 \cdot (36\Delta^2)^2$, we have

$$(3\Delta/\gamma)\sqrt{\rho(\frac{2048t^8\alpha}{\delta^4} + \eta)}$$

$$= (36\Delta^2)\sqrt{(\frac{1024t^8(10t)^{2t}(k-2)^{2t-4}(2t-3)\ln c}{p \cdot \delta^4 \cdot k^{2t-3}} + 3\Delta(2t-3)/c)}$$

$$\leq \sqrt{\frac{\epsilon^2}{2} + \frac{\epsilon^2}{4}} = \frac{\sqrt{3}}{2}\epsilon. \tag{6}$$

Assume that $val(\varPhi) = 1 - \epsilon$, let $\mu = (1 - \frac{\sqrt{3}}{2})\epsilon$, put it together with (6) we have $1 - \epsilon \leq 1 - \mu - (3\Delta/\gamma)\sqrt{\rho(\frac{2048t^8\alpha}{\delta^4} + \eta)}$, which means $weak\text{-}val(L) < \delta$ holds.

Let $C = \frac{(36\Delta^2)^2 4096t^8(10t)^{2t}(2t-3)\ln(c)}{\epsilon^2}$, then we have $p \geq \frac{C}{\delta^4 k}$ implies $weak\text{-}val(L) < \delta$. This ends up the proof.

Next we consider the reduction from label cover to max k-coverage. There are merely consequences of known reductions from label cover to max k-coverage in [5] and [10]. However, the previous work did not examine how the weak value δ affects the soundness of the reduction. The lemma that provides the completeness and soundness of the reduction is stated and proved below:

Lemma 2. *Let* $L = (U, V, \{\Sigma_u\}_{u \in U}, \{\Sigma_v\}_{v \in V}, \pi)$ *be the label cover instance, where* $|U| = k$, *and bipartite graph* (U, V) *is bi-regular with right degree* t. *There exist a reduction from label cover to max k-coverage that provides the completeness and soundness results below:*

- *(Completeness) if* $val(L) = 1$ *then* $OPT(\mathbf{U}) = 1$;
- *(Soundness) if* $weak\text{-}val(L) < \delta$ *then* $OPT(\mathbf{U}) \leq \delta + (1 - \delta)(1 - (1 - 1/t)^t)$.

Proof. Given a label cover instance $L = (U, V, \{\Sigma_u\}_{u \in U}, \{\Sigma_v\}_{v \in V}, \pi)$. Define set cover instance $(S, \mathbf{U})$ as follows:

- $\mathbf{U} = \{U_{(v,s_1,s_2,\ldots,s_t)} = (v, s_1, s_2, \ldots, s_{|\Sigma_v|}) : v \in V, s_1, s_2 \ldots, s|\Sigma_v| \in [t]\}$
- $S = \{S_{u,\sigma} = (u, \sigma) : u \in U, \sigma \in [|\Sigma_u|]\}$
- For $v \in V$, denote $N(v) = \{u_1, u_2, \ldots, u_t\}$ be the neighbor of V in the label cover instance L. for $i \in [t]$, if $S_{u_i,\sigma}$, $U_{(v,s_1,s_2,\ldots,s_{|\Sigma_v|})}$ satisfies $\pi_{u_i}(\sigma) = s_i$, then $S_{u_i,\sigma}$ covers $U_{(v,s_1,s_2,\ldots,s_{|\Sigma_v|})}$ (Figs. 1 and 2).

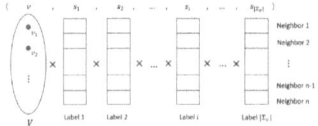

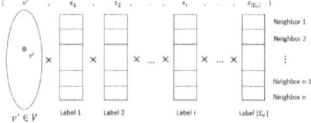

Fig. 1. The cover $\mathbf{U}$.   **Fig. 2.** The subset $\mathbf{U}_{v'} \subseteq \mathbf{U}$.

Let $N(S_{u,\sigma}) \subseteq \mathbf{U}$ be the set of elements covered by $S_{u,\sigma}$, and define $\mathbf{U}_{v'} = \{U_{(v,s_1,s_2,\ldots,s_{|\Sigma_v|})}\} \in \mathbf{U} : v = v'\}$. We have the following conclusions.

(i) For every $v' \in V$ where $N(v') = \{u_1, u_2, \ldots, u_t\}$, only for every $i, j \in [t]$, $\pi_{u_i}(\sigma_i) = \pi_{u_j}(\sigma_j)$ then $S' = \{S_{u_1,\sigma_1}, S_{u_2,\sigma_2}, \ldots, S_{u_t,\sigma_t}\}$ covers all fractions of $\mathbf{U}_{v'}$ (see Fig. 3).

(ii) Otherwise, if $\pi_{u_i}(\sigma_{a_i}) \neq \pi_{u_j}(\sigma_{b_j})$, then for any $v' \in V$ that satisfies $\{u_i, u_j\} \subseteq N(v')$, $N(S_{u_i,\sigma_{a_i}}) \cap \mathbf{U}_{v'}$ and $N(S_{u_j,\sigma_{b_j}}) \cap \mathbf{U}_{v'}$ has $1/t$ fraction of common elements (see Fig. 4).

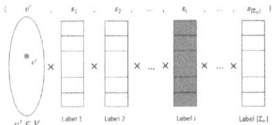

Fig. 3. The cover in $\mathbf{U}_{v'}$ with vertices with same label.   **Fig. 4.** The cover in $\mathbf{U}_{v'}$ with vertices with different label.

Completeness:

if $val(L) = 1$, then there exists a label $\sigma = \{\sigma_{u_1}, \sigma_{u_2} \ldots \sigma_{u_k}\}$ that for every $v_a \in V$, $N(v_a) = \{u_{a_1}, u_{a_2}, \ldots, u_{a_t}\}$, each $i \neq j$ satisfies $\pi_{u_{a_i}}(\sigma_{a_i}) = \pi_{u_{a_j}}(\sigma_{a_j})$.

By (i), we have for every $v_a \in V$, $\{S_{u_{a_1},\sigma_{u_{a_1}}}, \ldots, S_{u_{a_k},\sigma_{u_{a_k}}}\}$ covers all $\mathbf{U}_{v_a}$.

That means, $\{S_{u_1,\sigma_{u_1}}, \ldots, S_{u_k,\sigma_{u_k}}\}$ covers all the fractions of $\mathbf{U}$.

Soundness:

If $weak\text{-}val(L) < \delta$, let $S' = \{S_{u_1,\sigma_1}, S_{u_2,\sigma_2}, \ldots, S_{u_k,\sigma_k}\}$ be the subset of S with size k. By the definition of weak value, there exists $V' \subseteq V$ with size at least $(1 - \delta)|V|$ that for every $v \in V'$, if u_i, u_j satisfies $u_i \neq u_j \in N(v)$, then we have $\pi_{u_i}(\sigma_i) \neq \pi_{u_j}(\sigma_j)$.

Consider the coverage of each $\mathbf{U}_{v'}$ for every $v' \in V$. We have for every $S_{u,\sigma} \in S$, if $u \in N(v)$ then $|N(S_{u_i,\sigma_{a_i}}) \cap \mathbf{U}_{v'}| = |\mathbf{U}_{v'}|/t$. Also by (ii) we have, if $\{u_i, u_j\} \subseteq N(v')$, then $N(S_{u_i,\sigma_{a_i}}) \cap \mathbf{U}_{v'}$ and $N(S_{u_j,\sigma_{b_j}}) \cap \mathbf{U}_{v'}$ has $1/t$ fraction of common elements.

That means, denote c to be the number of set in S' that has neighbor in $\mathbf{U}_{v'}$, then only $1 - (1 - 1/t)^c$ fraction of $\mathbf{U}_{v'}$ can be covered by S'.

For $S' \subseteq S$, suppose that $|V'| = (1 - \delta)|V|$, Let $V' = \{v_1, v_2, ...v_{(1-\delta)|V|}\}$. Let $c_1, c_2...c_{(1-\delta)|V|}$ be the number of set in S' that has neighbors in $\mathbf{U}_{v_1}, \mathbf{U}_{v_2}, ..., \mathbf{U}_{v_{(1-\delta)|V|}}$, and let $c_0 = \sum_{v \in V \setminus V'} c_v$, and c_v be the the number of set in S' that has neighbors in $\mathbf{U}_v$.

We have the label cover instance L is bi-regular with right degree t, so each $u \in U$ has $t|V|/k$ neighbors in V. That means, for every $S_{u_i,\sigma_i} \in S$ there exists a collection of subsets $\{\mathbf{U}_v \subseteq U : v \in N(u_i)\}$ with size $t|V|/k$ that, only for every $\mathbf{U}_{v'} \in \{\mathbf{U}_v \subseteq U : v \in N(u_i)\}$, S_{u_i,σ_i} has neighbors in $\mathbf{U}_{v'}$. We have $|S'| = k$, then by the definitions we have $\sum_{i=0}^{(1-\delta)|V|} c_i = |V|t$.

The cover of S' in $\mathbf{U}$ can be expressed as the sum of the cover in two mutually exclusive subsets: $\{U_{(v,s_1,s_2,...,s_{|\Sigma_v|})} \in \bigcup \mathbf{U}_v : v \in V \setminus V'\}$ and $\{U_{(v,s_1,s_2,...,s_{|\Sigma_v|})} \in \bigcup \mathbf{U}_v : v \in V'\}$. Consider the cover in each subset separately, we have the following:

- Suppose for every $S_{u_i,\sigma_i}, S_{u_j,\sigma_j} \in S'$, there are no common neighbors in $\{U_{(v,s_1,s_2,...,s_t)} \in \bigcup \mathbf{U}_v : v \in V \setminus V'\}$, which maximizes the cover of S'. Then the cover of S' in $\{U_{(v,s_1,s_2,...,s_t)} \in \bigcup \mathbf{U}_v : v \in V \setminus V'\}$ can be written as $\frac{\min\{c_0, \delta|V| \cdot t\} \cdot |\mathbf{U}|}{|V|t}$.

- The cover of S' in $\mathbf{U}_{v_i}$ can be written as $(1 - (1 - 1/t)^{c_i}) \cdot \frac{|\mathbf{U}|}{|V|}$, so the cover of S' in $\{U_{(v,s_1,s_2,...,s_t)} \in \bigcup \mathbf{U}_v : v \in V \setminus V'\}$ can be written as $\sum_{i=1}^{(1-\delta)|V|}(1 - (1 - 1/t)^{c_i})) \cdot \frac{|\mathbf{U}|}{|V|}$.

Above all, if $weak\text{-}val(L) < \delta$ then the fraction of $\mathbf{U}$ covered by S' is at most

$$\max_{\{c_0,c_1,...,c_{(1-\delta)|V|} | \sum_{i=0}^{(1-\delta)|V|} c_i = |V|t)\}} \{(\frac{\min\{c_0, \delta|V| \cdot t\}}{t} + \sum_{i=1}^{(1-\delta)|V|} (1 - (1 - 1/t)^{c_i}))/|V|\}$$

$$\leq \max_{\{c_0,c_1,...,c_{(1-\delta)|V|} | \sum_{i=0}^{(1-\delta)|V|} c_i = |V|t)\}} \{(\frac{\min\{c_0, \delta|V| \cdot t\}}{t} + (1-\delta)|V| \cdot (1 - (1-1/t)^{\frac{\sum_{i=1}^{(1-\delta)|V|} c_i}{(1-\delta)|V|}}))/|V|\}$$

$$= \max_{\{c_0 \in [0, |V|t]\}} \{(\frac{\min\{c_0, \delta|V| \cdot t\}}{t} + (1 - \delta)|V| \cdot (1 - (1 - 1/t)^{\frac{|V|t-c_0}{(1-\delta)|V|}}))/|V|\}$$

$$\leq \delta + (1 - \delta)(1 - (1 - 1/t)^t).$$

Plug the results together and we can get the proof of Theorem 6.

Proof (Proof of Theorem 6).

The size of label cover instance in [10] is $N = k \cdot 2^{O(pm)}$, so the algorithm for δ-gap label cover runs in $T(k) \cdot N^{O(\frac{\phi k}{p})}$-time would violate Gap-ETH, where ϕ is a constant that satisfies given a 3-SAT case Φ, no $O(2^{\phi n})$ time algorithm can distinguish between $OPT(\Phi) = 1$ and $OPT(\Phi) < \epsilon$. By Theorem 7, let $p \geq \frac{C}{\delta^4}$, where $C = C(\epsilon, \Delta, t)$ is a constant. This implies that for any constant t, there exists a constant $R_\delta(t) = R(t, \epsilon, \phi, \Delta)$ that no $T(k) \cdot N^{\frac{\delta^4 k}{R_\delta(t)}}$ algorithm can distinguish label cover instance L within $val(L) = 1$ and $weak\text{-}val(L) \leq \delta$.

The size of max k-coverage instance is also $N = k \cdot 2^{O(pm)}$. From Lemma 2 and the reduction in [10], we have no $T(k) \cdot N^{\frac{\delta^4 k}{R_\delta(t)}}$ algorithm can distinguish max k-coverage instance $(S, \mathbf{U})$ within $OPT(\mathbf{U}) = 1$ and $OPT(\mathbf{U}) \leq \delta + (1 - \delta)(1 - (1 - 1/t)^t)$. Let $\lambda = \frac{\delta}{(1 - 1/t)^t}$, we have $\delta + (1 - \delta)(1 - (1 - 1/t)^t) = 1 - (1 - 1/t)^t + \lambda$, so there exist $R(t)$ that no $T(k) \cdot N^{\frac{\lambda^4 k}{R(t)}}$ running time algorithm can approximate max k-coverage within $1 - (1 - 1/t)^t + \lambda$, where $R(t) = O(t^t)$ is a fixed function of t. This ends up the proof.

4.2   Lower Bound for Max k-Coverage When λ is Smaller

We previously proved that when δ is a constant, there is no $\delta + (1 - \delta)(1 - (1 - 1/t)^t)$-approximation algorithm for max k-coverage in $T(k) \cdot N^{\frac{\delta^4 k}{R_\delta(t)}}$ running time. Since $t \ll k$, the parameter t matters the result when δ is smaller. To handle a smaller δ, we need to consider the influence of parameter t in the result.

Theorem 8. *Let $t(k)$ be a function of k. Assuming Gap-ETH, there is no $1 - 1/e + O(1/t(k))$-approximation algorithm for max k-coverage in $T(k) \cdot N^{o(\frac{k}{t(k)^4 f(t(k))})}$ running time, where $f(t) = t^8(10t)^{2t}(2t - 3)$.*

Proof. Assume that $t = t(k)$ be a function of k, and $f(t) = t^8(10t)^{2t}(2t - 3) = \Omega(R(t))$. By the proof of Theorem 6 we have the algorithm for max k-coverage with a running time of $T(k) \cdot N^{o(\frac{\delta^4 k}{f(t(k))})}$ would violate Gap-ETH.

The soundness of the max k-coverage needs to consider the influence by the parameter k. Using Taylor expansion we have

$$(1 - \frac{1}{t(k)})^{t(k)} = \frac{1}{e} - \frac{1}{2t(k)} - o(\frac{1}{t(k)}).$$

So the approximation ratio of the algorithms can be written as $\delta + (1 - \delta)(1 - \frac{1}{e} + \frac{1}{2t(k)} + o(\frac{1}{t(k)}))$. Let $\delta = O(\frac{1}{t(k)})$, then the approximation ratio of the algorithm can be simplified as

$$(\delta + (1 - \delta)(1 - \frac{1}{e} + \frac{1}{2t(k)} + o(\frac{1}{t(k)})))$$

$$= (1 - \frac{1}{e} + \frac{1}{2t(k)} + o(\frac{1}{t(k)})) + O(\frac{1}{t(k)})(\frac{1}{e} - \frac{1}{2t(k)} - o(\frac{1}{t(k)}))$$

$$= 1 - \frac{1}{e} + O(\frac{1}{t(k)}),$$

and the running time of the algorithm can be written as $T(k) \cdot N^{o(\frac{k}{t(k)^4 f(t(k))})}$. That is to say, there is no $1 - \frac{1}{e} + O(\frac{1}{t(k)})$-approximation algorithm for max k-coverage in $T(k) \cdot N^{o(\frac{k}{t(k)^4 f(t(k))})}$ running time.

Otherwise let $\delta = \Omega(\frac{1}{t(k)})$, let δ be the parameter, the relationship between δ and k can be rewritten as $\frac{1}{t(k)} = O(\delta)$. Then the approximation ratio of the algorithm can be simplified as

$$(\delta + (1-\delta)(1 - \frac{1}{e} + \frac{1}{2t(k)} + o(\frac{1}{t(k)}))$$

$$= (1 - \frac{1}{e} + O(\delta) + o(\delta)) + \delta(\frac{1}{e} - O(\delta) - o(\delta))$$

$$= 1 - \frac{1}{e} + O(\frac{1}{\delta}).$$

The running time of the algorithm is $T(k) \cdot N^{o(\frac{\delta^4 k}{f(t(k))})}$. Let $\delta = \frac{1}{t'(k)}$, then we have $\frac{1}{t'(k)} = \Omega(\frac{1}{t(k)})$. This implies that there is no $1 - \frac{1}{e} + O(\frac{1}{t'(k)})$-approximation algorithm for max k-coverage in $T(k) \cdot N^{o(\frac{k}{t'(k)^4 f(t'(k))})}$ running time, which is longer than $T(k) \cdot N^{o(\frac{k}{t'(k)^4 f(t'(k))})}$. That means, no $T(k) \cdot N^{o(\frac{k}{t(k)^4 f(t(k))})}$-time algorithm can approximate max k-coverage within a ratio of $1 - \frac{1}{e} + O(\frac{1}{t(k)})$.

4.3 Putting Things Together

Now we are ready to prove Theorem 1 and Theorem 2. Our hardness result for max k-coverage follows from Theorem 6 and Theorem 8.

Proof (Proof of Theorem 1). Suppose for the sake of contradiction that $\alpha(\lambda) \leq o(\lambda^4)$. Then there exists a $(1 - \frac{1}{e} + \lambda)$-approximation algorithm for max k-coverage running in $o(N^{\lambda^4 k})$ time. However, Theorem 6 implies that that no $T(k) \cdot N^{(\frac{\lambda^4}{f(t)})k}$ running time algorithm can approximate max k-coverage within $1 - (1 - 1/t)^t + \lambda$. Let t be a large constant, then there is no $1 - \frac{1}{e} + \lambda_0$-approximation algorithm for max k-coverage running in $o(N^{(\lambda_0 - c)^4 k})$ time, where $c = 1/e - (1 - 1/t)^t$ is a small constant, and $\lambda_0 = 1/e - (1 - 1/t)^t + \lambda = c + \lambda$. Since c is a small constant, $o(N^{(\lambda_0 - c)^4 k}) = o(N^{\lambda^4 k})$, which means no $1 - \frac{1}{e} + \lambda$-approximation algorithm for max k-coverage running in $o(N^{\lambda^4 k})$ time. This contradicts $\alpha(\lambda) \leq o(\lambda^4)$.

Proof (Proof of Theorem 2). Suppose for the sake of contradiction that $\alpha(\lambda) \leq o(\lambda^{\frac{1}{\lambda}})$. Then there exists a $(1 - \frac{1}{e} + \lambda)$-approximation algorithm for max k-coverage running in $o(N^{\lambda^{\frac{1}{\lambda}} k})$ time, where $\lambda = \lambda(k)$. Theorem 8 implies that no $T(k) \cdot N^{o(\frac{k}{t(k)^4 f(t(k))})}$-time algorithm can approximate max k-coverage within a ratio of $1 - \frac{1}{e} + O(\frac{1}{t(k)})$, where $f(t(k)) = O(t(k)^{t(k)})$. Let $\lambda = \frac{1}{t(k)}$, then according Theorem 8, no $1 - \frac{1}{e} + O(\lambda)$-approximation algorithm for max k-coverage running in $N^{o(\frac{k}{t(k)^4 f(t(k))})} = N^{o(\lambda^{\frac{1}{\lambda}} \cdot k)}$ time, which contradicts $\alpha(\lambda) \leq o(\lambda^{\frac{1}{\lambda}})$.

5 Conclusion

In this paper, we present an approximation algorithm for max k-coverage. We show that the optimal exponent α of the γ-approximation satisfies $\alpha \leq$

$(\gamma - 1 + 1/e)/\gamma$. We also derive a lower bound for the running time of any γ-approximation algorithm for max k-coverage, which is $O(N^{\lambda k})$ for some constant λ.

However, assuming plausible fine-grained hardness assumptions, the fastest γ-approximation algorithm of max k-coverage remains open. Our result of the problem is under the Gap-ETH assumption, and the lower bound of max k-coverage is not tight enough. Here are some open problems that came up from our result.

- The reduction in [10] uses an agreement testing theorem that does not impose any requirements on the function S. Can we provide a stronger agreement testing theorem with a function that has restrictions on it? With a stronger agreement testing theorem, we may provide a stronger lower bound for max k-coverage running in $O(N^{\lambda k})$ time?
- The reduction in [10] starts from Gap-ETH. Can we obtain lower bound for max k-coverage using a non-gap assumption like ETH?

References

1. Arora, S., Lund, C., Motwani, R., Sudan, M., Szegedy, M.: Proof verification and the hardness of approximation problems. J. ACM (JACM) **45**(3), 501–555 (1998)
2. Bringmann, K., Cassis, A., Fischer, N., Künnemann, M.: A structural investigation of the approximability of polynomial-time problems. In: Bojanczyk, M., Merelli, E., Woodruff, D.P. (eds.) 49th International Colloquium on Automata, Languages, and Programming, ICALP 2022, July 4-8, 2022, Paris, France. LIPIcs, vol. 229, pp. 30:1–30:20. Schloss Dagstuhl - Leibniz-Zentrum für Informatik (2022). https://doi.org/10.4230/LIPIcs.ICALP.2022.30
3. Dikstein, Y., Dinur, I.: Agreement testing theorems on layered set systems. In: 2019 IEEE 60th Annual Symposium on Foundations of Computer Science (FOCS), pp. 1495–1524. IEEE (2019)
4. Eisenbrand, F., Grandoni, F.: On the complexity of fixed parameter clique and dominating set. Theoret. Comput. Sci. **326**(1–3), 57–67 (2004)
5. Feige, U.: A threshold of ln n for approximating set cover. J. ACM (JACM) **45**(4), 634–652 (1998)
6. Feldmann, A.E., Lee, E., Manurangsi, P.: A survey on approximation in parameterized complexity: hardness and algorithms. Algorithms **13**(6), 146 (2020)
7. Hochba, D.S.: Approximation algorithms for NP-hard problems. ACM SIGACT News **28**(2), 40–52 (1997)
8. Hochbaum, D.S., Pathria, A.: Analysis of the greedy approach in problems of maximum k-coverage. Naval Res. Logist. (NRL) **45**(6), 615–627 (1998)
9. Lin, B., Ren, X., Sun, Y., Wang, X.: Constant approximating parameterized k-setcover is w [2]-hard. In: Proceedings of the 2023 Annual ACM-SIAM Symposium on Discrete Algorithms (SODA), pp. 3305–3316. SIAM (2023)
10. Manurangsi, P.: Tight running time lower bounds for strong inapproximability of maximum k-coverage, unique set cover and related problems (via t-wise agreement testing theorem). In: Proceedings of the Fourteenth Annual ACM-SIAM Symposium on Discrete Algorithms, pp. 62–81. SIAM (2020)

11. Nemhauser, G.L., Wolsey, L.A., Fisher, M.L.: An analysis of approximations for maximizing submodular set functions-i. Math. Program. **14**, 265–294 (1978)

12. Puatracscu, M., Williams, R.: On the possibility of faster SAT algorithms. In: Charikar, M. (ed.) Proceedings of the Twenty-First Annual ACM-SIAM Symposium on Discrete Algorithms, SODA 2010, Austin, Texas, USA, 17–19 January 2010, pp. 1065–1075. SIAM (2010). https://doi.org/10.1137/1.9781611973075.86

13. Sviridenko, M.: A note on maximizing a submodular set function subject to a knapsack constraint. Oper. Res. Lett. **32**(1), 41–43 (2004)

Nested and Interleaved Ticketing
for Multiple Travelers

Dongyu Lv[1], Yizhi Song[2], and Chao Xu[1(✉)]

[1] University of Electronic Science and Technology of China, Chengdu, China
the.chao.xu@gmail.com
[2] City University of Hong Kong, Hong Kong, China
yizhisong2-c@my.cityu.edu.hk

Abstract. Nested ticketing represents a prevalent practice in airline bookings, employed to bypass certain airline ticketing regulations with the aim of reducing costs on multiple round-trip tickets. We consider the computational complexity of nested ticketing, and it is 'covert' version, which we call interleaved ticketing. Consider multiple agents living in different locations and a single client who demands that one agent be present each week. The objective for the company is to schedule these agents and arrange their flight itineraries to minimize the cost. We show that when nested ticketing is allowed, the problem is NP-hard when there are at least two agents. We also show there exists a polynomial-time algorithm if only interleaved ticketing is allowed, and the number of airlines is bounded.

Keywords: Airline booking ploys · Matching · Dynamic programming

1 Introduction

Within commercial air transport, travelers may adopt a wide range of strategies to save on travel costs. Airline booking ploys are ways to circumvent airline ticket rules in order to spend less on the ticket [9]. One common case is nested ticketing, also known as back-to-back ticketing, bracketing and cross-ticketing.

Airlines often charge more for round-trip tickets that only stay the weekdays (mid-week) to target business travelers who prefer to return before the weekend. Conversely, tickets that include a weekend stay are usually cheaper, aiming at leisure travelers. Nested ticketing allows one to potentially save money on multiple round trips between two locations. Such an opportunity arises when a mid-week round-trip ticket is more expensive than the combination of two round-trip tickets involving a stay over the weekend. By buying two round-trip tickets and only using one flight leg per mid-week round-trip ticket, the mid-week round-trip ticket can be simulated at a lower fare.

Although these days airlines use much more sophisticated methods for ticket pricing [6,12,17], the strategy still holds: nested ticketing expands the possible combination of round trip tickets, and allow potential savings.

B. Li et al. (Eds.): IJTCS-FAW 2024, LNCS 14752, pp. 94–105, 2025.
https://doi.org/10.1007/978-981-97-7752-5_7

Previous studies for nested ticketing focus more on economic impact, e.g., price fairness for customers [3], or the legal and moral issues [4]. None of these studies the computational difficulty of employing such strategy.

Other common booking ploys like hidden city ticketing was considered extensively in [4,5,13], discussing their fairness and issues. We recommend the reader the report by U.S. Government Accountability Office for detailed information [10]. The pricing's impact on revenue was studied in [11,16].

For many airlines, nested ticketing that circumventing fare rules is explicitly forbidden, this includes all three major US airlines American Airlines [2], Delta [7] and United [15].

In order to be compliant, one can execute nested ticketing *across different airlines*. That is, for a single passenger, for each individual airline, the stay time of the round trips must not overlap. We call this *interleaved ticketing*. Interleaved ticketing has been known to the frequent flyer community at large (e.g. [1]), but no academic studies we are aware of. See Fig. 1 for an example of nested ticketing and interleaved ticketing with their money savings.

In this paper, we begin by examining an empirical scenario and subsequently model optimizing the cost of buying some round trip tickets as matching problems. We frame the problem as identifying the minimum weight matching satisfying some label and color constraints. Moreover, we provide a detailed discussion of various specific scenarios arising from two key conditions: (1) whether there is a single traveler or multiple travelers, and (2) whether nested ticketing within a single airline is allowed or forbidden.

Our Contribution. We show that the optimizing the round-trip itinerary of two people allowing nested ticketing is NP-hard. In contrast, for fixed number of airlines, one obtain polynomial time solvability if we only allow interleaved ticketing.

2   Preliminaries

Let $[n] = \{1, \ldots, n\}$ to denote all positive integers up to n. We consider multigraphs where there can be parallel edges and self-loops.

A *matching* is a set of disjoint edges. The matching is *perfect* if every vertex is contained in some edge of the matching. The bipartite perfect matching (BPM) problem is to find a perfect matching in a bipartite (multi)graph.

2.1   Problem Setup

We also establish three progressively more general scenarios.

The first scenario, referred to as the *solo scenario*, involves determining the optimal set of tickets for an individual who requires multiple round trips to the same city, a situation commonly faced by business travelers.

In the next scenario, we introduce the *dual scenario*, which considers a couple in a long-distance relationship, residing in separate home cities. They aim to visit

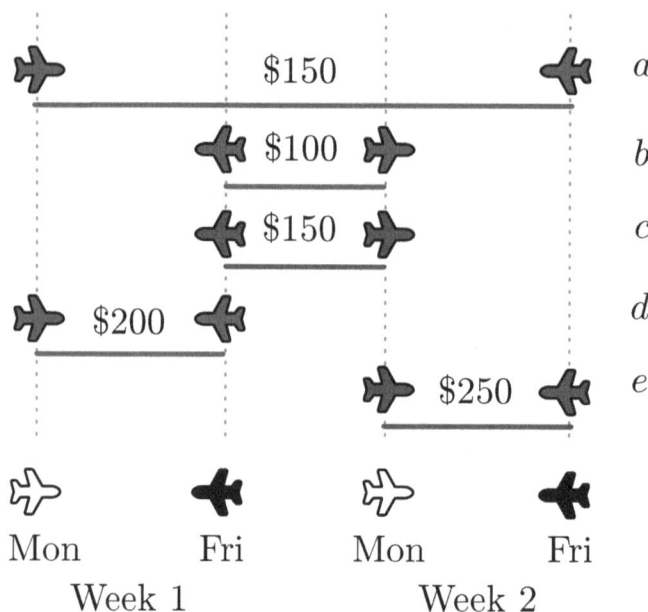

Fig. 1. Consider a two-week travel schedule where an individual based at A needs to travel to B and return to A on Monday and Friday, respectively. In this context, ✈ represents a flight from A to B, and ✈ symbolizes a flight from B to A. There are five round-trip tickets available: tickets a, b, and d are operated by airline 1, while tickets c and e are by airline 2. Tickets a, b, and c are relatively more affordable, as they include a weekend stay. For a standard round-trip covering the two weeks, one would typically choose tickets d and e, totaling a cost of \$450. However, a traveler can utilize nested ticketing strategies by purchasing round-trip tickets a and b instead, at a total cost of \$250. It is important to note that tickets a and b are nested within airline 1. To avoid nesting within any single airline, employing interleaved ticketing, the most cost-effective solution is to purchase round-trip tickets a and c, resulting in a total expense of \$300.

each other every weekend, with one partner flying to the other's city each Friday and returning on Sunday. This scenario is an extension of the solo scenario, as it can be reduced to the solo case by assigning an infinite cost to the second person's ticket. Additionally, within the dual scenario, we define a *fair dual scenario* where each individual is required to fly an equal number of times, assuming their meetings are an even number.

The *general scenario*, accommodating multiple travelers. Imagine a company with a client in city C_0 and p agents based in cities $C_1, \ldots, C_p$. The client requires the presence of some agent each week for the next n weeks, who can be different each week. Every Monday, an agent travels to the client's city, returning on Friday. The company's objective is to strategically schedule agents' visits and purchase tickets in a way that minimizes the total cost of travel.

One can observe when $p = 1$, the general scenario is the solo scenario.

It is somewhat less apparent, but when $p = 2$, the general scenario is the dual scenario. In the dual scenario, two individuals travel to each other's cities, rather than to a separate third location. Nonetheless, in the general scenario, the specific destination of each agent's travel is not the primary concern. The crucial aspect is that if an agent does travel, their departure and return city remain consistent, aligning with the conditions of the dual scenario.

The general scenario where only the simplest ticket buying strategy is allowed (buy a round trip ticket for each week) is a simple assignment problem. The objective is to assign each week to an agent such that the cost, defined as the round-trip expense for that agent in a given week, is minimized.

This is a trivial assignment problem. A simple greedy algorithm works: for each week i, we determine which agent incurs the lowest cost to fly and assign that agent to the week. This process is repeated for each subsequent week.

When considering nested ticketing, the problem can still be framed as an assignment problem, albeit with an altered cost structure. The task remains to assign specific weeks to agents. However, in this context, the cost of each assignment is contingent upon the *specific set of weeks* allocated to an agent. Once an assignment is established, the cost for each agent can be calculated using an algorithm designed to solve the solo scenario.

In the next section, we model our scenarios as graph matching problems with extra constraints. As a preview, we show the computational complexity of solving different scenarios in Fig. 2.

Travelers	1	2	$p \geq 3$
Nested ticketing	$n^{2+o(1)}$	NP-hard	NP-hard
Interleaved ticketing	$O(n^k)$	$O(n^{2k})$	$O(n^{pk})$

Fig. 2. The computational complexity of finding the optimum cost solution to the general scenario if nested ticketing or interleaved ticketing is allowed, for k airlines, where k is a constant.

2.2 Model

Consider the general scenario. For a schedule spanning n weeks, we build a multigraph comprising $2n$ vertices. The edges of this graph are labeled by a function $\ell : E \longrightarrow [p]$, where each label corresponds to a specific agent. In this representation, the $(2i - 1)$-th vertex signifies the outbound flight (departing from the agent's location) in the i-th week, while the $(2i)$-th vertex denotes the inbound flight (returning to the agent's location) in the same week. Here, i ranges from 1 to n, inclusively.

In a labelled graph where the vertices are $[2n]$, a matching is *homogeneous* if $(2i - 1)$ and $(2i)$ are covered by edges of the same label, for each i. A round-trip ticket for agent q, originating in the i-th week and returning in the j-th week with a cost of c, is modeled as an undirected edge connecting vertices $(2i - 1)$ and $(2j)$, bearing the label q and assigned the weight c. In a similar vein, a one-way ticket is represented by a self-loop. Note the graph is a bipartite graph. A homogeneous matching make sure the in and out flight of a week is taken by the same agent. Perfect matching forces each flight has to be taken. In order to minimize the total cost of the tickets, it translates to finding a homogeneous perfect matching of minimum weight.

Therefore, we are interested in solving the following problem

Problem 1 (Minimum weight homogeneous BPM ($\textbf{HM}(k)$)).
Input: A bipartite graph $G = ([2n], E)$ where the bipartition is all the even vertices and all the odd vertices. There is a label function on the edges $\ell : E \longrightarrow [k]$, and there is a weight function $w : E \longrightarrow \mathbb{R}$.
Output: A homogeneous perfect matching of minimum weight.

Now, if we impose a restriction to avoid nested ticketing for each airline, this can be modeled by introducing a color, to each edge. Specifically, for each ticket, we assign a color corresponding to its respective airline.

We say a matching M of $G = ([2n], E)$ is *linear*, if for any two edges of the same color $\{a, b\}, \{c, d\} \in M$, we have $[a, b] \cap [c, d] = \emptyset$. Here $[a, b]$ means the real interval from a to b.

To minimize the total cost of tickets in the general scenario that also does not allow nested ticketing (but allows interleaved ticketing), is equivalent to the following algorithmic problem.

Problem 2 (Minimum weight linear homogenous BPM ($\textbf{LM}(k, p)$)
Input: A bipartite graph $G = ([2n], E)$, an edge weight function $w : E \longrightarrow \mathbb{R}$, edge label function $\ell : E \longrightarrow [p]$ and edge color function $c : E \longrightarrow [k]$.
Output: A linear homogenous perfect matching M of minimum weight.

Finally, we want to add an additional problem just to handle the fair dual scenario, allowing only interleaved ticketing.

Problem 3 (Minimum weight fair linear homogenous BPM ($\textbf{FM}(k)$).
Input: A bipartite graph $G = ([2n], E)$ where n is an even number, an edge weight function $w : E \longrightarrow \mathbb{R}$, edge label function $\ell : E \longrightarrow [2]$ and edge color function $c : E \longrightarrow [k]$.

Output: A linear homogenous perfect matching M, so either label 0 or label 1 vertices appears exactly $n/2$ times. Under previous constraints, the weight is minimized.

In a solution, we say a vertex has label (color) k, if it is contained in an edge of label (color) k. In a solution, a vertex is the start vertex, if it is not the larger numbered vertex. Otherwise, it is the end vertex. Note in particular, this means a vertex with a self-loop is an end vertex.

3   Minimum Weight Homogeneous BPM

We set up the problem with a bipartite graph $G = ([2n], E)$. For simplicity, it is more helpful to consider the vertices partitions to be $U = \{u_1, \ldots, u_n\}$ and

$V = \{v_1, \ldots, v_n\}$, where u_i and v_i are supposed to be contained in edges of the same label. We use $G[U]$ to mean the *induced subgraph* of G on the vertex set U.

HM(1) is the standard minimum weight bipartite perfect matching problem, which can be solved in polynomial time [14]. Next, we will show **HM**(2) is a NP-hard problem. Since **HM**(i) is no harder than **HM**(j) for $i \leq j$, it proves **HM**(k) is NP-hard for all $k \geq 2$.

For the reduction, we need a few definitions. A matching is called a *rainbow matching*, if each edge in the matching has a different color.

Problem 4 (At-most-2 Rainbow BPM (A2RBPM).
Input: A bipartite graph $G = (V, E)$ with vertex bipartition A and B. The edges are partitioned into color classes $E_1, \ldots, E_m$, such that $|E_i| = 2$ for each i. Also, the edges in each E_i are disjoint.

Output: A perfect matching M that is a rainbow matching.

A2RBPM is known to be NP-complete [8]. We will reduce **A2RBPM** to **HM**(2) (Fig. 3).

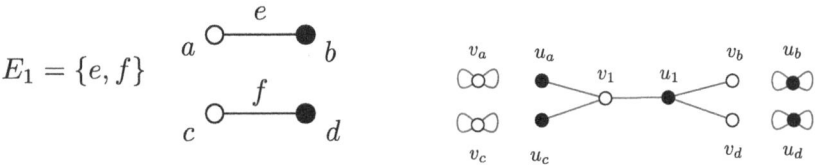

$E_1 = \{e, f\}$

(a) The color class E_1 for the rainbow matching problem.

(b) The transformed gadget.

Fig. 3. The reduction gadget.

Theorem 1. HM(2) *is NP-Hard.*

Proof. We begin the description of the reduction to **A2RBPM**. Let $G = (A \cup B, E)$, $E_1, \ldots, E_m$ be an instance of **A2RBPM**. We want to construct an instance to **HM**(2), with input $H = (U \cup V, F)$, label function ℓ and weight function w, such that G has a rainbow perfect matching if and only if the H has a homogenous matching. Note our weight function will just be the all 0 weight function.

The idea is each color class in G transforms into two vertices in H.

For each vertex $a \in A$, we construct two vertices $u_a \in U$ and $v_a \in V$. v_a is there only for formality. We add self-loops of each label to v_a. Similarly, for each vertex $b \in B$, we construct $u_b \in U$ and $v_b \in V$ where u_b has a self-loop of each label. We call v_a and u_b dummy vertices.

For each color class $E_i = \{e, f\}$, we create vertices u_i and v_i and add to U and V, respectively. We add an edge between u_i and v_i for each label. Assuming that $e = \{a, b\}$ and $f = \{c, d\}$. We add edges $u_a v_i$ and $v_b u_i$ of label 0. We also add edges $u_c v_i$ and $v_d u_i$ of label 1. This forms a single gadget. This reduction takes polynomial time.

We show that a rainbow matching in G exists if and only if a homogeneous perfect matching in H exists.

Without loss of generality, assume the rainbow matching in G using edges $a_i b_i \in E_i$ for i from 1 to n. This would imply a homogeneous perfect matching in H as follows: u_i and v_i is matched to u_{a_i}, and v_{b_i}, and has the same label (either 0 or 1 depending on construction). By our construction, u_a and v_b are matched for $a \in A$ and $b \in B$. For all color classes E_j, we add $u_j v_j$ into the matching. Finally, for each dummy vertex, add the self-loop with the corresponding vertex's label.

On the other hand, suppose H has a homogeneous perfect matching M. We construct a rainbow perfect matching M' in G. We look at each i where $u_i v_i \notin M$. Because M is a perfect matching, then u_i matched with v_a and v_i matched with u_b for some a and b. We add the edge $e = ab$ to M'. M' is a rainbow matching since the edges do not have the same color. M' is a perfect matching, since if u_a is matched in M for $a \in A$, then a is also matched in M'. It holds true for v_b where $b \in B$.

Hence, this shows **A2RBPM** reduces to **HM**(2), and therefore **HM**(2) is NP-hard.

4   Minimum Weight Linear Homogeneous BPM

In this section, we show a polynomial time algorithm for **LM**(k, p) for fixed k and p.

4.1   Single Label

For ease of understanding, we start our discussion with the scenario involving only a single label. In this case, the primary focus is on colors rather than labels. We build a solution through dynamic programming.

Imagine building the solution incrementally by sequentially inspecting vertices from 1 to n. At the i-th step, the task is to determine the color of the vertex and decide whether this vertex represents the start or end of an edge of that color. We then move to the next vertex and continue this process until a complete solution is formed. Every solution can be constructed in this manner: one simply needs to take any given solution and follow its pattern, performing the construction sequentially from vertices 1 to n. What information is required for each decision? As an example, if we chose the vertex to have the color red, if the last red vertex is not an end vertex of an edge, then the current vertex has to be an end vertex of a red edge.

Consequently, when assigning a specific color to a vertex, it is imperative to know the location of the previous vertex of the same color and whether it is an end vertex.

Therefore, we are able to derive the required dynamic programming formulation. This information can be effectively encoded using two vectors, A and B. Here, A_i represents the index of the preceding vertex with color i, while B_i is assigned the value of 1 if the preceding vertex of color i is a start vertex, and 0 if otherwise.

Formally, let $A \in ([2n] \cup \{\infty\})^k$. We say A is *valid*, if all the non-infinity elements in A are distinct. Let $\mathcal{A}$ be all the valid vectors. We let $\mathcal{A}_m$ to be all valid vectors where the maximum element is m. $B \in \{0,1\}^k$. Define $\mathcal{M}(A,B)$ to be the set of feasible solutions such that the last vertex before m of color i is A_i, and there is an edge $\{A_i, t\}$ for some $t > i$ if and only if $B_i = 0$. Note if there is no vertex before m of color i, A_i would be ∞. Define $D[A,B,m] = \min_{M \in \mathcal{M}(A,B)} w(M[V_m])$, where V_m are the first m vertices.

The solution to our problem is precisely $\min\{D[A,B,2n] \mid A \in \mathcal{A}_{2n}, B \in \{0,1\}^k\}$. We describe the recurrence relation, where A is a valid vector with $\max(A) = m$, and $A_i = m$.

First, in order to construct a solution captured by $D[A,B,m]$, we can look at what is happening before, at $D[A', B', m-1]$. Observe here $A'_j = A_j$ for all $j \neq i$, and $B'_j = B_j$ for all $j \neq i$. Therefore, we just have to know about A'_i and B'_i. We consider two different cases. If $B_i = 1$, then this implies m is an end vertex of color i, and we can ask where is the start vertex. Again, there are two cases: either the start vertex is m, or the start vertex is before m. In the first case, we can see A'_i to be anything other than the positive elements in A'_j and $B'_i = 1$. In the second case, we would mark A'_i to be any valid number, and set $B'_i = 0$. Subsequently, if $B_i = 0$, this situation becomes straightforward. We simply choose A'_i to be any valid number and then set $B'_i = 1$.

The base case is when $m = 0$, which we would define it to be 0. The pseudocode can be seen in Fig. 4.

Observe that the number of states in the dynamic program initially appears to be $O(n^k 2^k n)$. However, the last parameter is actually determined by the first vector, reducing the number of states to $O(n^k 2^k)$. To compute a single state, we require $O(n)$ time. Thus, the total running time would be $O(n^{k+1} 2^k)$. A more refined analysis, however, allows us to eliminate a factor of n.

Consider a graph representing the states in the dynamic program. There is an edge from state $(A', B', m-1)$ to state (A, B, m) if $D[A,B,m]$ calls $D[A',B',m-1]$. We assert that this graph has at most $O(kn^k 2^k)$ edges. Indeed, for a given $(A', B', m-1)$, there are at most $2k$ different (A, B, m) that depend on it. This is because A can differ from A' in only one coordinate, and if different, that coordinate must be m. Thus, the number of possible A is at most k. Similarly, B and B' can differ in at most one position, which must align with the same position of A, allowing at most 2 choices. As such, the number of outgoing edges from $(A', B', m-1)$ is at most $2k$. Consequently, the dynamic program operates in $O(kn^k 2^k)$ time. For a fixed k, the runtime is $O(n^k)$, leading us to the following theorem.

Theorem 2. *$LM(k,1)$ can be solved in $O(n^k)$ time for a graph on n vertices for fixed k.*

```
D[A, B, m]:
  if m = 0 :
      return 0
  i ← the index where A_i = m
  A' ← A
  B' ← B
  X ← ∅
  N ← [m − 1] \ {A_j|j ∈ [k]} ∪ {∞}
  for n' in N:
      A'_i ← n'
      if B_i = 0:
          B'_i ← 1
          X ← X ∪ {D[A', B', m − 1]}
      if B_i = 1:
          B'_i ← 0
          X ← X ∪ {D[A', B', m − 1] + w(e(A'_i, m, i))}
          B'_i ← 1
          X ← X ∪ {D[A', B', m − 1] + w(e(m, m, i))}
  return min(X)
```

Fig. 4. Pseudocode to find the optimal value. Assume $e(a, b, i)$ returns the edge between a and b with color i.

4.2 Multiple Labels

The idea is basically identical to the single label version. However, we now need to concurrently track both the color and label of each vertex. We define the ordered pair comprising the color and label of a vertex as the vertex's signature. For instance, a vertex with label i and color j possesses the signature (i, j).

In place of the vectors A and B, we now require matrices A and B. Here, $A_{i,j}$ represents the last vertex that holds the signature (i, j), while matrix B indicates whether the last vertex with signature (i, j) is an end vertex. The complete pseudocode for this dynamic programming approach is provided in Fig. 5. The primary modification from the single label version is the enforcement that the label of vertex $(2i − 1)$ must match that of vertex $(2i)$. Utilizing a similar process, we arrive at our main theorem.

Theorem 3. $LM(k, p)$ can be solved in $O(n^{kp})$ time for a graph on n vertices when k and p are fixed.

Remark 1. If we define k_q to be the number of possible colors when fixing the label to be q, then the running time would be $O(n^{\sum_{q=1}^{p} k_q})$.

Furthermore, it is straightforward to observe that $FM(k)$ can be resolved by additionally recording the number of occurrences of label 0 in the solution. This modification results in an increase in the running time by an additional factor of $O(n)$ compared to $LM(k, 2)$. Consequently, we establish the following theorem.

Theorem 4. $FM(k)$ can be solved in $O(n^{2k+1})$ time for fixed k.

$D[A, B, m]$:
  if $m = 0$:
    return 0
  $i, q \leftarrow$ the index where $A_{i,q} = m$
  $A' \leftarrow A$
  $B' \leftarrow B$
  $X \leftarrow \emptyset$
  ⟪*Make sure the labels are the same.*⟫
  if $i\%2 = 0$ and $m - 1 \notin \{A_{j,q} | j \in [k]\}$:
    $N \leftarrow \{m - 1\}$
  else:
    $N \leftarrow [m - 1] \setminus \{A_{j,q'} | j \in [k], q' \in [p]\} \cup \{\infty\}$
  for n' in N:
    $A'_{i,q} \leftarrow n'$
    if $B_{i,q} = 0$:
      $B'_{i,q} \leftarrow 1$
      $X \leftarrow X \cup \{D[A', B', m - 1]\}$
    if $B_{i,q} = 1$:
      $B'_{i,q} \leftarrow 0$
      $X \leftarrow X \cup \{D[A', B', m - 1] + w(e(A'_{i,q}, m, i, q))\}$
      $B'_{i,q} \leftarrow 1$
      $X \leftarrow X \cup \{D[A', B', m - 1] + w(e(m, m, i, q))\}$
  return $\min(X)$

Fig. 5. Pseudocode to find the optimal value for multiple labels. Assuming $e(a, b, i, q)$ returns the edge between a and b with color i and label q.

5   Experiments

We acquired a dataset of airfare prices from Google Flights, which includes the ticket information from New York to Los Angeles for ten weeks starting from March 11, 2024 (Monday). During the data collection, we limited the airlines to three commonly seen carriers on this route, namely American, United, and Delta, and the tickets were specifically for flights departing on Monday and returning on Friday.

For comparative analysis, we conducted three distinct experimental sets. In the first strategy, termed 'simple round trip', we purchased round-trip tickets on a weekly basis without allowing for any nesting. In the second strategy, we permitted interleaved ticketing. The third and final strategy allowed for nested ticketing. The outcomes of these different strategies are detailed in Table 1.

As indicated in Table 1, for purchasing tickets across consecutive weeks of future travel, interleaved ticketing typically offers a savings of about 10% compared to the simple round trip strategy. While nested ticketing may be more cost-effective, this method is prohibited by airlines. With only a marginal price difference compared to nested ticketing, interleaved ticketing emerges as the more advantageous ticket-purchasing strategy for compliance with airline policies.

Table 1. The cost for solo scenario, in US dollars.

Price	Week Count				
	4 weeks	5 weeks	6 weeks	7 weeks	8 weeks
Simple Round Trip	1352	1687	2022	2366	2701
Interleaved Ticketing	1221	1501	1836	2137	2424
Nested Ticketing	1193	1464	1798	2098	2387

Table 2. The cost for dual scenario, in US dollars.

Price	Week Count				
	4 weeks	5 weeks	6 weeks	7 weeks	8 weeks
Simple Round Trip	1297	1632	1967	2311	2646
Interleaved Ticketing	1183	1463	1753	2043	2378
Nested Ticketing	1183	1463	1753	2043	2378

In the dual scenario, further experiments were conducted and comparatively analyzed. The results of these experiments are displayed in Table 2.

The findings align with our expectations. Similar to the solo scenario, interleaved ticketing in the dual scenario yields savings of over 10% compared to simple round trips. Interestingly, for our specific data, nested ticketing is not more cost-effective than interleaved ticketing, despite differences in flight selections under varying constraints. This suggests that the optimal solution might inherently exclude any nesting within airlines.

When considering fairness between two individuals, we documented the results under a specific scenario where both parties fly an equal number of times over a six-week period. The results are as follows: the cost for a simple round trip totaled $1967, while both interleaved and nested ticketing strategies resulted in a cost of $1764, approximately 10% less expensive.

Acknowledgements. Chao like to thank Lingyu Xu for the inspiration of the problem, Naonori Kakimura for helpful pointers related to constrained matchings, and 美卡论坛, where he learned about airline booking ploys.

References

1. Back-to-back tickets. to do or not to do? (1999). https://www.flyertalk.com/forum/milesbuzz/938-back-back-tickets-do-not-do.html. Accessed 14 Jan 2024
2. Agent International Passenger Rules Airline Tariff Publishing Company and Fares. Tariff no. aa1 (2023). https://www.aa.com/content/images/tariff/american-airlines-general-rules-of-the-international-tariff.pdf. Accessed 14 Jan 2024
3. Aslani, S., Modarres, M., Sibdari, S.: On the fairness of airlines' ticket pricing as a result of revenue management techniques. J. Air Transp. Manag. **40**, 56–64 (2014)

4. Bischoff, G., Maertens, S., Grimme, W.: Airline pricing strategies versus consumer rights. Transp. J. **50**(3), 232–250 (2011)
5. Cook, G.N., Billig, B.G.: Airline Operations and Management: A Management Textbook, 2nd edn. Routledge, London (2023)
6. Currie, C.S.M., Cheng, R.C.H., Smith, H.K.: Dynamic pricing of airline tickets with competition. J. Oper. Res. Soc. **59**(8), 1026–1037 (2008)
7. Delta. Delta domestic general rules tariff (2023). https://www.delta.com/content/dam/delta-www/pdfs/dl-dgr-master-22jun23.pdf. Accessed 14 Apr 2024
8. Le, V.B., Pfender, F.: Complexity results for rainbow matchings. Theoret. Comput. Sci. **524**, 27–33 (2014)
9. Meire, S., Derudder, B.: Pirating the skies? A review of airline booking ploys. Res. Transp. Bus. Manag. **43**, 100721 (2022)
10. U.S. Government Accountability Office. Aviation competition: Restricting airline ticketing rules unlikely to help consumers. Report to Congressional Committees (2001)
11. Jaelynn, O., Huh, W.: Hidden city travel and its impact on airfare: the case with competing airlines. Transp. Res. Part B: Methodol. **156**, 101–109 (2022)
12. Otero, D.F., Akhavan-Tabatabaei, R.: A stochastic dynamic pricing model for the multiclass problems in the airline industry. Eur. J. Oper. Res. **242**(1), 188–200 (2015)
13. Rakowski, J.J.: Does the consumer have an obligation to cooperate with price discrimination? Bus. Ethics Q. **14**(2), 263–274 (2004)
14. Schrijver, A.: Combinatorial Optimization. Algorithms and Combinatorics, 2003rd edn. Springer, Berlin (2002)
15. Inc. United Airlines. Contract of carriage document (2023). https://www.united.com/en/us/fly/contract-of-carriage.html. Accessed 14 Jan 2024
16. Wang, Z., Ye, Y.: Hidden-city ticketing: the cause and impact. Transp. Sci. **50**(1), 288–305 (2016)
17. Williams, K.: Dynamic airline pricing and seat availability. Cowles Foundation discussion paper (2020)

Longest $(k]$-Tuple Common Substrings

Tiantian Li[1], Haitao Jiang[1], Lusheng Wang[2], and Daming Zhu[1(✉)]

[1] School of Computer Science and Technology, Shandong University, Qingdao, China
{htjiang,dmzhu}@sdu.edu.cn
[2] Department of Computer Science, City University of Hong Kong, Hong Kong, China
cswangl@cityu.edu.hk

Abstract. A $(k]$-*tuple common substring* (abbr. $(k]$-CSS) is a common subsequence of two or more given strings including at most k common substrings. The complexity of finding a longest $(k]$-CSS of 2 strings is still open. We present a dynamic programming algorithm for finding a longest $(k]$-CSS of two strings in $O(kn_1n_2)$ time and space where n_1 and n_2 are the given string lengths. To breakthrough the quadratic space complexity, we present an algorithm for finding a longest $(k]$-CSS in $O(kn_1n_2)$ time and $O(n_1 + kn_2)$ space.

Keywords: Algorithm · Complexity · Common substring

1 Introduction

String matching has greatly influenced computer science and play an essential role in various real-world problems such as genome data analysis and data compression [8,9]. Finding common subsequences among given strings is a representative and attractive topic in string matching [6]. A longest common subsequence (abbr. LCS) of two strings has been known able to be found in quadratic time and space and even better, in quadratic time and linear space [7,13]. In terms of the time complexity bound, an LCS of two strings of length n can be found in $O(\frac{n^2}{\log n})$ time if the alphabet is finite [11] whereas finding an LCS of two strings in $O(n^{2-\epsilon})$ time refutes the popular Strong Exponential Time Hypothesis [1].

Since LCS is usually short of measuring string similarity caused by consecutively matching letters, Benson et al. introduced LCS_t and $LCS_{\geq t}$, a type of longest common subsequences of two strings in which all common substrings are of length t or at least t of two strings [3,4]. LCS_t and $LCS_{\geq t}$ can both be found in $O(n^2)$ time and $O(tn)$ space if the input strings are of length n [14].

Common substrings are meaningful to characterize *conserved areas* in bio-sequences [5,10]. In many situations, conservative areas in bio-sequences are asked with not only order-consistent but also number limited common substrings. In eukaryotes for example, a gene that codes for a protein is typically made up of (4 to 20) alternating exons and introns where exons contribute to the protein codes whereas introns do not [2]. A cDNA (a string) consists only of the concatenation of exons in a gene.

To reflect characteristics of conservative areas such as cDNA, a common subsequence of m (≥ 2) strings with at most k common substrings, called a $(k]$-*tuple common substring* (abbr. $(k]$-CSS), has been proposed [10]. It is NP-Hard to find a longest

B. Li et al. (Eds.): IJTCS-FAW 2024, LNCS 14752, pp. 106–114, 2025.
https://doi.org/10.1007/978-981-97-7752-5_8

$(k]$-CSS of m strings if k and m are under no restriction of their values [12]. A longest $(k]$-CSS of m strings can be found in $O(mn^k)$ time and $O(kmn)$ space where n is the length sum of given strings. This implies finding a longest $(k]$-CSS of m strings is computationally easy, if k is a constant [10]. However, the complexity of finding a longest $(k]$-CSS where there is no constant limitation on k of two strings is still open. Dynamic programming might be a primary idea one can think of for finding a longest $(k]$-CSS of two strings. It has not been seen doing well if one wishes to ask only one recursive formula for expressing the length of a longest $(k]$-CSS of two strings as that has been done for finding an LCS or an LCS_t.

We propose two recursive formulae of a system to express the longest $(k]$-CSS length of two strings and present a dynamic programming algorithm to find a longest $(k]$-CSS of two strings in $O(kn_1 n_2)$ time and space where n_1 and n_2 are lengths of the given strings. As we know, this is the first time for using two parallel formulae instead of one to express the length of a LCS-marked common subsequence of two strings for algorithm design. To break through the quadratic space limitation of this algorithm, we present to improve this algorithm to run in $O(kn_1 n_2)$ time and $O(n_1 + kn_2)$ space (in Sect. 3).

2   Preliminaries

Let Σ denote a finite alphabet. A symbol string is referred to as *on* Σ if all symbols in the string are in Σ. The number of symbols in a string s is referred to as the *length* of s and will be denoted by $|s|$. For a string $s = s[1] \ldots s[n]$ on Σ, we denote by $s[i, j]$ where $1 \le i \le j \le n$ the consecutive substring $s[i] \ldots s[j]$ of s. Two strings $s_1 = s_1[1, k]$ and $s_2 = s_2[1, k]$ are referred to as *identical* if $s_1[i] = s_2[i]$ for $i \in [1, k]$. We denote it as $s_1 = s_2$ for two strings s_1 and s_2 to be identical. The concatenation of two strings s_1 and s_2 is $s_1 \parallel s_2$ where $\parallel$ will be used as an operator to concatenate two strings all through. Let $s = s[1, n]$ and $\lambda = \lambda[1, l]$ be two strings on Σ with $l \le n$. The substring $s[t, t + l - 1]$ for $t \in [1, n - l + 1]$ is referred to as an *occurrence* of λ if $\lambda = s[t, t + l - 1]$. A *begin-point* (resp. *endpoint*) of λ in s is t (resp. $t + l - 1$), if $s[t, t + l - 1]$ is an occurrence of λ in s. Let $\lambda_1, \ldots, \lambda_l$ be a sequence of non-null strings on Σ. We refer to $s[\alpha[1], \alpha[1] + |\lambda_1| - 1], \ldots, s[\alpha[l], \alpha[l] + |\lambda_l| - 1]$, a sequence of respective occurrences of $\lambda_1, \ldots, \lambda_l$ in s as an *occurrence* of the sequence $\lambda_1, \ldots, \lambda_l$ in s, if $\alpha[i] + |\lambda_i| \le \alpha[i+1]$ for $i \in [1, l - 1]$.

Let s_1 and s_2 denote two strings on Σ and $S = \{s_1, s_2\}$. We refer to a sequence of $l \ge 1$ non-null strings $\lambda_1, \ldots, \lambda_l$ on Σ as an *l-tuple common substring* of S, if there are respective occurrences of the sequence $\lambda_1, \ldots, \lambda_l$ in both s_1 and s_2. An l-tuple common substring will be abbreviated as an l-CSS. A 1-CSS of S is a *common substring* of S. We shall denote by $\lambda_1 \oplus \lambda_2 \oplus \ldots \oplus \lambda_l$ the l-CSS of S constituted by the sequence of common substrings $\lambda_1, \ldots, \lambda_l$ of S. Let $\Lambda = \lambda_1 \oplus \lambda_2 \oplus \ldots \oplus \lambda_l$ denote an l-CSS of S. Then λ_t is the t-th *member* in Λ for $t \in [1, l]$ and λ_1 and λ_l are the *first* and the *last* in Λ. *The length* of an l-CSS refers to the length sum of all members in the l-CSS. We shall say λ_t for $t \in [1, l]$ is *with a begin-point* at $\alpha[t]$ in s_1 (resp. s_2), if $s_1[\alpha[t], \alpha[t] + |\lambda_t| - 1] = \lambda_t$ is the t-th member in an occurrence of Λ in s_1 (resp. s_2). If in s_1 (resp. s_2), λ_t is with a begin-point at $\alpha[t]$, then it is with an endpoint at $\alpha[t] + |\lambda_i| - 1$. Let Λ_1 and Λ_2

denote an l_1-CSS and an l_2-CSS of S. If the last member of Λ_1 is with an endpoint at less than a begin-point of the first member of Λ_2 in both s_1 and s_2, we shall denote by $\Lambda_1 \oplus \Lambda_2$ the $(l_1 + l_2)$-CSS of S combined from Λ_1 and Λ_2.

An l-CSS of S with $l \leq k$ will also be referred to as a $(k]$-CSS of S. A $(k]$-CSS of S is *longest*, if its length is no less than that of any $(k]$-CSS of S. The problem of longest $(k]$-CSS of two strings is formulated as follows.

Definition 1. *Longest* $(k]$*-CSS of two strings (abbr. $LCSS_k$): Given a set* $S = \{s_1, s_2\}$ *and an integer* k, *where* s_1 *and* s_2 *are two strings on* Σ. *Find a longest* $(k]$*-CSS of* S *and an occurrence of it in both* s_1 *and* s_2.

3   Longest $(k]$-CSS of Two Strings

Assume $s_1 = s_1[1, n_1]$ and $s_2 = s_2[1, n_2]$ are two strings on Σ. Let $S = \{s_1, s_2\}$ and a given integer k constitute an instance of $LCSS_k$. First, we present a dynamic programming algorithm for $LCSS_k$ whose time and space complexity are $O(kn_1n_2)$. In our experience, an $O(kn_1n_2)$ space algorithm cannot work to solve $LCSS_k$ on a personal computer with memory limited to 16 GB even if $n_1 + n_2 = 30,000$ and $k = 2$. To overcome this barrier, we present a dynamic programming plus divide-and-conquer algorithm for $LCSS_k$ whose time and space complexity are $O(kn_1n_2)$ and $O(n_1 + kn_2)$ respectively.

3.1   A Dynamic Programming Algorithm

To get a longest $(k]$-CSS of S by dynamic programming, we present two recursive formulae of a system expressing the length of a longest $(t]$-CSS of $\{s_1[1, i], s_2[1, j]\}$ from two aspects for $i \in [0, n_1]$, $j \in [0, n_2]$ and $t \in [0, k]$.

Let $A(i, j, t, 1)$ denote the set of $(t]$-CSS of $\{s_1[1, i], s_2[1, j]\}$ whose last member is with endpoints at i in s_1 and j in s_2 respectively. We denote by $F(i, j, t, 1)$ the length of a longest $(t]$-CSS over all in $A(i, j, t, 1)$. Let $A(i, j, t, 0)$ denote the set of $(t]$-CSS of $\{s_1[1, i], s_2[1, j]\}$ whose last member is with an endpoint at less than i in s_1 or less than j in s_2. We denote by $F(i, j, t, 0)$ the length of a longest $(t]$-CSS over all in $A(i, j, t, 0)$.

Since $A(i, j, t, 1) \cup A(i, j, t, 0)$ is the set of $(t]$-CSS of $\{s_1[1, i], s_2[1, j]\}$, the length of a longest $(t]$-CSS of $\{s_1[1, i], s_2[1, j]\}$ is $\max\{F(i, j, t, 1), F(i, j, t, 0)\}$. Let us see how to recursively express $F(i, j, t, 0)$ and $F(i, j, t, 1)$ for $i \in [0, n_1], j \in [0, n_2]$ and $t \in [0, k]$. If $i = 0$ or $j = 0$ or $t = 0$, then let $F(i, j, t, 0) = 0$ and $F(i, j, t, 1) = 0$.

Lemma 1.

$$F(i, j, t, 0) = \begin{cases} 0 & if\ i \cdot j \cdot t = 0; \\ \max_{q \in \{0,1\}} \begin{cases} F(i-1, j, t, q) \\ F(i, j-1, t, q) \end{cases} & otherwise. \end{cases} \tag{1}$$

Proof. Since $A(i, j, t, 0) = A(i-1, j, t, 1) \cup A(i, j-1, t, 1) \cup A(i-1, j, t, 0) \cup A(i, j-1, t, 0)$ for $i > 0, j > 0$ and $t > 0$, $F(i, j, t, 0) = \max\{F(i-1, j, t, 0), F(i-1, j, t, 1), F(i, j-1, t, 0), F(i, j-1, t, 1)\}$. The recursive expression for $F(i, j, t, 0)$ is given in Formula (1).

We proceed to get $F(i, j, t, 1)$ recursively. If $s_1[i] = s_2[j]$, then $F(i, j, t, 1) > 0$. If $s_1[i] \neq s_2[j]$, let us appoint $F(i, j, t, 1) = 0$.

Lemma 2.

$$F(i, j, t, 1) = \begin{cases} 1 + \max \begin{cases} F(i-1, j-1, t, 1) \\ F(i-1, j-1, t-1, 0) \end{cases} & \text{if } s_1[i] = s_2[j]; \\ 0 & \text{otherwise.} \end{cases} \qquad (2)$$

Proof. Since the last members of all (t)-CSS in $A(i, j, t, 1)$ are with endpoints at i in s_1 and j in s_2, $s_1[i] = s_2[j]$. Since $F(i, j, t, 1)$ is the length of a longest one in $A(i, j, t, 1)$, $F(i, j, t, 1)$ can be obtained from the following two sources.

If $s_1[i]$ is an extension of the last member of a longest (t)-CSS in $A(i-1, j-1, t, 1)$, then $F(i, j, t, 1) = F(i-1, j-1, t, 1) + 1$. Otherwise, $s_1[i]$ is the first symbol of the last member of a longest (t)-CSS in $A(i, j, t, 1)$, then $s_1[i-1] \neq s_2[j-1]$. In this situation, $F(i, j, t, 1) = F(i-1, j-1, t-1, 0) + 1$. Because of this, $F(i, j, t, 1)$ is expressed recursively by Formula (2).

A dynamic programming algorithm arises from Formulae (1) and (2) to obtain $F(i, j, t, 1)$ and $F(i, j, t, 0)$ for $i \in [0, n_1]$, $j \in [0, n_2]$ and $t \in [0, k]$. The length of a longest (k)-CSS of S is $\max\{F(n_1, n_2, k, 1), F(n_1, n_2, k, 0)\}$. The endpoints of members in a longest (k)-CSS of S in s_1 and s_2 can be obtained along a traceback path from $F(n_1, n_2, k, 1)$ or $F(n_1, n_2, k, 0)$ to some $F(i, j, t, q)$ with $i \cdot j \cdot t = 0$ and be stored in a $2 \cdot k$ array. As long as the endpoints of members in a longest (k)-CSS in s_1 and s_2 are reported, all members in a longest (k)-CSS can be identified in linear time of the size of input strings.

It needs to compute $O(kn_1n_2)$ cells of $F(i, j, t, 0)$ and $F(i, j, t, 1)$ by dynamic programming. It takes $O(1)$ time to get the value of a cell by Formulae (1) and (2). The time and space complexity of the dynamic programming algorithm for $LCSS_k$ is $O(kn_1n_2)$. This implies that $LCSS_k$ is polynomial-time solvable without restriction on k, even though the longest (k)-CSS problem for m strings is NP-Hard without restriction on both m and k.

Theorem 1. *$LCSS_k$ can be solved by dynamic programming in $O(kn_1n_2)$ time and $O(kn_1n_2)$ space, where n_1 and n_2 are lengths of the two input strings.*

3.2 Dynamic Programming Meets Divide-and-Conquer

In this subsection, we manage to improve dynamic programming algorithm so that $LCSS_k$ is solved in $O(kn_1n_2)$ time and $O(n_1 + kn_2)$ space.

If $n_1 \cdot n_2 \cdot k = 0$, then there is no need to find a longest (k)-CSS of $\{s_1, s_2\}$. If $n_1 = 1$ or $n_2 = 1$ or $k = 1$, then a longest 1-CSS of $\{s_1, s_2\}$ is a satisfactory solution that can be found in linear time. Assume $n_1 > 1$, $n_2 > 1$ and $k > 1$ below. To solve $LCSS_k$ by divide-and-conquer, we start with dividing $LCSS_k$ into smaller subproblems.

a) Divide $LCSS_k$ into smaller subproblems. We propose to cut s_1 and s_2 into the same number of substrings that constitute subproblems of $LCSS_k$, then find solutions of the subproblems and combine them into a longest $(k]$-CSS of S.

For a string $s = s[1] \ldots s[n]$ on Σ, let $\overline{s[i,j]}$ with $1 \le i \le j \le n$ denote $s[j] \ldots s[i]$, the inversion of $s[i,j]$. For an arbitrary $(k]$-CSS $\varLambda = \lambda_1 \oplus \ldots \oplus \lambda_l$ of S with $l \le k$, let $\overline{\varLambda} = \overline{\lambda_l} \oplus \ldots \oplus \overline{\lambda_1}$, that is a $(k]$-CSS of $\{\overline{s_1}, \overline{s_2}\}$.

For $i \in [1, n_1 + 1]$, $j \in [1, n_2 + 1]$ and $t \in [0, k]$, let $F^*(i, j, t, 1)$ denote the longest length of $(t]$-CSS of $\{s_1[i, n_1], s_2[j, n_2]\}$ whose last member is with endpoints at $n_1 - i + 1$ in $s_1[i, n_1]$ and $n_2 - j + 1$ in $s_2[j, n_2]$ and $F^*(i, j, t, 0)$ the longest length of $(t]$-CSS of $\{s_1[i, n_1], s_2[j, n_2]\}$ whose last member is with an endpoint at less than $n_1 - i + 1$ in $s_1[i, n_1]$ or less than $n_2 - j + 1$ in $s_2[j, n_2]$. Let $F^*(i, j, t, 1) = F^*(i, j, t, 0) = 0$ if $i = n_1 + 1$ or $j = n_2 + 1$ or $t = 0$.

To represent a $(k]$-CSS of S that is a combination of solutions of smaller subproblems of $LCSS_k$, let us specialize two forms of $(k]$-CSS of S that are combined from a multiple tuple common substring of $\{s_1[1, i], s_2[1, j]\}$ and a multiple tuple common substring of $\{s_1[i + 1, n_1], s_2[j + 1, n_2]\}$ for $i \in [0, n_1]$ and $j \in [0, n_2]$.

Form 1: For $t \in [0, k]$, let $\varLambda_1$ be a longest $(t]$-CSS of $\{s_1[1, i], s_2[1, j]\}$ and $\varLambda_2$ a longest $(k - t]$-CSS of $\{s_1[i + 1, n_1], s_2[j + 1, n_2]\}$. Then $\varLambda_1 \oplus \varLambda_2$ is a $(k]$-CSS of S.

Form 2: For $t \in [1, k]$, let $\varLambda_1$ be a longest $(t]$-CSS of $\{s_1[1, i], s_2[1, j]\}$ in $A(i, j, t, 1)$ and $\varLambda_2$ a longest $(k - t + 1]$-CSS of $\{s_1[i + 1, n_1], s_2[j + 1, n_2]\}$ whose last member is with endpoints at $n_1 - i$ in $s_1[i + 1, n_1]$ and $n_2 - j$ in $s_2[j + 1, n_2]$. Then $\varLambda_1 \oplus \overline{\varLambda_2}$ is a $(k]$-CSS of S.

Let $D(i, j, t)$ denote the length of a longest $(t]$-CSS of $\{s_1[1, i], s_2[1, j]\}$ and $D^*(i, j, t)$ the length of a longest $(t]$-CSS of $\{s_1[i, n_1], s_2[j, n_2]\}$. Then $D(i, j, t) = \max\{F(i, j, t, 1), F(i, j, t, 0)\}$ and $D^*(i, j, t) = \max\{F^*(i, j, t, 1), F^*(i, j, t, 0)\}$. It follows that $D(i, j, t) = 0$ if $i \cdot j \cdot t = 0$ and $D^*(i, j, t) = 0$ if $i = n_1 + 1$ or $j = n_2 + 1$ or $t = 0$.

For $i \in [1, n_1]$, $j \in [0, n_2]$ and $t \in [0, k]$, let $M_1(i, j, t)$ be the length of a $(k]$-CSS of S in Form 1, then $M_1(i, j, t) = D(i, j, t) + D^*(i + 1, j + 1, k - t)$. For $i \in [1, n_1]$, $j \in [0, n_2]$ and $t \in [1, k]$, let $M_2(i, j, t)$ be the length of a $(k]$-CSS of S in Form 2, then $M_2(i, j, t) = F(i, j, t, 1) + F^*(i + 1, j + 1, k - t + 1, 1)$. For an arbitrary $\hat{i} \in [1, n_1]$, the following formula holds

$$D(n_1, n_2, k) = \max\{ \max_{\substack{j \in [0, n_2] \\ t \in [0, k]}} M_1(\hat{i}, j, t), \ \max_{\substack{j \in [0, n_2] \\ t \in [1, k]}} M_2(\hat{i}, j, t)\} \tag{3}$$

By Formula (3), for a given $\hat{i} \in [1, n_1]$ there are $\hat{j} \in [0, n_2]$ and $\hat{t} \in [0, k]$ such that $D(n_1, n_2, k) = M_1(\hat{i}, \hat{j}, \hat{t})$ or there are $\hat{j} \in [0, n_2]$ and $\hat{t} \in [1, k]$ such that $D(n_1, n_2, k) = M_2(\hat{i}, \hat{j}, \hat{t})$. For a given $\hat{i}$, if $D(n_1, n_2, k) = M_2(\hat{i}, \hat{j}, \hat{t})$, then subproblems of $LCSS_k$ arise to ask for a longest $(t]$-CSS and a longest $(k - t + 1]$-CSS whose last members should be with fixed endpoints in s_1 and s_2. To design an algorithm for $LCSS_k$ that can avoid such subproblems, we re-voice the second form of a $(k]$-CSS of S.

Let $G(i, j)$ denote the length of the longest common suffix of $s_1[1, i]$ and $s_2[1, j]$. Then $G(i, j) = \max\{r \mid s_1[i - r + 1, i] = s_2[j - r + 1, j]\}$. If $s_1[i] = s_2[j]$, then $G(i,$

$j) > 0$. If $s_1[i] \neq s_2[j]$, then we appoint $G(i,j) = 0$. The following formula is to obtain $G(i,j)$ recursively:

$$G(i,j) = \begin{cases} G(i-1,j-1) + 1 & \text{if } s_1[i] = s_2[j]; \\ 0 & \text{otherwise.} \end{cases} \quad (4)$$

Let $G^*(i,j)$ denote the length of the longest common suffix of $\overline{s_1[i,n_1]}$ and $\overline{s_2[j,n_2]}$ and $G^*(i,j) = 0$ if $i = n_1 + 1$ or $j = n_2 + 1$ or $s_1[i] \neq s_2[j]$.

One can verify $D(i-1,j-1,t) + 1 \geq D(i,j,t)$ holds for $i \geq 1, j \geq 1$. Hence $M_2(i,j,t)$ is further expressed as

$$\begin{aligned} M_2(i,j,t) = {} & D(i - G(i,j), j - G(i,j), t-1) + G(i,j) + G^*(i+1,j+1) \\ & + D^*(i + G^*(i+1,j+1) + 1, j + G^*(i+1,j+1) + 1, k-t). \end{aligned} \quad (5)$$

Therefore if there is a longest (k)-CSS of S combined from a (t)-CSS of $\{s_1[1,i], s_2[1,j]\}$ and a $(k - t + 1)$-CSS of $\{\overline{s_1[i+1,n_1]}, \overline{s_2[j+1,n_2]}\}$ for $i \in [0,n_1]$ and $j \in [0,n_2]$ in Form 2, then $G(i,j) > 0$ and $G^*(i+1,j+1) > 0$ and, there is a longest (k)-CSS of S with length $M_2(i,j,t)$ in the following manner:

Form 2$^+$: For $t \in [1,k]$, let Λ_1 be a longest $(t-1)$-CSS of $\{s_1[1, i - G(i,j)], s_2[1, j - G(i,j)]\}$ and Λ_2 a longest $(k-t)$-CSS of $\{\overline{s_1[i^*,n_1]}, \overline{s_2[j^*,n_2]}\}$ where $i^* = i + G^*(i+1,j+1) + 1$ and $j^* = j + G^*(i+1,j+1) + 1$. Then $\Lambda_1 \oplus s_1[i - G(i,j) + 1, i^* - 1] \oplus \Lambda_2$ is a (k)-CSS of S.

By Formula (3) and the above analysis, one can always divide $LCSS_k$ into subproblems whose solutions can be combined into a solution of $LCSS_k$, a longest (k)-CSS of S in Form 1 or Form 2$^+$. For a given $\hat{i}$, if $\hat{j}$ and $\hat{t}$ have been known with $D(n_1, n_2, k) = M_1(\hat{i}, \hat{j}, \hat{t})$, then a longest (k)-CSS of S in Form 1 can be combined from a solution of $LCSS_{\hat{t}}$ for $\{s_1[1,\hat{i}], s_2[1,\hat{j}]\}$ and a solution of $LCSS_{k-\hat{t}}$ for $\{\overline{s_1[\hat{i}+1,n_1]}, \overline{s_2[\hat{j}+1,n_2]}\}$. If $D(n_1, n_2, k) = M_2(\hat{i}, \hat{j}, \hat{t})$, then a longest (k)-CSS of S in Form 2$^+$ can be combined from a solution of $LCSS_{\hat{t}-1}$ for $\{s_1[1,\hat{i} - G(\hat{i},\hat{j})], s_2[1,\hat{j} - G(\hat{i},\hat{j})]\}$ and a solution of $LCSS_{k-\hat{t}}$ for $\{\overline{s_1[\hat{i}^*,n_1]}, \overline{s_2[\hat{j}^*,n_2]}\}$ where $\hat{i}^* = \hat{i} + G^*(\hat{i}+1,\hat{j}+1) + 1$ and $\hat{j}^* = \hat{j} + G^*(\hat{i}+1,\hat{j}+1) + 1$ together with $s_1[\hat{i} - G(\hat{i},\hat{j}) + 1, \hat{i}^* - 1]$. The way to get $\hat{j}$ and $\hat{t}$ for a given $\hat{i}$ will be left for the next subsubsection.

b) The algorithm. To identify $\hat{j}$ and $\hat{t}$ for a given $\hat{i}$ for dividing $LCSS_k$ into smaller subproblems, it follows from Formula (3) and Form 1 and Form 2$^+$ that $D(\hat{i}, j, t)$, $F(\hat{i}, j, t, 1)$, $G(\hat{i}, \hat{j})$, $D^*(\hat{i}+1, j+1, k-t)$, $F^*(\hat{i}+1, j+1, t, 1)$ and $G^*(\hat{i}+1, \hat{j}+1)$ should be ready for use.

For a given $i \in [1, n_1]$, it follows from Formulae (1), (2) and (4) that $F(i, j', t', q')$ and $G(i, j')$ for all $j' \in [0, n_2], t' \in [0, k]$ and $q' \in [0, 1]$ can be obtained from all $F(i - 1, j, t, q), F(i, j-1, t, q)$ and $G(i-1, j)$ for all $j \in [0, n_2], t \in [0, k]$ and $q \in [0, 1]$. Hence $D(\hat{i}, j, t), F(\hat{i}, j, t, 1)$ and $G(\hat{i}, \hat{j})$ can be obtained via dynamic programming algorithm in an iterative manner: Use a $2 \cdot (n_2 + 1) \cdot (k + 1) \cdot 2$ matrix and a $2 \cdot (n_2 + 1)$ matrix to maintain $F(i, j, t, q), F(i-1, j, t, q), G(i, j)$ and $G(i-1, j)$ insistently in execution of the nested loops for i, j and t during the dynamic programming algorithm. Then $F(\hat{i}, j, t, 0/1)$ and $G(\hat{i}, \hat{j})$ will be identified at the end of the $\hat{i}^{th}$ loop for i from which, $D(\hat{i}, j, t) = \max\{F(\hat{i}, j, t, 1), F(\hat{i}, j, t, 0)\}$ can be obtained for $j \in [0, n_2]$ and $t \in [0, k]$. Given $\hat{i}$, we denote by $\mathcal{A}(\hat{i}, k, S)$ a subroutine to obtain $F(\hat{i}, j, t, 1)$, $G(\hat{i}, j)$ and $D(\hat{i}, j, t)$ in the way as stated above.

The space complexity of $\mathcal{A}(\hat{i}, k, S)$ is $O(n_1 + kn_2)$ because there are $O(kn_2)$ cells in all matrices used by $\mathcal{A}(\hat{i}, k, S)$. The time complexity of $\mathcal{A}(\hat{i}, k, S)$ is $O(\hat{k}in_2)$ because one cell in matrices used by the subroutine is accessed once in each of the $\hat{i}$ loops. Using $F(\hat{i}, j, t, 1)$, $G(\hat{i}, j)$ and $D(\hat{i}, j, t)$ reported by $\mathcal{A}(\hat{i}, k, S)$ for a given $\hat{i}$, $M_1(\hat{i}, j, t)$ and $M_2(\hat{i}, j, t)$ can be obtained in $O(kn_2)$ time for $j \in [0, n_2]$ and $t \in [0, k]$. Then a longest (k)-CSS of S can be found by divide-and-conquer technique as Algorithm 1.

ïž£

Algorithm 1. $D\&C(S, k)$

Require: $S = \{s_1[1, n_1], s_2[1, n_2]\}$, an integer $k \geq 1$.
Ensure: The occurrences of a longest (k)-CSS of S in s_1 and s_2.
1: If $n_1 = 1$ or $n_2 = 1$ or $k = 1$ then return a longest 1-CSS of S.
2: Let $\hat{i} = \lfloor \frac{n_1}{2} \rfloor$. Invoke $\mathcal{A}(\hat{i}, k, S)$ to obtain $D(\hat{i}, j, t)$, $F(\hat{i}, j, t, 1)$ and $G(\hat{i}, j)$, invoke $\mathcal{A}(n_1 - \hat{i}, k, \overline{S})$ to obtain $D^*(\hat{i} + 1, j + 1, t)$, $F^*(\hat{i} + 1, j + 1, t, 1)$ and $G^*(\hat{i} + 1, j + 1)$ for $j \in [0, n_2]$ and $t \in [0, k]$ where $\overline{S} = \{\overline{s_1}, \overline{s_2}\}$.
3: Obtain $M_1(\hat{i}, j, t)$ and $M_2(\hat{i}, j, t)$, identify $\hat{j} \in [0, n_2]$ and $\hat{t} \in [0, k]$ according to Formula (3) such that $M_1(\hat{i}, \hat{j}, \hat{t}) = D(n_1, n_2, k)$ or $M_2(\hat{i}, \hat{j}, \hat{t}) = D(n_1, n_2, k)$.
4: If $M_1(\hat{i}, \hat{j}, \hat{t}) = D(n_1, n_2, k)$, then invoke $D\&C(\{s_1[1, \hat{i}], s_2[1, \hat{j}]\}, \hat{t})$ to get Λ_1, invoke $D\&C(\{\{s_1[\hat{i} + 1, n_1], s_2[\hat{j} + 1, n_2]\}, k - \hat{t})$ to get Λ_2. Return $\Lambda_1 \oplus \overline{\Lambda_2}$.
5: If $M_2(\hat{i}, \hat{j}, \hat{t}) = D(n_1, n_2, k)$, then invoke $D\&C(\{s_1[1, \hat{i} - G(\hat{i}, \hat{j})], s_2[1, \hat{j} - G(\hat{i}, \hat{j})]\}, \hat{t} - 1)$ to get Λ_1, invoke $D\&C(\{\{s_1[\hat{i}^*, n_1], s_2[\hat{j}^*, n_2]\}, k - \hat{t})$ to get Λ_2 where $\hat{i}^* = \hat{i} + G^*(\hat{i} + 1, \hat{j} + 1) + 1$ and $\hat{j}^* = \hat{j} + G^*(\hat{i} + 1, \hat{j} + 1) + 1$. Return $\Lambda_1 \oplus s_1[\hat{i} - G(\hat{i}, \hat{j}) + 1, \hat{i}^* - 1] \oplus \overline{\Lambda_2}$.

Since all matrices used in this algorithm contain $O(kn_2)$ cells, the space complexity of the algorithm is $O(n_1 + kn_2)$. By the following theorem, we answer the most concerned question about the time complexity of the algorithm.

Theorem 2. *Algorithm 1 takes $O(kn_1 n_2)$ time to return a longest (k)-CSS of $S = \{s_1[1, n_1], s_2[1, n_2]\}$.*

Proof. Let $T(n_1, n_2, k)$ denote the time complexity of Algorithm 1 with input S. By identifying $\hat{j}$ and $\hat{t}$ for a given $\hat{i}$, $LCSS_k$ is divided into two subproblems whose solutions can be combined into a solution of $LCSS_k$, a longest (k)-CSS of S in Form 1 or Form 2^+. Since solving those subproblems for a longest (k)-CSS in Form 2^+ has no higher time complexity than that in Form 1, we only analyze the bound of $T(n_1, n_2, k)$ in the situation where a longest (k)-CSS of S in Form 1 is in all recursions of the algorithm.

Since it takes $O(kn_1 n_2)$ time to invoke $\mathcal{A}(\hat{i}, k, S)$ and $O(kn_2)$ additional time to get some $\hat{j}$ and $\hat{t}$ with $M_1(\hat{i}, \hat{j}, \hat{t}) = D(n_1, n_2, k)$ or $M_2(\hat{i}, \hat{j}, \hat{t}) = D(n_1, n_2, k)$,

$T(n_1, n_2, k)$ satisfies Formula (6) where c_0 and c_1 are constants.

$$T(n_1, n_2, k) = \begin{cases} T(\lfloor \frac{n_1}{2} \rfloor, \hat{j}, \hat{t}) + T(\lceil \frac{n_1}{2} \rceil, n_2 - \hat{j}, k - \hat{t}) + c_1 k n_1 n_2 & \text{if } n_1 \geq 2; \\ c_0 n_2 & \text{otherwise.} \end{cases} \quad (6)$$

We prove by induction that $T(n_1, n_2, k) \leq (c_0 + 2c_1)k n_1 n_2$. If $n_1 = 2$, then $T(2, n_2, k) \leq (c_0 + 2c_1)k n_2$. Assume $T(n_1, n_2, k) \leq (c_0 + 2c_1)k n_1 n_2$ for $n_1 \in [2, m_0 - 1]$, where m_0 is an integer with $m_0 \geq 3$. For $n_1 = m_0$,

$$\begin{aligned} T(m_0, n_2, k) &= (c_0 + 2c_1)(\frac{m_0 \hat{j} \hat{t}}{2} + \frac{m_0(n_2 - \hat{j})(k - \hat{t})}{2}) + c_1 k m_0 n_2 \\ &\leq k[\frac{(c_0 + 2c_1)m_0 n_2}{2} + c_1 m_0 n_2] \quad (7) \\ &\leq (c_0 + 2c_1)k m_0 n_2. \end{aligned}$$

It follows by induction on n_1 that $T(n_1, n_2, k) \leq (c_0 + 2c_1)k n_1 n_2$.

4   Conclusion

It is significant to design approximation algorithms for $LCSS_k$ of m strings. We wonder whether there exist effective algorithms for $LCSS_k$ of two or more strings if the (k]-CSS are asked with length limited substrings, from below or above.

Acknowledgments. This work is supported by grant from the NSF of China (NSFC: 62272272).

References

1. Abboud, A., Backurs, A., Williams, V.V.: Tight hardness results for LCS and other sequence similarity measures. In: Foundations of Computer Science, pp. 59–78. IEEE (2015)
2. Baker, T.A., Bell, S.P., Gann, A., Levine, M., Losick, R., Inglis, C.: Molecular Biology of the Gene. Pearson (2008)
3. Benson, G., Levy, A., Maimoni, S., Noifeld, D., Shalom, B.R.: LCSk: a refined similarity measure. Theor. Comput. Sci. **638**, 11–26 (2016)
4. Benson, G., Levy, A., Shalom, B.R.: Longest common subsequence in k length substrings. In: Similarity Search and Applications: 6th International Conference, pp. 257–265. Springer (2013)
5. Fremin, B.J., Bhatt, A.S.: Comparative genomics identifies thousands of candidate structured RNAs in human microbiomes. Genome Biol. **22**(1), 100–100 (2021)
6. Gusfield, D.: Algorithms on Strings, Trees, and Sequences: Computer Science and Computational Biology. Cambridge University Press, Cambridge (1997)
7. Hirschberg, D.S.: A linear space algorithm for computing maximal common subsequences. Commun. ACM **18**(6), 341–343 (1975)
8. Hládek, D., Staš, J., Pleva, M.: Survey of automatic spelling correction. Electronics **9**(10), 1670 (2020)
9. Kociumaka, T., Radoszewski, J., Starikovskaya, T.: Longest common substring with approximately k mismatches. Algorithmica **81**(6), 2633–2652 (2019)

10. Li, T., Zhu, D., Jiang, H., Feng, H., Cui, X.: Longest k-tuple common sub-strings. In: 2022 IEEE International Conference on Bioinformatics and Biomedicine (BIBM), pp. 63–66. IEEE (2022)
11. Masek, W.J., Paterson, M.S.: A faster algorithm computing string edit distances. J. Comput. Syst. Sci. **20**(1), 18–31 (1980)
12. Michael, M., Nicolas, F., Ukkonen, E.: On the complexity of finding gapped motifs. J. Discrete Algorithms **8**(2), 131–142 (2010)
13. Wagner, R.A., Fischer, M.J.: The string-to-string correction problem. J. ACM (JACM) **21**(1), 168–173 (1974)
14. Zhu, D., Wang, L., Wang, T., Wang, X.: A space efficient algorithm for the longest common subsequence in k-length substrings. Theor. Comput. Sci. **687**, 79–92 (2017)

Scheduling Two Types of Jobs
with Minimum Makespan

Song Cao and Kai Jin$^{(\boxtimes)}$

Shenzhen Campus of Sun Yat-sen University, Shenzhen, Guangdong, China
caos6@mail2.sysu.edu.cn, jink8@mail.sysu.edu.cn

Abstract. We consider scheduling two types of jobs (A-job and B-job) to p machines and minimizing their makespan. A group of same type of jobs processed consecutively by a machine is called a batch. For machine v, processing x A-jobs in a batch takes $k_v^A x^2$ time units for a given speed k_v^A, and processing x B-jobs in a batch takes $k_v^B x^2$ time units for a given speed k_v^B. We give an $O(n^2 p \log(n))$ algorithm based on dynamic programming and binary search for solving this problem, where n denotes the maximal number of A-jobs and B-jobs to be distributed to the machines.

Keywords: Machine scheduling · Makespan · Dynamic programming · Binary search

1 Introduction

We consider scheduling two types of jobs A and B to machines and minimizing their makespan, i.e., the maximum time completing all jobs. Suppose n_a A-jobs and n_b B-jobs have to be scheduled, and can be distributed to p machines. A group of A-jobs processed consecutively by some machine is called an *A-batch*, and a group of B-jobs processed consecutively by some machine is called a *B-batch*. A batch refers to an A-batch or a B-batch. A machine processing a batch continuously will overheat over time, resulting in decreased performance. This leads to a quadratic relationship between processing time and the number of jobs in a batch. For the v-th machine, processing x A-jobs in a batch takes $k_v^A x^2$ time units, and processing x B-jobs in a batch takes $k_v^B x^2$ time units, where k_v^A, k_v^B are given speed. Moreover, for the v-th machine, assume that it takes t_v^A extra time units to process any A-batch (the overhead to switch the status to processing A-jobs), and t_v^B extra time units to process any B-batch (the overhead to switch the status to processing B-jobs). No empty batch is allowed, and two neighbouring batches of any machine must be of different types. Denote $n = \max(n_a, n_b)$.

This research was supported by Department of Science and Technology of Guangdong Province (Project No. 2021QN02X239) and Shenzhen Science and Technology Program (Grant No. 202206193000001, 20220817175048002).

© The Author(s), under exclusive license to Springer Nature Singapore Pte Ltd. 2025
B. Li et al. (Eds.): IJTCS-FAW 2024, LNCS 14752, pp. 115–122, 2025.
https://doi.org/10.1007/978-981-97-7752-5_9

This paper gives an $O(n^2 p \log(n))$ time algorithm for solving the above problem based on binary search. Given a parameter L, we have to determine whether there is a scheduling with makespan not exceeding L time units.

For convenience, we use a pair (a, b) to indicate the task-combination formed by a A-jobs and b B-jobs (it is always assumed that $a \geq 0$ and $b \geq 0$). Let B denote the set consisting of all b for which the machines can finish the task-combination (n_a, b) within L time units. Clearly, the makespan can be smaller or equal to L time units if and only if $n_b \in B$. We prove that the set B is an interval, whose two endpoints can be computed by dynamic programming that takes $O(n^2 p)$ time. This leads to our final algorithm that runs in $O(n^2 p \log(n))$ time. The major challenge for designing the algorithm lies in proving the theorem that says B is an interval, for which a lot of analysis are shown in this paper.

1.1   Related Works

Machine scheduling problem is widely studied for its practical importance. Consider scheduling jobs $J_1, \ldots, J_n$ on parallel machines $M_1, \ldots, M_m$ and minimizing the makespan. Preemption(job splitting), precedence relation and release date are not considered for each job. The processing time that machine M_i processes job J_j is denoted as p_{ij}. Classified by machine environment, when $p_{ij} = p_j$, it's called identical parallel machine; when $p_{ij} = p_j / s_i$ for a given speed s_i of machine M_i, it's called uniform parallel machine; and the general case, when $p_{ij} = p_j / s_{ij}$, it's called unrelated parallel machine. According to the three-field classification note [1], the above three cases can be denoted as $P||C_{max}$, $Q||C_{max}$ and $R||C_{max}$. The problem we solve in this paper is a variation of $R||C_{max}$.

$R||C_{max}$ is NP-hard since its simple version P||Cmax has been proved NP-hard [2]. Studies on these problems focus on finding polynomial ρ-approximation algorithm ($\rho > 1$), that is, polynomial time algorithm which generates solution at most ρ times the optimal solution [1]. Lenstra et al. [3] proved that for $R||C_{max}$, the worst-case ratio for approximation is at least $3/2$ unless $P = NP$. And they present a polynomial 2-approximation algorithm. Ghirardi et al. [4] provides an heuristic called Recovering Beam Search, which generates approximate solutions in $O(n^3 m)$ time.

Different variations of unrelated parallel machine problem have been studied. Consider minimizing L_p norm of unrelated machines' completion time, Alon et al. [5] provides an $(1+\varepsilon)$-approximation scheme, that is, a family of algorithms $\{A_\varepsilon\}$ ($\varepsilon > 0$) such that, each $\{A_\varepsilon\}$ is $1 + \varepsilon$-approximation, and its running time may depend on input and ε [1]. Azar et al. [6] provides a 2-approximation algorithm for $p > 1$, and $\sqrt{2}$-approximation for $p = 2$. And Kumar et al. [7] improves it with a better-than-2 approximation algorithm for all $p \geq 1$. Im and Li [8] consider weighted completion time with job release date. They improve a 2-approximation to 1.8786-approximation for both preemptive and non-preemptive problems. For $R||C_{max}$, when matrix rank formed by job processing time is small, Bhaskara et al. [9] show that it admits an QPTAS for rank 2, and is APX-hard for rank 4. Chen et al. [10] continue this research and prove it APX-hard for rank 3. Deng et al. [11] consider a more general problem called GLB problem, where the load of each machine is a norm, and minimizes generalized makespan(also a norm). They provide a polynomial $O(\log n)$-approximation algorithm. Im and Li [8] give a 1.45-approximation algorithm for minimizing total weighted completion time, and a $\sqrt{4/3}$-approximation for minimizing L_2-norm. Bamas et al. [12] obtain a polynomial $(2 - 1/n^\varepsilon)$-approximation for two-value makespan minimization, and a polynomial $(1.75 + \varepsilon)$-approximation for the restricted assignment case.

The organization of this paper is as follows. Section 2 considers the one machine case and proves several important observations for this case. Section 3 considers the multiple machines case and gives our main algorithm.

2 The One Machine Case

Consider any fixed machine v with parameters k_v^A, k_v^B, t_v^A and t_v^B. Denote by $f(a, b)$ the minimum time units for this machine to finish task-combination (a, b), and denote by S_v the set of task-combinations that can be finished by machine v within L time units (where L is a bound that will be specified by a binary search in the main algorithm). Formally, $S_v = \{(a, b) \mid f(a, b) \leq L\}$.

A crucial property of S_v is described in the following theorem.

Theorem 1. *For fixed $a \geq 0$, those b for which $(a, b) \in S_v$ are consecutive.*

To prove this theorem, we need some knowledge about $f(a, b)$ as shown below.

Definition 1. *For $a \geq 0$ and $s \geq 0$, define $cost^A(a, s)$ to be the minimum time units for machine v to process a A-jobs in s batches (here, we allow empty batch length and hence $cost^A(a, s)$ is well-defined even for the case $a < s$). Formally,*

$$cost^A(a, s) = min_{\substack{x_1 + \ldots + x_s = a \\ x_1 \geq 0, \ldots, x_s \geq 0}} \{s \cdot t_v^A + (x_1^2 + \ldots + x_s^2) \cdot k_v^A\}.$$

Note that for $s = 0$, we have $cost^A(0, 0) = 0$ and $cost^A(a, 0) = \infty$ for $a > 0$. Similarly, define

$$cost^B(b, s) = min_{\substack{x_1 + \ldots + x_s = b \\ x_1 \geq 0, \ldots, x_s \geq 0}} \{s \cdot t_v^B + (x_1^2 + \ldots + x_s^2) \cdot k_v^B\}.$$

For $a \geq 0$ and $s > 0$, the following equation holds (trivial proof omitted).

$$cost^A(a, s) = s \cdot t_v^A + \left[(a \bmod s)(\left[\frac{a}{s}\right] + 1)^2 + (s - (a \bmod s)) \left[\frac{a}{s}\right]^2 \right] \cdot k_v^A. \quad (1)$$

According to (1), for fixed $s \geq 1$, function $cost^A(a, s)$ of variable a increases on $[0, \infty)$. Similarly, function $cost^B(b, s)$ of variable b increases on $[0, \infty)$.

Lemma 1 [13]. $f(a, b) = min_{0 \leq s \leq min(b, a+1)} cost(a, b, s)$, *where*

$$cost(a, b, s) = cost^B(b, s) + mcost^A(a, s), \quad (2)$$

and

$$mcost^A(a, s) = \min_{\substack{s' \in \{s-1, s, s+1\}, s' \geq 0}} cost^A(a, s'). \quad (3)$$

Lemma 1 is not that obvious – An empty batch is allowed in the definition of $cost^A$ and $cost^B$, yet **not** allowed in processing the task-combination (a, b).

Lemma 2 [13]. *For $a \geq 0$, all the functions below are increasing functions of s:*

1. $cost^A(a, s+1) - cost^A(a, s)$ *(namely, the difference function of $cost^A(a, s)$);*
2. $cost^B(a, s+1) - cost^B(a, s)$ *(namely, the difference function of $cost^B(a, s)$);*
3. $mcost^A(a, s + 1) - mcost^A(a, s)$ *(namely, the difference function of $mcost^A(a, s)$).*

In other words, $cost^A(a, s), cost^B(a, s), mcost^A(a, s)$ are convex functions of s.

Lemma 3. [13]

1. *For fixed $a \geq 0$ and $b \geq 0$, function $cost(a, b, s+1) - cost(a, b, s)$ of s (namely, the difference function of $cost(a, b, s)$) is increasing.*
2. *For fixed $s \geq 0$ and $a \geq 0$, function $cost(a, b, s)$ of b is increasing.*
3. *For fixed $s \geq 0$ and $b \geq 0$, function $cost(a, b, s)$ of a is increasing.*
4. *Function $cost(x, x + 1, x + 1)$ of x is increasing.*

We are ready to prove Theorem 1. Suppose $a \geq 0$ is fixed in the following.

We introduce a table M with $n_b + 1$ rows and $a + 2$ columns to prove Theorem 1: for (b, s) where $0 \leq b \leq n_b$, and $0 \leq s \leq a + 1$, define

$$M(b, s) = \begin{cases} [cost(a, b, s) \leq L], & s \leq b; \\ - \text{ (leave it as undefined)}, & s > b, \end{cases} \quad (4)$$

where $[\cdot]$ denotes the Iverson bracket. See Table 1 for an example. Rows in M are indexed with 0 to n_b. Columns in M are indexed with 0 to $a + 1$.

Table 1. An example of M, where $n_b = 4$ and $a = 2$.

b	s			
	0	1	2	3
0	1	–	–	–
1	1	1	–	–
2	0	0	1	–
3	0	0	1	0
4	0	0	0	0

Proof (of Theorem 1). Assume $a \geq 0$ is fixed. Recall that this theorem states that $\{b \mid f(a, b) \leq L\}$ are consecutive. Following Lemma 1, $f(a, b) \leq L$ if and only if the b-th row in M has a 1. Hence it reduces to showing that (i) the rows in M with at least one 1 are consecutive.

We now point out two facts about table $M(b, s)$.

Fact 1. If $M(b, s) = 1$ and $s \leq b - 1$, then $M(b - 1, s) = 1$.

Proof Since $M(b, s) = 1$, we know $cost(a, b, s) \leq L$. By Lemma 3.2 we know $cost(a, b - 1, s) \leq cost(a, b, s) \leq L$. So $M(b - 1, s) = 1$. □

Fact 2. If $M(s, s) = 1$ and $M(s + 1, s + 1) = 0$, then $M(s', s') = 0$ for $s' > s + 1$.

Proof Since $M(s, s) = 1$ and $M(s + 1, s + 1) = 0$, we know $cost(a, s, s) \leq L < cost(a, s + 1, s + 1)$. So,

$$cost(a, s, s) < cost(a, s + 1, s + 1). \tag{5}$$

Observe that $cost(a, s', s') = cost^B(s', s') + mcost^A(a, s') = s't_v^B + s'k_v^B + mcost^A(a, s')$. Following Lemma 2, function $cost(a, s', s')$ of s' is convex. Further since (5), $cost(a, s', s')$ is strictly increasing when $s' \geq s + 1$. Therefore $cost(a, s', s') > cost(a, s + 1, s + 1) > L$, which means $M(s', s') = 0$. □

According to Fact 2, the 1's in the diagonal of M are consecutive. According to Fact 1, the 1's in any particular column are consecutive and the lowest 1 (if any) in that column always appears at the diagonal of M. Together, we obtain argument (i). □

2.1 More Properties of S_v

For $0 \leq a \leq n_a, 0 \leq b \leq n_b$, define $F^{(v)}(a, b) = [(a, b) \in S_v] = [f(a, b) \leq L]$. Abbreviate $F^{(v)}$ as F when v is clear. This table represents S_v. See Table 2 for an illustration. Whereas Theorem 1 shows the consecutiveness of 1's within any row of F, the next theorem shows properties of F between its consecutive rows.

Theorem 2. *For any fixed $a \geq 0$,*

1. *If some row of F are all 0's, so is the next row of F.*
2. *If both row a and row $a + 1$ contain 1, at least one of the following holds:*
   (1) there is b such that $F(a, b) = F(a + 1, b) = 1$.
   (2) $F(a, a + 1) = F(a + 1, a + 2) = 1$.

Table 2. An example of $F^{(v)}$, which represents S_v.

b \ a	0	1	2	3
0	1	1	1	1
1	0	1	0	0
2	0	0	0	0
3	0	0	0	0

We introduce two subregions of F. The area in which $b \leq a + 1$ is referred to as *Area 1*. The area in which $b \geq a + 1$ is referred to as *Area 2*.

Lemma 4. [13]

1. For $(a, b), (a + 1, b)$ in Area 1, $F(a, b) = 0$ implies $F(a + 1, b) = 0$.
2. For $(a, b), (a, b + 1)$ in Area 2, $F(a, b) = 0$ implies $F(a, b + 1) = 0$.
3. If $F(a, a + 1) = 0$, then $F(a + 1, a + 2) = 0$.
4. If $F(a + 1, a + 1) = 0$ and $F(a + 1, a + 2) = 1$, then $F(a, a + 1) = 1$.

We can now proof Theorem 2.

Proof (of Theorem 2). 1. Suppose row a are all 0's. By Lemma 4.1 and 4.3, we know $F(a+1, i) = 0$ for all $i \leq a+2$. Further according to Lemma 4.2, row $a+1$ are all 0's.

2. If there exists $b \leq a + 1$, such that $F(a + 1, b) = 1$, we know $F(a, b)$ and $F(a + 1, b)$ are both in *Area 1*. So by Lemma 4.1 we know $F(a, b) = 1$.

Otherwise, $F(a + 1, b) = 0$ for all $b \leq a + 1$. So there exists $b_0 > a + 1$ such that $F(a+1, b_0) = 1$. By Lemma 4.2 we know $F(a+1, a+2) = 1$. So by Lemma 4.4 we know $F(a, a + 1) = 1$.   □

3   Multiple Machines Case

Below we give an $O(n^2 p \log(n))$ algorithm based on dynamic programming and binary search, and Subsect. 3.2 proves its running time.

3.1   Algorithm

For convenience, let $b_v(a) = \{b \geq 0 \mid f(a, b) \leq L\}$. Given finite set $A, B \subseteq \mathbb{Z}$, we define $A + B = \{a + b \mid a \in A, b \in B\}$ as their *sumset*.

Definition 2. *For $v \geq 0$, $a \geq 0$ and $L \geq 0$, define $dp(v, a)$ to be the set of amount b of B-jobs, such that the first v machines can finish the task-combination (a, b) within L time units. Formally,*

$$dp(v, a) = \bigcup_{\substack{x_1 + \ldots + x_v = a \\ x_1 \geq 0, \ldots, x_v \geq 0}} (b_1(x_1) + \ldots + b_v(x_v)).$$

We can immediately get recurrence formula for $dp(v, a)$ by enumerating the number of A-jobs finished by the last machine:

$$dp(v, a) = \bigcup_{0 \leq a' \leq a} (dp(v - 1, a - a') + b_v(a')). \tag{6}$$

We obtain the answer by binary searching the smallest time units L that makes $dp(p, n_a)$ containing n_b. The following theorem assures that $dp(p, n_a)$ can be calculated in $O(n^2 p)$ time. The upper bound of L is $O(n^2)$ by scheduling all jobs to one machine. Therefore the answer can be calculated in $O(n^2 p \log(n))$ time.

Theorem 3. *Set $dp(v, a)$ is an interval.*

Suppose Theorem 3 holds. Let $dp(v, a) = [dp(v, a).l, dp(v, a).r]$. Theorem 1 states that $b_v(a)$ is an interval. Let $b_v(a) = [b_v(a).l, b_v(a).r]$. And (6) can be simplified as:

$$dp(v, a).l = min_{0 \leq a' \leq a} \left(dp(v - 1, a - a').l + b_v(a').l \right),$$

$$dp(v, a).r = max_{0 \leq a' \leq a} \left(dp(v - 1, a - a').r + b_v(a').r \right).$$

3.2   Proof of Theorem 3

To prove this theorem, we need some knowledge about $dp(v, a)$ as shown below.

Lemma 5 [13]. *For any $v \geq 1$ and $a \geq 0$,*

1. *If $dp(v, a) = \emptyset$, then $dp(v, a + \delta) = \emptyset$ for all $\delta > 0$.*
2. *If $dp(v, a) \neq \emptyset$ and $dp(v, a + 1) \neq \emptyset$, at least one of the following holds:*
   (1) $dp(v, a) \cap dp(v, a + 1) \neq \emptyset$.
   (2) there exists $b_0 \in dp(v, a)$ and $b_0 + 1 \in dp(v, a + 1)$.

We can now prove Theorem 3.

Proof (of Theorem 3). We prove it by induction on v. When $v = 1$, we have $dp(1, a) = b_1(a)$. The conclusion holds by Theorem 1.

Suppose conclusion holds for $v - 1$. Lemma 5.1 states that non-empty $dp(v, a)$ is consecutive when a changes. Theorem 2.1 states that non-empty $b_v(a)$ is consecutive when a changes. So (6) can be written as

$$dp(v, a) = \bigcup_{a' \in [l, r]} (dp(v - 1, a - a') + b_v(a')),$$

where $dp(v - 1, a - a') + b_v(a') \neq \emptyset$ for all $a' \in [l, r]$.

To prove $dp(v, a)$ an interval, it reduces to prove that for all $a' \in [l, r - 1]$, $dp(v-1, a-a')+b_v(a') \cup dp(v-1, a-(a'+1))+b_v(a'+1)$ is an interval. According to Theorem 2.2 and Lemma 5.2, there are four cases we need to discuss.

Case 1. $b_v(a') \cap b_v(a' + 1) \neq \emptyset$; $dp(v - 1, a - a') \cap dp(v - 1, a - (a' + 1)) \neq \emptyset$
The intersection of $dp(v-1, a-a')+b_v(a')$ and $dp(v-1, a-(a'+1))+b_v(a'+1)$ is non-empty. By conclusion hypothesis and Theorem 1, their union is an interval.

Case 2. $a' + 1 \in b_v(a'), a' + 2 \in b_v(a' + 1)$;
$dp(v - 1, a - a') \cap dp(v - 1, a - (a' + 1)) \neq \emptyset$
Suppose $b_0 \in dp(v - 1, a - a') \cap dp(v - 1, a - (a' + 1))$, we have $b_0 + a' + 1 \in dp(v - 1, a - a') + b_v(a')$, and $b_0 + a' + 2 \in dp(v - 1, a - (a' + 1)) + b_v(a' + 1)$. By conclusion hypothesis and Theorem 1, their union is an interval.

Case 3. $b_v(a') \cap b_v(a' + 1) \neq \emptyset$;
there exists $b_0 \in dp(v - 1, a - (a' + 1)), b_0 + 1 \in dp(v - 1, a - a')$
Proof is same as in case 2.

Case 4. $a' + 1 \in b_v(a'), a' + 2 \in b_v(a' + 1)$;
there exists $b_0 \in dp(v - 1, a - (a' + 1)), b_0 + 1 \in dp(v - 1, a - a')$

Then $b_0 + a' + 2 \in dp(v - 1, a - a') + b_v(a')$, and $b_0 + a' + 2 \in dp(v - 1, a - (a' + 1)) + b_v(a' + 1)$. So their intersection is non-empty. By conclusion hypothesis and Theorem 1, their union is an interval. □

References

1. Lawler, E.L., Lenstra, J.K., Rinnooy Kan, A.H.G., Shmoys, D.B.: Sequencing and scheduling: algorithms and complexity. In: Handbooks in Operations Research and Management Science, vol. 4, pp. 445–522 (1993)
2. Garey, M.R., Johnson, D.S.: "Strong" np-completeness results: motivation, examples, and implications. J. ACM (JACM) **25**(3), 499–508 (1978)
3. Lenstra, J.K., Shmoys, D.B., Tardos, É.: Approximation algorithms for scheduling unrelated parallel machines. Math. Program. **46**, 259–271 (1990)
4. Ghirardi, M., Potts, C.N.: Makespan minimization for scheduling unrelated parallel machines: a recovering beam search approach. Eur. J. Oper. Res. **165**(2), 457–467 (2005)
5. Alon, N., Azar, Y., Woeginger, G.J., Yadid, T.: Approximation schemes for scheduling. In: SODA, pp. 493–500. Citeseer (1997)
6. Azar, Y., Epstein, A.: Convex programming for scheduling unrelated parallel machines. In: Proceedings of the Thirty-Seventh Annual ACM Symposium on Theory of Computing, pp. 331–337 (2005)
7. Kumar, V.S.A., Marathe, M.V.: Approximation algorithms for scheduling on multiple machines. In: 46th Annual IEEE Symposium on Foundations of Computer Science (FOCS 2005), pp. 254–263. IEEE (2005)
8. Im, S., Li, S.: Better unrelated machine scheduling for weighted completion time via random offsets from non-uniform distributions. In: 2016 IEEE 57th Annual Symposium on Foundations of Computer Science (FOCS), pp. 138–147. IEEE (2016)
9. Bhaskara, A., Krishnaswamy, R., Talwar, K., Wieder, U.: Minimum makespan scheduling with low rank processing times. In: Proceedings of the Twenty-Fourth Annual ACM-SIAM Symposium on Discrete Algorithms, pp. 937–947. SIAM (2013)
10. Chen, L., Marx, D., Ye, D., Zhang, G.: Parameterized and approximation results for scheduling with a low rank processing time matrix. In: 34th Symposium on Theoretical Aspects of Computer Science (STACS 2017). Schloss-Dagstuhl-Leibniz Zentrum für Informatik (2017)
11. Deng, S., Li, J., Rabani, Y.: Generalized unrelated machine scheduling problem. In: Proceedings of the 2023 Annual ACM-SIAM Symposium on Discrete Algorithms (SODA), pp. 2898–2916. SIAM (2023)
12. Bamas, É., Lindermayr, A., Megow, N., Rohwedder, L., Schlöter, J.: Santa Claus meets makespan and matroids: algorithms and reductions. In: Proceedings of the 2024 Annual ACM-SIAM Symposium on Discrete Algorithms (SODA), pp. 2829–2860. SIAM (2024)
13. Cao, S., Jin, K.: Scheduling two types of jobs with minimum makespan (2024)

Blockchain Technology for Digital Asset Ownership

Jasmine Siu Lee Lam[1]([✉]) [iD] and Kee Wei Lee[2]

[1] Technical University of Denmark, 2800 Kongens Lyngby, Denmark
`jasmlam@dtu.dk`

[2] Nanyang Technological University, 50 Nanyang Avenue, Singapore 639798, Singapore

Abstract. Blockchain technology has developed tremendously as the world becomes increasingly digitalised. Today, in addition to recording and transmitting currencies, blockchains are also used for other assets such as digital collectibles, showing blockchains' growing potential in digital asset ownership. This study adopts content analysis of the state-of-the-art literature as well as a case-study approach to analyse the unique potential and challenges of blockchains to digital asset ownership. A major potential according to the findings is the opportunity to monetise and unlock newer digital assets. Under-developed intellectual property laws and regulations remain a key challenge. Discussions and recommendations are presented along with real-world cases.

Keywords: Blockchain · Digital asset · Virtual asset · Non-Fungible Tokens

1 Introduction

Blockchain is a distributed ledger which handles transactions in a secured network. The open and trust-less nature of blockchain, along with the rise in stablecoins, has also led to the rise in decentralised financial services which could be broadly defined as "a variety of financial products, services, activities, and arrangements supported by smart contract-enabled distributed ledger technology." (President's Working Group on Financial Markets (PWG), Federal Deposit Insurance Corporation (FDIC) and Office of the Comptroller of the Currency (OCC), 2022, p. 9). Although initially gaining popularity as a currency, blockchain technology is no longer limited to just cryptocurrencies and has instead branched out massively to include other various assets due to the technology's rapid advancement. Today, while blockchains are still used to record and transmit currencies, such as in fundraising campaigns (Egkolfopoulou, 2022), they are also used for other assets like digital collectibles (Pu & Lam, 2023), showing blockchains' growing potential in digital asset ownership. Regions such as Dubai, UAE, have even begun adopting laws governing virtual assets and set up regulatory agencies overseeing the sector (Reuters, 2022).

As the world continues to rapidly digitalise, an expanding portion of economic activity revolves around the digital world. With the growth of the digital world, also known as the metaverse, the role of digital goods represented by Non-Fungible Tokens

B. Li et al. (Eds.): IJTCS-FAW 2024, LNCS 14752, pp. 123–129, 2025.
https://doi.org/10.1007/978-981-97-7752-5_10

(NFTs) also increases (Hammi et al., 2023). With blockchain laying the foundation for the next version of the internet, business owners and users increasingly face multiple challenges such as general cyber vulnerabilities (WEF, 2022). Given the increasingly important role blockchain technology plays in the future society, it is therefore important that any potential and drawback be studied carefully.

Corresponding to the rapid growth of blockchain, there are increasing research interests in this area. Tokens (NFTs) to represent ownership of digital assets has led to increased adoption by different organisations. Digital assets include data, digital art, intellectual property, game items represented via NFTs, etc. Experimentation with this new technology reveals the different benefits and limitations of such an approach. Shatkovskaya et al. (2018) investigated how a system of tracking intellectual property can be developed and increasingly commercialised to become an asset class. Establishing a link between the digital asset and the blockchain opens up various possibilities for the asset to be monetised, traded, and managed. Pérez-Solà & Herrera-Joancomartí (2020) has shown how artists can monetise their work through royalties and enforce the rights to their works via the blockchain. Due to the ability to protect asset ownership and rights, blockchain has also been proposed as a solution to promoting the data economy. Studies such as those from Yu & Zhao (2019) recommended the adoption of blockchain to overcome some of the common problems encountered in data marketplaces. These studies examined beyond merely asset ownership and open the possibilities for many other interactions involving these assets. Castonguay & Stein Smith (2020) on the other hand, found that, in practice, blockchain and cryptocurrencies are more likely to encounter malfeasance, fraud, and manipulation than is commonly understood. Valeonti et al. (2021) noted how NFTs can be applied to represent digital collectibles and explored their potential and limitations as an asset to help organisations raise funds. Ante (2023) analysed NFTs' 14 biggest submarkets, the overall trend and state of the market. The conclusion was that the NFT market till 2021 was immature. We see more works in proposing new approaches and frameworks to advance blockchain technology to facilitate digital asset ownership in recent years. For example, Solouki & Bamakan (2022) introduced dynamic NFTs which are more responsive to changing circumstances. Identifying that intellectual property rights require better protection, Ferro et al. (2023) proposed an approach to digital rights management by using smart contracts.

Digital assets are different from physical assets in many ways. The unique features of blockchain-based digital asset ownership have not been specifically examined in the literature. This paper will contribute to narrowing this identified knowledge gap. The study focuses specifically on blockchain technology's impact on digital asset ownership and adopts content analysis of the state-of-the-art literature as well as a case-study approach to analyse the potential and challenges of blockchains to digital asset ownership.

This paper endeavors to provide insights to better understand and design future systems for integrating blockchain-based digital assets into the economy. This will assist technology developers and policymakers to determine what and how much to regulate to promote innovation and integrate blockchain into the digital economy whilst keeping disruption and costs to a minimum. The following sections will present the case study methodology, discuss potential and challenges surrounding digital assets and draw a conclusion.

2 Methodology

This study's methodology consists of two steps. In the first step, content analysis of the literature and market information in the subject matter is performed. This step on one hand provides a fundamental understanding of digital asset ownership and the application of blockchain technology. On the other hand, current information from industry sources such as market data and technical reports are examined to ensure the timeliness of our research. The second step is a case study approach to analysing digital asset as an asset class managed by blockchain. The case study approach allows us to explore a phenomenon in-depth, in its natural context, especially if the line between context and phenomenon is blurry (Yin, 2009). This approach has been shown to be helpful in inductively identifying and coming up with new factors or constructs (Starman, 2013). Case studies are also able to analyse complex situations that might have multiple variables (Starman, 2013) which is apt in studying the complexities surrounding blockchain technology for asset ownership. Taking reference from Pu & Lam (2023) and Vanelslander et al. (2019), multiple case studies are used as this offers insights from different practices and perspectives. However, cases are limited to allow for adequate time for analysis and to avoid the potential pitfall of over-collecting (Crowe et al., 2011). Potential cases are selected with reference to the literature and most updated repository from library and internet. All the screened cases are existing real-world companies and projects. Cross validation of information is conducted by checking different sources, for example, company websites, technical reports, and third-party commentary pieces. We have completed four case studies and draw inferences from them, while this paper will cite key examples from two of them, namely, IBM and Samsung.

3 Results and Discussion

The most distinctive characteristic of digital assets is being virtual and intangible. This makes digital asset ownership more mobile and transferable, but at the same, less visible and controllable. For instance, audiovisual media such as entertainment videos and animations can be created, stored, and transacted digitally. Blockchain technology plays an increasingly active role to unlock the value of digital assets. Following the research process as presented in the above section, the unique potential and challenges of blockchain technology for digital asset ownership are derived. The findings are summarised in Table 1.

In terms of potential and the largest advantage, blockchain is able to monetise and unlocks newer digital assets. Unlike physical assets, digital assets can nowadays "live" on the internet or any digital systems and are much more easily integrated into the blockchain. Yet like physical assets, digital assets that were once illiquid can increasingly be tokenised and thus be unlocked. There have been numerous examples of digital pieces that were easily tokenised and thus sold, helping organisations such as museums improve funding (Valeonti et al., 2021). Data stored on servers are another example of a newly "unlocked" asset which can be commercialised (Grabis et al., 2020). This could also extend to other forms of intellectual property such as music, art, (Shatkovskaya et al., 2018) or any other form of digital collectibles. There have been cases of commercial companies jumping on the bandwagon such as Samsung's digital store in Decentraland, a blockchain-powered Metaverse world (Samsung, 2024). In particular, the case of Samsung builds on the established Ethereum blockchain with smart contract functionality, hence the potential of tokenising digital assets is realised more effectively. Many new products, such as smart home products, can be created and sold. Samsung claims that the digital store enables new relationships with its customers, thus generates business opportunities.

Table 1. Unique potential and challenges of blockchain technology for digital asset ownership.

Asset class	Potential	Challenges
Digital	- Monetises and unlocks newer assets - Greater and fined-grained access control over the asset	- Intellectual property laws and regulations - Limited scalability and data storage on the blockchain

Due to the presence of smart contracts and automation, digital assets also benefit from increased access control. What used to be just an image, or a video can now, with the help of NFTs, become unique on the blockchain, introducing scarcity (Valeonti et al., 2021). Owners of these virtual assets can now have greater control of the asset as it is now represented by an NFT token. Transactions through the blockchain allow one to transfer, sell, loan, borrow against the asset, or even fractionalise the underlying asset with many real-world cases having already done so (Genç, 2023; QUIROZ-GUTIERREZ, 2022). In both past studies and real-world examples, proposed models built on the blockchain can even enable the automatic sharing of profit through royalties (OpenSea, 2022; Pérez-Solà & Herrera-Joancomartí, 2020). These greater access control not only applies to just works like art or music. Studies have also been exploring the benefit of using the blockchain for digital marketplaces for data used in machine learning algorithms (Baranwal Somy et al., 2019). Yu & Zhao (2019) proposed a data marketplace that could help enhance security, improve traceability, and protect the rights and privacy of data providers. The potential is well demonstrated by the case of Orion open source blockchain database developed by IBM (IBM, 2023). The greater control over the digital asset and the trustless nature of the blockchain helps foster a more secure and private marketplace which is ideal for sensitive data, especially in the digital age where access to private information is sometimes abused. A distinction of this blockchain apart from others lies in the feature that it is a centralised blockchain, not decentralised. This is made possible by integrating a cryptography-based layer on top of a classical database. Hence, benefits and functionalities brought by blockchain are added on those of a classical database.

However, challenges and limitations of using blockchain for digital asset ownership also exist. Digital assets are susceptible to the difficulties in governance. There are gaps especially when it comes to laws and regulations. The ownership of a huge number of new digital assets enabled by blockchain has become a problem as regulations are not equipped to regulate these assets. This is a common theme noted in many of the cases studied such as by Shatkovskaya et al. (2018) and Valeonti et al. (2021). One point of uncertainty is how copyrights are allocated during transactions concerning intellectual property represented by NFTs (Valeonti et al., 2021). This will be a major area requiring improvement, not only for a particular case but the entire field of blockchain. Policy makers are recommended to accelerate the amendments of laws and regulations to incorporate digital asset ownership.

Another area of concern is the scalability of the blockchain to handle and store information of these digital assets. One area of concern for data is the limit to storing the data directly on-chain (Grabis et al., 2020) as in some cases off-chain storage has been shown to be more lightweight (Baranwal Somy et al., 2019). These limitations will vary according to the type of blockchain designed and used. Until sufficient legal clarity and technological advances have been made, they will remain bottlenecks to the ownership of digital assets. To address this challenge, Orion open source blockchain database developed by IBM combines classical database functionalities with blockchain features. The aim is to offer a higher level of scalability (IBM, 2023). Therefore, with advanced design of the blockchain architecture, it is possible to mitigate the drawback of scalability.

4 Conclusion

Blockchain technology has developed tremendously as the world becomes increasingly digitalised. The potential of blockchain for asset ownership is numerous but one should not neglect its drawbacks. This paper analysed digital asset ownership facilitated by blockchain with special attention to the unique potential and challenges when applicable to digital assets. The key findings are summarised here. The top two areas of potential are: 1) the opportunity to monetise and unlock newer assets, 2) greater and fined-grained access control over the digital asset. The two major challenges are: 1) under-developed intellectual property laws and regulations, 2) limitation challenges in scalability and data storage on the blockchain. We discussed these areas of potential and challenges along with real-world cases.

This paper presents a contribution which assists users, policymakers, and other interested parties to gain a better overview of the implications of blockchain technology in digital asset ownership. This could potentially help broad-based planning of policies and regulations to facilitate the adoption of blockchain. Blockchain is a rapidly emerging field, and it is hoped that innovation can continue to flourish while regulations help to keep users operating in a secure environment. Because under-developed intellectual property laws and regulations remain a key barrier to digital asset ownership, this is a recommended future research direction in addition to technology advancement of blockchain's functionality.

Disclosure of Interests.   The authors have no competing interests to declare that are relevant to the content of this article.

References

Ante, L.: Non-fungible token (NFT) markets on the Ethereum blockchain: temporal development, cointegration and interrelations. Econ. Innov. New Technol. **32**(8), 1216–1234 (2023)

Baranwal Somy, N., et al.: Ownership preserving AI market places using blockchain. In: 2019 IEEE International Conference on Blockchain (Blockchain) (2019). https://doi.org/10.1109/blockchain.2019.00029

Castonguay, J.J., Stein Smith, S.: Digital assets and blockchain: hackable, fraudulent, or just misunderstood? Account. Perspect. **19**(4), 363–387 (2020)

Crowe, S., Cresswell, K., Robertson, A., Huby, G., Avery, A., Sheikh, A.: The case study approach. BMC Med. Res. Methodol. **11**(1) (2011). https://doi.org/10.1186/1471-2288-11-100

Egkolfopoulou, M.: Crypto's war test leaves future of money debate wide open. Bloomberg (2022). https://www.bloomberg.com/news/articles/2022-03-10/crypto-s-test-since-russian-invasion-leaves-future-of-money-debate-wide-open. Accessed 5 Mar 2024

Ferro, E., et al.: Digital assets rights management through smart legal contracts and smart contracts. Blockchain: Res. Appl. **4**(3), 100142 (2023)

Genç, E.: How can you share an NFT? Fractional NFTs explained (2023). www.coindesk.com. https://www.coindesk.com/learn/how-can-you-share-an-nft-fractional-nfts-explained/. Accessed 5 Dec 2023

Grabis, J., Stankovski, V., Zarins, R.: Blockchain enabled distributed storage and sharing of personal data assets. In: 2020 IEEE 36th International Conference on Data Engineering Workshops (ICDEW) (2020). https://doi.org/10.1109/icdew49219.2020.00-13

Hammi, B., Zeadally, S., Perez, A.J.: Non-fungible tokens: a review. IEEE Internet Things Mag. **6**(1), 46–50 (2023)

IBM: The Orion blockchain database: empowering multi-party data governance (2023). https://www.ibm.com/blog/the-orion-blockchain-database-empowering-multi-party-data-governance/

OpenSea: 10. Setting fees on secondary sales. OpenSea Developer Documentation (2022). https://docs.opensea.io/docs/10-setting-fees-on-secondary-sales. Accessed 3 Mar 2024

Pérez-Solà, C., Herrera-Joancomartí, J.: BArt: trading digital contents through digital assets. Concurr. Comput. Pract. Exp. **32**(12), e5490 (2020)

Pu, S., Lam, J.S.L.: The benefits of blockchain for digital certificates: a multiple case study analysis. Technol. Soc. **72**, 102176 (2023)

President's Working Group on Financial Markets (PWG), Federal Deposit Insurance Corporation (FDIC), & Office of the Comptroller of the Currency (OCC): Report on Stablecoins. In U.S. Department of the Treasury (2022). https://home.treasury.gov/system/files/136/StableCoinReport_Nov1_508.pdf

Quiroz-Gutierrez, M.: Someone got a $1.25 million loan by using NFTs as collateral. Fortune (2022). https://fortune.com/2022/02/07/nft-collateral-for-million-dollar-loans/#:~:text=Borrowers%20who%20use%20the%20site. Accessed 5 Mar 2024

Reuters: Dubai adopts first virtual asset law, establishes regulator. Reuters (2022). https://www.reuters.com/world/middle-east/dubai-adopts-first-law-governing-virtual-assets-ruler-2022-03-09/. Accessed 5 Dec 2023

Samsung: Create, collect and connect in the Samsung metaverse (2024). https://www.samsung.com/us/explore/sustainability/create-collect-and-connect-in-the-metaverse/

Shatkovskaya, T.V., Shumilina, A.B., Nebratenko, G.G., Isakova, J.I., Sapozhnikova, E.Y.: Impact of technological blockchain paradigm on the movement of intellectual property in the digital space. Eur. Res. Stud. J. **XXI**(Special Issue 1), 397–406 (2018). https://doi.org/10.35808/ersj/1190

Solouki, M., Bamakan, S.M.H.: An in-depth insight at digital ownership through dynamic NFTs. Procedia Comput. Sci. **214**, 875–882 (2022)

Starman, A.B.: The case study as a type of qualitative research. J. Contemp. Educ. Stud. 28–43 (2013)

Valeonti, F., Bikakis, A., Terras, M., Speed, C., Hudson-Smith, A., Chalkias, K.: Crypto collectibles, museum funding and OpenGLAM: challenges, opportunities and the potential of non-fungible tokens (NFTs). Appl. Sci. **11**(21), 9931 (2021). https://doi.org/10.3390/app112 19931

Vanelslander, T., et al.: A serving innovation typology: mapping port-related innovations. Transp. Rev. **39**(5), 611–629 (2019)

WEF: The global risks report 2022 (2022)

Yin, R.K.: Case Study Research: Design and Methods. SAGE (2009)

Yu, B., Zhao, H.: Research on the construction of big data trading platform in China. In: Proceedings of the 2019 4th International Conference on Intelligent Information Technology - ICIIT 2019 (2019). https://doi.org/10.1145/3321454.3321474

On the Optimal Mixing Problem of Approximate Nash Equilibria in Bimatrix Games

Xiaotie Deng[1] , Dongchen Li[2] , and Hanyu Li[1(✉)]

[1] CFCS, School of Computer Science, Peking University, Beijing 100871, China
{xiaotie,lhydave}@pku.edu.cn
[2] Department of Computer Science, The University of Hong Kong,
Pokfulam, Hong Kong
dongchen.li@connect.hku.hk

Abstract. This paper introduces the optimal mixing problem, a natural extension of the computation of approximate Nash Equilibria (NE) in bimatrix games. The problem focuses on determining the optimal convex combination of given strategies that minimizes the approximation (i.e., regret) in NE computation. We develop algorithms for the exact and approximate optimal mixing problems and present new complexity results that bridge both practical and theoretical aspects of NE computation. Practically, our algorithms can be used to enhance and integrate arbitrary existing constant-approximate NE algorithms, offering a powerful tool for the design of approximate NE algorithms. Theoretically, these algorithms allow us to explore the implications of support restrictions on approximate NE and derive the upper-bound separations between approximate NE and exact NE. Consequently, this work contributes to theoretical understandings of the computational complexity of approximate NE under various constraints and practical improvements in multi-agent reinforcement learning (MARL) and other fields where NE computation is involved.

1 Introduction

The problem of approximate *Nash equilibrium* (NE) computation is interesting and fundamental from both theoretical and pragmatic perspectives. Theoretically, approximate NE builds bridges between several important complexity classes related to TFNP [23], especially PPAD [2,6]. Practically, an approximate NE solver is a core component in multi-agent reinforcement learning (MARL) [14,15,22,25], which has been successfully applied to train machine agents that can defeat the top human players in electronic games [28,33].

To define the approximate NE problem, consider a two-player bimatrix game with payoff matrices $R \in [0,1]^{m \times n}$ and $C \in [0,1]^{m \times n}$. Define

$$f(x,y) = \max\{f_R(x,y), f_C(x,y)\} \geq 0$$

with $f_R(x, y) = \max\{Ry\} - x^\mathsf{T} Ry$ and $f_C(x, y) = \max\{C^T x\} - x^\mathsf{T} Cy$. Intuitively, in the game given by payoff matrices R, C, when the two players select x, y respectively, $f(x, y)$ is a measure of their willingness to unilaterally deviate from the current strategy. Following [9,31], the goal of NE computation in bimatrix games can be written as:

$$\arg\min_{x,y} f(x, y) \quad \text{s.t.} \quad x \in \Delta_m, y \in \Delta_n. \tag{1}$$

In the literature [20], $f(x, y)$ is called the *approximation* of (x, y). A strategy profile (x, y) is an NE if $f(x, y) = 0$ and an ϵ-NE if $f(x, y) \leq \epsilon$. Since $f \geq 0$ and NE always exists [26], the solution of (1) must be an NE.

1.1   The Optimal Mixing Problem

In this paper, we propose the *optimal mixing problem* of approximate NE, which is a natural extension of the approximate NE computation. Given s, t strategies $x_1, \ldots, x_s$ and $y_1, \ldots, y_t$ of the two players, the (s, t)-optimal mixing problem is:

$$\arg\min_{\alpha,\beta} f(\alpha_1 x_1 + \cdots + \alpha_s x_s, \beta_1 y_1 + \cdots + \beta_t y_t) \quad \text{s.t.} \quad \alpha \in \Delta_s, \beta \in \Delta_t. \tag{2}$$

Intuitively, the problem seeks the convex combination of $x_1, \ldots, x_s$ and $y_1, \ldots, y_t$ that has the minimum approximation. We also define the ϵ-*optimal* (s, t)-*mixing problem* by allowing an ϵ additive tolerance in the objective, i.e., the output is required to have approximation no more than $f^* + \epsilon$, where f^* is the optimal value of (2).

The approximate and exact optimal mixing problem naturally extends the NE computation: when the input of the approximate and exact optimal mixing problem is $e_1, \ldots, e_m$ and $e_1, \ldots, e_n$, the standard basis of $\mathbb{R}^m$ and $\mathbb{R}^n$, it is exactly (1).

The motivation of the optimal mixing problem is twofold.

- **Algorithm design and analysis**:
  To guarantee a certain approximation bound, current polynomial-time algorithms for approximate NE all follow a *search-and-mix* method [20]. Such methods can be divided into two polynomial-time phases. In the search phase, an algorithm computes a fixed number of strategies of each player. In the mixing phase, the algorithm then make *specific* convex combinations of the selected strategies and outputs the one with the minimum f value. However, such a design paradigm has several limitations:
    - **Different approaches seldom integrate.** The search phases of these algorithms follow very different and even incomparable approaches, e.g., gradient descent [9,31], linear programming [7], and zero-sum game [1,5]. However, it is worth considering whether an algorithm with better approximation bounds can be designed by combining these different approaches. The optimal mixing problem, which allows to mix any strategies, offers a unified framework for such a combination.

- **Overemphasis on worst cases.** To guarantee an approximation bound, all mixing phase design in the literature only focuses on the worst-case instances [20], which are rare compared to other instances [3,12,20,32]. However, such focuses may hinder the practical usefulness of these algorithms. It is natural to directly find the *optimal* convex combination for every instance, which is essentially an optimal mixing problem.

- **Computational complexity of approximate NE under support restrictions.** It was shown in [4,13] that adding certain natural requirements increases the complexity of NE computation from PPAD-complete [2] to NP-complete. However, we can easily find a requirement to falsify this conclusion on approximate NE (which is a more computationally appropriate notion[1]). For example, on a certain support, deciding the existence of NE is NP-complete [4]. In contrast, by shifting the probability of the strategy profile a little, we can show that for any support, there is always an ϵ-NE on it.

  From above observations, we know that the complexity of approximate NE computation could be very different from that of exact NE computation if we put certain restrictions on the solution. Thus, we need to reexamine the effect of such restriction over approximate NE. The optimal mixing problem provides a unified framework to study such restrictions, especially the support restriction.

1.2 Our Contributions

Bearing these motivations, we develop algorithms for the approximate and exact optimal mixing problems. Informally, we have that:

Theorem 1 (Main results). *When $s \leq 2, t \leq 3$ or $s = 1, t = \text{poly}(n)$, there is algorithm solving any optimal mixing problem in $\text{poly}(m, n)$ time. Moreover, there exists an algorithm solving any ϵ-optimal (s,t)-mixing problem in time*

$$\text{poly}(m, n, s, t) \left(e + \frac{e}{\sqrt{\epsilon/2}} \right)^{s+t} \text{ and space } \text{poly}(m, n, s, t).$$

This theorem provides various implications, as is described below.

- **Algorithm design and analysis:**
  - **Integration of different approaches:** Our algorithms can be used to combine *arbitrary* strategies computed by different approaches. Thus, the combined strategy fuses various advantages from these different approaches.
  - **Instance-optimal mixing phases:** With these algorithms, we propose general approximate and exact polynomial-time algorithms for instance-optimal mixing phases. There is no need to design the mixing phase ad hoc in the future. Interestingly, [10] shows that there is an automatic way

---

[1] According to [21], approximate NE is a more realistic solution concept by involving bounded rationality.

to derive the approximation bound for any search-and-mix algorithm, even when using our algorithms as the mixing phase. Thus, together with our algorithms, the mixing phase design is fully automated.[2]

- **Computational complexity of approximate NE under support restrictions**:
  - **Restrictions enlarge the complexity**: As is shown in Table 1, our algorithms establish upper bounds for finding the best approximate NE over certain support. Table 1 shows the upper bound complexity of approximate NE with support restrictions (the left column, where the support of their approximate NE must be over certain pure strategies) is significantly larger than that of approximate NE without support restrictions (the right column). This is consistent with the cases in exact NE computation [4, 13].
  - **Approximate NE with restrictions could be more difficult than exact NE**: The upper bound results shows some counter-intuitive separation between the complexity of approximate NE and that of exact NE. The complexity of deciding the existence of $1/n^{O(1)}$-NE under support restrictions is $2^{O(n \log n)}$. Surprisingly, this is even higher than the complexity for *exact* NE under support restrictions (which is $2^{O(n)}$ by support enumeration [27]). People usually think that approximate NE is easier than exact NE. However, when there is no existence guarantee (due to support restrictions), things could be different: to show the non-existence of $1/n^{O(1)}$-NE is harder than to show the non-existence of exact NE.

Table 1. This table shows the results of ϵ-optimal (s,t)-mixing problem with distinct pure-strategy input. Without loss of generality, we assume $m = n$ and $s \leq t$. The results are presented in "lower bound; upper bound" pairs if the two bounds are not matched. Our new results are colored in blue.

ϵ	s, t	
	$\Theta(n) \leq n$	n
$> 1/3$	$?; 2^{O(n)}$	$\text{poly}(n)$ [9]
$\in [\epsilon^*, 1/3]$	$?; 2^{O(n)}$	$?; n^{O(\log n / \epsilon^2)}$ [21]
$< \epsilon^*, \text{const}$	quasi-poly$(n)^4$ [30]; $2^{O(n)}$	quasi-poly$(n)^4$ [30]; $n^{O(\log n / \epsilon^2)}$ [21]
$= 1/n^{O(1)}$	PPAD-hard [2]; $2^{O(n \log n)}$	PPAD-complete [2]; $2^{O(n)}$ [19]
$= 1/2^{O(n)}$	PPAD-hard [2]; $2^{O(n^2)}$	PPAD-complete [2]; $2^{O(n)}$ [19]
$= 0$ (exact)	FNP-hard [4, 13]; ?	PPAD-complete [2]; $2^{O(n)}$ [19]

---

[2] A perhaps surprising result given by [10] is that for each algorithm in the literature, the approximation bound with our algorithms as the mixing phase is the same as original ad hoc ones!.

Remark 1. One may note that in Theorem 1, we stop at $(2,3)$ for exact optimal mixing problem. There is a fair reason for this. In general, the optimal mixing problem has the form of a quadratically constrained quadratic program (QCQP). The exact solution is an element belonging to an algebraic variety. For $(2,3)$, this solution can be reduced to a solution of a univariate quintic equation, and can be radically solved. However, for $(3,3)$, it is even not clear how to reduce the problem to a univariate equation. Such a phenomenon is not unique in NE computation. For example, the exact NE of a 3-player game is also an algebraic variety, and it is still not clear how to solve it using a Turing machine [11].

1.3   Related Work

Complexity and Approximation of NE. The computational complexity of approximate NE has been extensively studied. Initially, Papadimitriou [29] introduces a general complexity class PPAD and shows that computing $1/2^n$-NE lies in PPAD. Later, computing $1/\operatorname{poly}(n)$-NE is shown to be PPAD-complete for k-player games with any fixed $k \geq 2$ [2,6]. Moreover, computing NE in two-player games is hard even in the smoothed meaning [2], or restricting the rank of game to constant [24]. These results establish the hardness of approximate NE computing with polynomial-small approximation. It is well-believed that computing NE could require exponential time (ETH for PPAD). See, e.g. [18,30].

For constant approximation, it seems to be easier than polynomial-small approximation. For any given $\epsilon > 0$, there is an algorithm finding an ϵ-NE [21] in $n^{O(\log n/\epsilon^2)}$ time (QPTAS). Assuming ETH for PPAD, Rubinstein [30] shows that there exists a constant $\epsilon^* > 0$ such that computing an ϵ^*-NE in a two-player $n \times n$ game requires $n^{\log^{1-o(1)} n}$ time. This matches the QPTAS result [21] up to $o(1)$ term.

The lower bound results on constant approximation lead to the study of the upper bound, i.e., seeking the minimum ϵ such that there exists a polynomial-time algorithm computing an ϵ-NE. Most literature focuses on two-player (bimatrix) games in the literature. A series of polynomial-time algorithm [1,2,7–9,17,31] have been proposed, with the approximation from the beginning of $3/4$ [17] to the state-of-the-art $1/3 + \delta$ [9]. For a more thorough introduction, see [20].

Notably, all the results above, including algorithms and hardness results, heavily rely on the existence of NE. In fact, if we want to find NE with certain natural requirements, such as strong NE, NE over a certain support, or with certain social welfare, such NE may not exist. Moreover, by using SAT to make reductions, [4,13] show that deciding the existence of such NE is NP-compete. The FNP-hardness result in Table 1 is a direct corollary of this reduction. However, to the best of our knowledge, there is no literature at all for similar discussions over approximate NE.

NE Computation in Practical Applications. Emerging from game theory, NE computation has been widely applied in many fields, including Internet economics, computer science, and machine learning. Most prominently, NE computing is a core component in many multi-agent reinforcement learning (MARL) algorithms, including PSRO [25], Nash-Q [14], Nash-VI [22], and Nash-V learning [15]. MARL has been successfully applied to train machine agents that can defeat the top human players in electronic games, including AlphaStar [33] in StarCraft II and OpenAI Five [28] in Dota 2. With such fruitful applications, it is demanding to design efficient algorithms for NE.

1.4 Paper Organization

This paper is organized as follows. In Sect. 2, we introduce the basic concepts and notations. In Sect. 3, we present the polynomial-time algorithms for the optimal mixing problem. In Sect. 4, we present an algorithm for the approximate optimal mixing problem. In Sect. 5, we show how to apply the optimal mixing problem to make an instance-optimal enhancement to the search-and-mix method in the literature.

2 Preliminaries

Asymptotic Notations. We use the standard asymptotic notations $O(\cdot)$ and $\Theta(\cdot)$ to describe the asymptotic behavior of functions. For two positive functions f and g, $f = O(g)$ means that there exists a constant $c > 0$ such that $f(n) \leq c \cdot g(n)$ for all sufficiently large n. $f = \Theta(g)$ means that $f = O(g)$ and $g = O(f)$.

Vectors and Matrices. Denote the n-dimensional Euclidean space by $\mathbb{R}^n$. The standard orthonormal basis of $\mathbb{R}^n$ is $e_1, \ldots, e_n$. Notation $[n] := \{1, \ldots, n\}$ represents an index set. For vector $v \in \mathbb{R}^n$, denote its ith item by v_i. For vector $u \in \mathbb{R}^n$, define the following operators: $\max\{u\} := \max\{u_1, \ldots, u_n\}$, $\min\{u\} := \min\{u_1, \ldots, u_n\}$. For two vectors $v, w \in \mathbb{R}^n$, notation $v \geq w$ represents that $v_i \geq w_i$ holds for every $i \in [n]$.

For an $m \times n$ matrix A, denote its ith row by A_i, its jth column by A^j, and its item at ith row jth column by A_{ij}. Its transpose is denoted by A^T.

Simplex and Convex Combinations. A standard $(n-1)$-simplex is the set $\Delta_n := \{\alpha \in \mathbb{R}^n : \alpha \geq 0 \text{ and } \sum_{i=1}^n \alpha_i = 1\}$. A simplex can be viewed as a probability space and its elements are probability vectors. For given elements $z_1, \ldots, z_w$ from $\mathbb{R}^n$, the set of their *convex combinations* is defined to be $\{\alpha_1 z_1 + \cdots + \alpha_w z_w : \alpha \in \Delta_w\}$, where any vector $\alpha \in \Delta_w$ determines a convex combination $\alpha_1 z_1 + \cdots + \alpha_w z_w$.

Games, Mixed Strategies, and Best Responses. We only focus on *bimatrix games*, in which there are two players. We refer to them as the row player and the column player. A game can be defined by a pair of payoff matrices R and C in $[0, 1]^{m \times n}$. When the row player chooses the ith row and the column player chooses the jth column, their payoffs are denoted by R_{ij} and C_{ij}, respectively.

For each player, a *(mixed) strategy* of the row (column) player is a vector $x \in \Delta_m$ ($y \in \Delta_n$). In particular, pure strategies are a specific pure strategy is chosen with a probability of 1. A *strategy profile* (x, y) refers to a pair of mixed strategies x and y from the row and column players, respectively. Given the strategy profile, the payoffs of the row player and the column player are $x^\mathsf{T} R y$ and $x^\mathsf{T} C y$, respectively. A *best response* against a strategy x from the row player is a mixed strategy of the column player that maximizes the expected payoff against x.

Approximate Nash Equilibria from the Optimization Viewpoint. We follow [31] to define ϵ-NE. First, define the regret of the row player and the column player as follows:

$$f_R(x, y) := \max\{Ry\} - x^\mathsf{T} R y \text{ and } f_C(x, y) := \max\{C^\mathsf{T} x\} - x^\mathsf{T} C y.$$

Define $f(x, y) := \max\{f_R(x, y), f_C(x, y)\}$. Then a strategy profile (x, y) is an ϵ-NE if $f(x, y) \leq \epsilon$. $f(x, y)$ is called the *approximation* of (x, y). Particularly, a strategy profile is an NE if it is a 0-NE. The minimum of f over $\Delta_m \times \Delta_n$ is always 0 by the existence of NE [26].

3   Polynomial-Time Algorithms for Optimal Mixing Problems

In this section, we propose polynomial-time algorithms for the optimal mixing problem. Recall that the optimal mixing problem is defined as follows.

Definition 1 (Optimal (s, t)-mixing problems). *An optimal (s, t)-mixing problem has the following input and output.*

- Input*: Bimatrix game (R, C), mixed strategies $x_1, \ldots, x_s$ of the row player and $y_1, \ldots, y_t$ of the column player.*
- Output*: Coefficients $\alpha^* \in \Delta_s$, $\beta^* \in \Delta_t$ that minimize*

$$f(\alpha_1 x_1 + \cdots + \alpha_s x_s, \beta_1 y_1 + \cdots + \beta_t y_t).$$

For convenience, we name an algorithm as an *optimal (s, t)-mixing algorithm* if it solves any optimal (s, t)-mixing problem.

A summary of the results in this section is presented in Theorem 2.

Theorem 2. *The following statements hold.*

1. *For any t, there exists an optimal $(1, t)$-mixing algorithm in $O(mnt + L(t, m))$ time, where $L(t, m)$ is the time complexity of solving a linear program with t variables and m constraints.*
2. *There exists an optimal $(2, 2)$-mixing algorithm in $O(mn)$ time.*
3. *There exists an optimal $(2, 3)$-mixing algorithm in $O(m^2(n + \log m) + n \log n)$ time.*

To obtain the results, we begin by scrutinizing the form of the problem. The objective function in Definition 1 can be expanded as follows:

$$\max\{ \max\{R(\beta_1 y_1 + \cdots + \beta_t y_t)\} - (\alpha_1 x_1 + \cdots + \alpha_s x_s)^\mathsf{T} R(\beta_1 y_1 + \cdots + \beta_t y_t),$$
$$\max\{C^\mathsf{T}(\alpha_1 x_1 + \cdots + \alpha_s x_s)\} - (\alpha_1 x_1 + \cdots + \alpha_s x_s)^\mathsf{T} C(\beta_1 y_1 + \cdots + \beta_t y_t)\}. \tag{3}$$

A direct observation is that we can suppose without loss of generality that $s \le t$. Otherwise, we simply exchange the positions of the players.

We first consider the simplest situation where $s = 1$. In this case, Δ_s is degenerated to a single point. (3) is degenerated to the following form:

$$\max\{ \max\{R(\beta_1 y_1 + \cdots + \beta_t y_t)\} - x_1^\mathsf{T} R(\beta_1 y_1 + \cdots + \beta_t y_t),$$
$$\max\{C^\mathsf{T} x_1\} - x_1^\mathsf{T} C(\beta_1 y_1 + \cdots + \beta_t y_t)\}. \tag{4}$$

This can further be expanded to:

$$\max\{\beta_1 (Ry_1)_1 + \cdots + \beta_t (Ry_t)_1 - \beta_1 (x_1^\mathsf{T} Ry_1) - \cdots - \beta_t (x_1^\mathsf{T} Ry_t), \ldots,$$
$$\beta_1 (Ry_1)_m + \cdots + \beta_t (Ry_t)_m - \beta_1 (x_1^\mathsf{T} Ry_1) - \cdots - \beta_t (x_1^\mathsf{T} Ry_t), \tag{5}$$
$$(C^\mathsf{T} x_1) - \beta_1 (x_1^\mathsf{T} Cy_1) - \cdots - \beta_t (x_1^\mathsf{T} Cy_t)\}.$$

Now, the objective function becomes the maximum of $m + 1$ functions being all linear in β, respectively. In addition, the constraint $\beta \in \Delta_t$ is also linear in β. Using a standard transformation, we can transform the problem into a linear program with $t + 1$ variables and $m + t + 2$ constraints. Since the linear program can be solved in polynomial time [16], we obtain a polynomial-time optimal $(1, t)$-mixing algorithm.

Then, we consider the general optimal (s, t)-mixing problem. In this case, all terms in (3) are non-degenerated. There are three major components in (3): inner maximum terms $\max\{R(\beta_1 y_1 + \cdots + \beta_t y_t)\}$ and $\max\{C^\mathsf{T}(\alpha_1 x_1 + \cdots + \alpha_s x_s)\}$, bilinear terms $(\alpha_1 x_1 + \cdots + \alpha_s x_s)^\mathsf{T} R(\beta_1 y_1 + \cdots + \beta_t y_t)$ and $(\alpha_1 x_1 + \cdots + \alpha_s x_s)^\mathsf{T} C(\beta_1 y_1 + \cdots + \beta_t y_t)$, and the outermost maximum operator (i.e., $\max\{f_R, f_C\}$). Different terms present different difficulties:

1. Inner maximum terms are piecewise-linear in β and α, respectively, thus convex but non-differentiable.
2. Bilinear terms are bilinear in β and α, thus differentiable but nonconvex.
3. The outermost maximum operator is non-differentiable.

Our solution is sketched below:

1. Since the inner maximum terms have a piecewise-linear structure, we can divide the problem into subproblems on each linear piece.
   - To determine the linear pieces, we resort to the famous *half-plane intersection problem* in computational geometry.
   - This overcomes the first difficulty.
2. For each subproblem, we derive necessary conditions for the global optima. Thus, by scanning all points satisfying the conditions, we can find a global optimum.

- To give the necessary conditions, we combines discrete geometry (linear-algebraic characterizations of polytopes) and optimization (KKT conditions).
- To scan these points, we formulate the conditions into various optimization problems (e.g., univariate quadratic programs and linear programs) and solve them.
- This overcomes the second and the third difficulty.

3. Finally, we show that there are polynomial number of linear pieces and in each linear piece a constant number of points to check. Thus, we can find the global optimum in polynomial time.

Since the algorithms are quite involved and the analysis is highly technical, we defer the details to the full version of this paper.

To demonstrate the main idea of the algorithms, below we sketch the implementation of the optimal $(2, 2)$-mixing algorithm over arbitrary strategies x_1, x_2 and y_1, y_2.

Denote the set of all possible convex combinations as $\mathcal{A} := \{(\alpha_1 x_1 + \alpha_2 x_2, \beta_1 y_1 + \beta_2 y_2) : \alpha, \beta \in \Delta_2\}$. Observe that the form of the function f_R over the mixing region $\mathcal{A}$ is:

$$f_R(\alpha_1 x_1 + \alpha_2 x_2, \beta_1 y_1 + \beta_2 y_2)$$
$$= \max\{R(\beta_1 y_1 + \beta_2 y_2)\} - (\alpha_1 x_1 + \alpha_2 x_2)^\mathsf{T} R(\beta_1 y_1 + \beta_2 y_2).$$

Note that $\beta_2 = 1 - \beta_1$, thus the max term can be written as $\max\{\beta_1 R(y_1 - y_2) + R y_2\}$. It has the form of the maximum of m linear functions about β_1, which is piecewise linear in β_1.

Now, we want to compute the exact form of the piecewise linear function, given by a sequence of breakpoints $0 = b_1 \leq \cdots \leq b_t = 1$ $(t \leq m+1)$ so that f is linear in β on each $[b_i, b_{i+1}]$. We need to compute the exact form of this problem, that is to compute the value of $R(y_1 - y_2)$ and $R y_2$ with time $O(mn)$. Then, this becomes a famous problem in computational geometry called the *envelope problem*, which can be solved in time $O(m \log m)$.

Similarly, we can compute the linear pieces given by breakpoints $0 = a_1 \leq \cdots \leq a_s = 1$ $(s \leq n+1)$ in time $O(n \log n)$. Therefore, on each grid $[a_i, a_{i+1}] \times [b_j, b_{j+1}], i \in [s], j \in [t]$, both f_R and f_C are linear in α and β, respectively.

Then, we minimize the objective function over each grid and compare the results to take the one with minimal f value. By doing so, we obtain the global minimum of f on region $\mathcal{A}$.

On each grid, the objective function is in the form of the maximum of two bilinear functions g_1 and g_2. However, it is still non-differentiable. We apply the KKT condition from continuous optimization to obtain the necessary optimal conditions for this problem. We can show that the minimum must be attained at the following three kinds of points:

1. Points where the partial derivative of g_1 or g_2 with respect to α or β is zero.
2. The four vertices of the grid.
3. Points where g_1 and g_2.

Now we show that the number of points to be checked of each kind is bounded by a constant. For the first kind, since the partial derivatives of bilinear functions g_1 and g_2 are linear, the problem finally reduces to a univariate linear program, which can be solved in constant time. For the second kind, there are only four points. Finally, for the third kind, we can solve the relation between α and β from $g_1 = g_2$ and substitute it into the objective function. Then, we obtain a univariate quadratic program, which can be easily minimized by checking at most six points.

In words, on each grid, we only need constant time to compute the minimum f. Thus, by scanning over all grids in $O(mn)$ time, we can compute the global minimum of f on $\mathcal{A}$. The total complexity is given by $O(mn+m\log m+n\log n) = O(mn)$.

4   An Algorithm for Approximate Optimal Mixing Problems

In this section, we present an algorithm solving any ϵ-optimal (s,t)-mixing problem. Recall that the ϵ-optimal (s,t)-mixing problem is defined as follows.

Definition 2 (ϵ-Optimal (s,t)-mixing problem). *Given $\epsilon > 0$, the ϵ-optimal (s,t)-mixing problem has the following input and output:*

- Input*: Bimatrix game (R,C), mixed strategies $x_1,\ldots,x_s$ of the row player and $y_1,\ldots,y_t$ of the column player, and an $\epsilon > 0$.*
- Output*: Coefficients $\tilde{\alpha} \in \Delta_s, \tilde{\beta} \in \Delta_t$ such that for any solution $\alpha^* \in \Delta_s, \beta^* \in \Delta_t$ to the optimal (s,t)-mixing problem, the following inequality holds:*

$$f(\tilde{\alpha}_1 x_1+\cdots+\tilde{\alpha}_s x_s, \tilde{\beta}_1 y_1+\cdots+\tilde{\beta}_t y_t) \leq f(\alpha_1^* x_1+\cdots+\alpha_s^* x_s, \beta_1^* y_1+\cdots+\beta_t^* y_t)+\epsilon.$$

The main result in this section is the following theorem.

Theorem 3. *There exists an algorithm solving any ϵ-optimal (s,t)-mixing problem in time* $\text{poly}(m,n,s,t)\left(e + \frac{e}{\sqrt{\epsilon/2}}\right)^{s+t}$ *and space* $\text{poly}(m,n,s,t)$.

Corollary 1. *For ϵ-optimal (s,t)-mixing problem, when s,t are constant, there exists an FPTAS; when $s,t = O(\log n)$, there exists a PTAS.*

We note that term $\left(e + \frac{e}{\sqrt{\epsilon/2}}\right)^{s+t}$ is unlikely to be improved to $\text{poly}(m,n,s,t,1/\epsilon)$. Otherwise, by taking all pure strategies as input, we can obtain an FPTAS for ϵ-NE, which is proved by Theorem 1.3 in [2] to be impossible unless PPAD $\subseteq$ P.

As is observed in Sect. 3, the objective function f contains the maximum terms and bilinear terms. The bilinear terms causes non-convexity, which makes it very hard to find the global minimum.

Based on such observations, our method adopts the following idea. First, we use small grids to cover the whole domain. Next, on each grid, we use linear functions to approximate the bilinear terms. After the linear approximation, the objective function becomes piecewise linear, which can be solved by linear programming. By a delicate selection of grid, we can ensure that the optimal solution of the linear approximation is close to the optimal solution of the original problem, thus giving a close enough approximation. Finally, we output the best solution among all grids. The full algorithm and the proof of Theorem 3 are given in the full version of this paper.

5   Applications to the Search-and-Mix Methods

In this section, we show how to apply the optimal mixing problem to make an instance-optimal enhancement to the search-and-mix methods in the literature.

As is summarized by [20], in the literature, polynomial-time algorithms for approximate NE follow a *search-and-mix* method. Such methods can be divided into two phases. In the search phase, an algorithm computes several strategies of each player in polynomial time. In the mixing phase, the algorithm then make convex combinations of the selected strategies into several strategy profiles and outputs the profile with the minimum f value. An illustration is presented in Fig. 1.

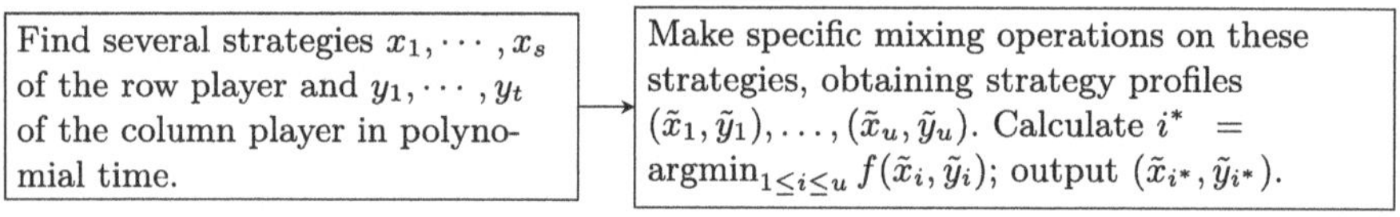

| Find several strategies $x_1, \cdots, x_s$ of the row player and $y_1, \cdots, y_t$ of the column player in polynomial time. | Make specific mixing operations on these strategies, obtaining strategy profiles $(\tilde{x}_1, \tilde{y}_1), \ldots, (\tilde{x}_u, \tilde{y}_u)$. Calculate $i^* = \operatorname{argmin}_{1 \le i \le u} f(\tilde{x}_i, \tilde{y}_i)$; output $(\tilde{x}_{i^*}, \tilde{y}_{i^*})$. |

Fig. 1. Procedure of the search-and-mix method in the literature

However, the mixing phases in the literature are *ad hoc*, since the mixing coefficients are selected specifically for corresponding search phase with certain properties. A typical example is as follows.

Example 1 (BBM algorithm [1]).

- *Search phase*: Compute an NE (x^*, y^*) of the zero-sum game $(R-C, C-R)^3$. Let $g_1 = f_R(x^*, y^*)$ and $g_2 = f_C(x^*, y^*)$. By symmetry, assume without loss of generality that $g_1 \ge g_2$. Then compute $r_1 \in \operatorname{br}_R(y^*)$ and $b_2 \in \operatorname{br}_C(r_1)$.
- *Mixing phase*: Mix strategies in the search phase and obtain strategy profiles (x^*, y^*) and $(r_1, (1 - \delta_2)y^* + \delta_2 b_2)$, where $\delta_2 = (1 - g_1)/(2 - g_1)$. Output the one with the smaller f value.

---

[3] Computing NE in zero-sum games can be modeled by a linear program and thus can be solved in polynomial time. See, e.g., [27].

Note that the mixing coefficient δ_2 is chosen specifically for this search phase to produce an optimal approximation bound of 0.38. If we choose δ_2 to be other values, for example, $1/2$, then it is not hard to show that the approximation guarantee will only be 0.5.

Now, we relate the optimal mixing problem to the search-and-mix methods in the literature. The traditional ad hoc designed mixing phases focus too much on the worst case and not useful in practice. However, from the perspective of our work, the mixing phase is essentially an optimal mixing problem. We can use the approximate and exact optimal mixing problem to design new mixing phases for the search-and-mix methods, which computes the *instance-optimal* mixing coefficients. The new procedure is presented in Fig. 2.

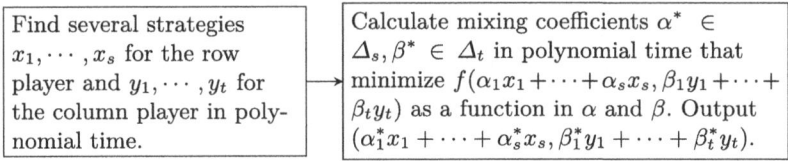

Fig. 2. The new procedure for the search-and-mix method

Our exact algorithm for the optimal mixing problem can cover the need of *all* mixing phases in the literature. Moreover, our approximation algorithm can be used for most future mixing phases. For any constant number of strategies in the search phase, our approximation algorithm is an FPTAS. Beyond that, when $s, t = O(\log n)$, our approximation algorithm is a PTAS and when $s, t = \text{poly}(\log n)$, our approximation algorithm is a QPTAS. They can all be used in the new polynomial-time procedure of the search-and-mix methods.

To conclude this section, as an example, we show how to design a new mixing phase for Example 1 using the optimal mixing problem.

Example 2 (New mixing phase for the BBM algorithm). Recall that in the search phase we obtain x^*, r_1 for the row player and y^*, b_2 for the column player. Then, the new mixing phase is to input payoff matrices R, C, row player's mixed strategies x^*, r_1, column player's mixed strategies y^*, b_2, solve the optimal $(2, 2)$-mixing problem to obtain α^*, β^* and then output $(\alpha^* x^* + (1 - \alpha^*) r_1, \beta^* y^* + (1 - \beta^*) b_2)$.

Acknowledgments. This work is supported the Natural Science Foundation of China (Grant No. 62172012) and Natural Science Foundation of China (Grant No. 6212290003).

References

1. Bosse, H., Byrka, J., Markakis, E.: New algorithms for approximate nash equilibria in bimatrix games. In: Deng, X., Graham, F.C. (eds.) WINE 2007. LNCS, vol. 4858,

pp. 17–29. Springer, Heidelberg (2007). https://doi.org/10.1007/978-3-540-77105-0_6

2. Chen, X., Deng, X., Teng, S.H.: Settling the complexity of computing two-player Nash equilibria. J. ACM **56**(3), 14:1–14:57 (2009). https://doi.org/10.1145/1516512.1516516

3. Chen, Z., Deng, X., Huang, W., Li, H., Li, Y.: On tightness of the tsaknakis-spirakis algorithm for approximate nash equilibrium. In: Caragiannis, I., Hansen, K.A. (eds.) SAGT 2021. LNCS, vol. 12885, pp. 97–111. Springer, Cham (2021). https://doi.org/10.1007/978-3-030-85947-3_7

4. Conitzer, V., Sandholm, T.: New complexity results about Nash equilibria. Games Econ. Behav. **63**(2), 621–641 (2008). https://doi.org/10.1016/j.geb.2008.02.015

5. Czumaj, A., Deligkas, A., Fasoulakis, M., Fearnley, J., Jurdziński, M., Savani, R.: Distributed methods for computing approximate equilibria. In: Cai, Y., Vetta, A. (eds.) WINE 2016. LNCS, vol. 10123, pp. 15–28. Springer, Heidelberg (2016). https://doi.org/10.1007/978-3-662-54110-4_2

6. Daskalakis, C., Goldberg, P.W., Papadimitriou, C.H.: The complexity of computing a Nash equilibrium. SIAM J. Comput. **39**(1), 195–259 (2009). https://doi.org/10.1137/070699652

7. Daskalakis, C., Mehta, A., Papadimitriou, C.: Progress in approximate nash equilibria. In: Proceedings of the 8th ACM Conference on Electronic Commerce, pp. 355–358. ACM, San Diego California USA (June 2007). https://doi.org/10.1145/1250910.1250962

8. Daskalakis, C., Mehta, A., Papadimitriou, C.: A note on approximate Nash equilibria. In: Spirakis, P., Mavronicolas, M., Kontogiannis, S. (eds.) WINE 2006. LNCS, vol. 4286, pp. 297–306. Springer, Heidelberg (2006). https://doi.org/10.1007/11944874_27

9. Deligkas, A., Fasoulakis, M., Markakis, E.: A polynomial-time algorithm for 1/3-approximate Nash equilibria in bimatrix games. In: Chechik, S., Navarro, G., Rotenberg, E., Herman, G. (eds.) 30th Annual European Symposium on Algorithms, ESA 2022, 5–9 September. LIPIcs, vol. 244, pp. 41:1–41:14. Schloss Dagstuhl - Leibniz-Zentrum für Informatik, Berlin/Potsdam, Germany (2022). https://doi.org/10.4230/LIPIcs.ESA.2022.41

10. Deng, X., Li, D., Li, H.: The Search-and-Mix Paradigm in Approximate Nash Equilibrium Algorithms (October 2023). https://doi.org/10.48550/arXiv.2310.08066

11. Etessami, K., Yannakakis, M.: On the complexity of Nash equilibria and other fixed points. SIAM J. Comput. **39**(6), 2531–2597 (2010). https://doi.org/10.1137/080720826

12. Fearnley, J., Igwe, T.P., Savani, R.: An empirical study of finding approximate equilibria in bimatrix games. In: Bampis, E. (ed.) SEA 2015. LNCS, vol. 9125, pp. 339–351. Springer, Cham (2015). https://doi.org/10.1007/978-3-319-20086-6_26

13. Gilboa, I., Zemel, E.: Nash and correlated equilibria: some complexity considerations. Games Econom. Behav. **1**(1), 80–93 (1989). https://doi.org/10.1016/0899-8256(89)90006-7

14. Hu, J., Wellman, M.P.: Nash Q-learning for general-sum stochastic games. J. Mach. Learn. Res. **4**(Nov), 1039–1069 (2003)

15. Jin, C., Liu, Q., Wang, Y., Yu, T.: V-Learning–a simple, efficient, decentralized algorithm for multiagent RL. In: ICLR 2022 Workshop on Gamification and Multiagent Solutions (2022)

16. Khachiyan, L.G.: Polynomial algorithms in linear programming. USSR Comput. Math. Math. Phys. **20**(1), 53–72 (1980). https://doi.org/10.1016/0041-5553(80)90061-0

17. Kontogiannis, S.C., Panagopoulou, P.N., Spirakis, P.G.: Polynomial algorithms for approximating Nash equilibria of bimatrix games. In: Spirakis, P., Mavronicolas, M., Kontogiannis, S. (eds.) WINE 2006. LNCS, vol. 4286, pp. 286–296. Springer, Heidelberg (2006). https://doi.org/10.1007/11944874_26

18. Kothari, P.K., Mehta, R.: Sum-of-squares meets Nash: lower bounds for finding any equilibrium. In: Proceedings of the 50th Annual ACM SIGACT Symposium on Theory of Computing, pp. 1241–1248. ACM, Los Angeles CA USA (June 2018). https://doi.org/10.1145/3188745.3188892

19. Lemke, C.E., Howson, J.T., Jr.: Equilibrium points of bimatrix games. J. Soc. Ind. Appl. Math. **12**(2), 413–423 (1964). https://doi.org/10.1137/0112033

20. Li, H., et al.: A survey on algorithms for Nash equilibria in finite normal-form games. Comput. Sci. Rev. **51**, 100613 (2024). https://doi.org/10.1016/j.cosrev.2023.100613

21. Lipton, R.J., Markakis, E., Mehta, A.: Playing large games using simple strategies. In: Proceedings of the 4th ACM Conference on Electronic Commerce, pp. 36–41 (2003)

22. Liu, Q., Yu, T., Bai, Y., Jin, C.: A sharp analysis of model-based reinforcement learning with self-play. In: International Conference on Machine Learning, pp. 7001–7010. PMLR (2021)

23. Megiddo, N., Papadimitriou, C.H.: On total functions, existence theorems and computational complexity. Theor. Comput. Sci. **81**(2), 317–324 (1991). https://doi.org/10.1016/0304-3975(91)90200-L

24. Mehta, R.: Constant rank bimatrix games are PPAD-hard. In: Proceedings of the Forty-Sixth Annual ACM Symposium on Theory of Computing, pp. 545–554. ACM, New York New York (May 2014). https://doi.org/10.1145/2591796.2591835

25. Muller, P., et al.: A generalized training approach for multiagent learning. In: ICLR, pp. 1–35. ICLR (2020)

26. Nash, J.: Non-cooperative games. Ann. Math. **54**(2), 286–295 (1951)

27. Nisan, N., Roughgarden, T., Tardos, E., Vazirani, V.V.: Algorithmic Game Theory. Cambridge university press, Cambridge (2007)

28. OpenAI, Berner, C., et al.: Dota 2 with Large Scale Deep Reinforcement Learning (December 2019)

29. Papadimitriou, C.H.: On the complexity of the parity argument and other inefficient proofs of existence. J. Comput. Syst. Sci. **48**(3), 498–532 (1994). https://doi.org/10.1016/S0022-0000(05)80063-7

30. Rubinstein, A.: Settling the complexity of computing approximate two-player nash equilibria. In: 2016 IEEE 57th Annual Symposium on Foundations of Computer Science (FOCS), pp. 258–265 (October 2016). https://doi.org/10.1109/FOCS.2016.35

31. Tsaknakis, H., Spirakis, P.G.: An optimization approach for approximate Nash equilibria. In: Deng, X., Graham, F.C. (eds.) WINE 2007. LNCS, vol. 4858, pp. 42–56. Springer, Heidelberg (2007). https://doi.org/10.1007/978-3-540-77105-0_8

32. Tsaknakis, H., Spirakis, P.G., Kanoulas, D.: Performance evaluation of a descent algorithm for bi-matrix games. In: Papadimitriou, C., Zhang, S. (eds.) WINE 2008. LNCS, vol. 5385, pp. 222–230. Springer, Heidelberg (2008). https://doi.org/10.1007/978-3-540-92185-1_29

33. Vinyals, O., et al.: Grandmaster level in StarCraft II using multi-agent reinforcement learning. Nature **575**(7782), 350–354 (2019). https://doi.org/10.1038/s41586-019-1724-z

Finding Fair and Efficient Allocations Under Budget Constraints

Yuanyuan Wang, Xin Chen$^{(\boxtimes)}$, Qizhi Fang, Qingqin Nong, and Wenjing Liu

Ocean University of China, Qingdao 266100, China
wyy8088@stu.ouc.edu.cn, {chenxin1403,qfang,qqnong,liuwj}@ouc.edu.cn

Abstract. We study the problem of how to fairly and efficiently allocate indivisible items (goods) to agents under budget constraints. Each item has a specific size, and each agent has a budget that limits the total size of the items received. To better explore efficiency, we introduce the concept of tightness, where all agents are tight. An agent is considered as tight if adding any unallocated item to her bundle would exceed her budget. Interestingly, we observe that every Pareto optimal (PO) allocation can be extended into a tight allocation while maintaining the values of the agents' bundles. We achieve an overall negative result for general even identical binary valuations: there exists no allocation meeting both tightness and envy-freeness up to any item (EFX), and even relaxing it to any desired approximate EFX still has been proved impossible. However, for single-valued valuations, we illustrate that an EFX and tight allocation always exist, and it can be computed using a polynomial algorithm. For identical single-valued valuations, we establish the existence of 1/2-EFX and PO allocations, with the approximation ratio being the best possible. To further our efforts to study fairness and efficiency, we introduce a relaxed concept of tightness, partial tightness (PT), in which only the unenvied agents are tight. We find that 1/2-EFX and PT allocations are achievable by providing a pseudo-polynomial time algorithm.

Keywords: Fair Allocation · Budget Constraints · EFX · PO

1 Introduction

The fair allocation of indivisible items has been a significant problem in mathematical economics and computer science [4,19,20]. The target is to "fairly" allocate a set of indivisible resources (such as matching courses in a school [7]) among multiple agents.

How to capture the definition of fairness and compute fair allocations is the central task in the fair allocation problem. Over the last decade, besides the classic notion of fairness, envy-freeness (EF), a number of attempts at relaxing the

This work is supported in part by the National Natural Science Foundation of China (Nos. 11871442, 12171444, 12201590, 12301418) and Natural Science Foundation of Shandong (No. ZR2022QA014).

B. Li et al. (Eds.): IJTCS-FAW 2024, LNCS 14752, pp. 144–158, 2025.
https://doi.org/10.1007/978-981-97-7752-5_12

rules of EF have been extensively studied, such as the prevalent envy-freeness up to one item (EF1, [16]), in which the envy can be eliminated by the hypothetical removal of some good in other's bundle, and envy-freeness up to any item (EFX, [8]), in which the envy can be eliminated by the hypothetical removal of any good in the other's bundle.

The majority of literature [1, 12, 17] consider the setting that every agent can receive an arbitrary bundle of indivisible items, which means that all possible allocations are feasible for the involved agents. However, in many real-world scenarios, the items may have sizes (or costs) and each agent can only receive items that the total size does not exceed her budget. These observations have recently led to an active line of work [3, 13, 20], which is focused on the existence and computation of fair allocations under budget constraints. Wu et al. [20] proposed two notions, *envy*, an agent i envies agent j if she can find a feasible subset (the size of the set does not exceed her budget) from j's bundles that are more valuable than her current bundle, EF, if any agent does not envy others. Accordingly, we propose a notion of EFX allocation, in which the envy of agent i can be eliminated by the hypothetical removal of any good in any i's feasible envious subset from j's bundle.

Due to the limitation of the budgets, a partition of m items may not be guaranteed for feasibility. Hence, in the budget-setting, leaving some items unallocated is permitted. Every agent gets an empty set that is budget-feasible, and it is already EF and EFX, but it is not efficient by any measurement of efficiency. Based on this situation, besides fairness, efficiency should also be taken into consideration. In this paper, we consider a new measurement of efficiency, *tightness*, where all agents are tight, to guarantee that each agent gets as many items as possible within her budget. Our goal is to explore the relationship of tightness and Pareto Optimality (PO), and find allocations that meet both EFX and tight (or PO).

1.1 Our Contributions

In this paper, we consider allocating m indivisible items to n agents under budget constraints. Each item has a nonnegative size, each agent has a positive budget and an additive valuation function $v_i(\cdot)$. We introduce tightness as a measurement of efficiency. An allocation A is tight if, for any agent i, her total size $s(A_i)$ is smaller than her budget, but the size $s(A_i \cup \{g\})$ exceeds the budget for any unallocated item g. We observe that every PO allocation can be extended into a tight allocation and maintain the values of agents' bundles, and strive to examine the current problem of allocations where both fairness and efficiency are met. Our main results are shown in Table 1.

The first part focuses on the existence of EFX and tight allocations. For general additive valuations, we prove that EFX and tight are incompatible, and there exists no α-EFX and tight allocation, even for the special cases of two agents with binary valuations, where $\alpha > 0$. However, for single-valued valuations, we devise a polynomial-time algorithm that computes an EFX and tight allocation.

Table 1. A summary of our main results. "−" indicates an open question.

	Settings	Upper Bounds	Lower Bounds
EFX & Tight	General	0	None
	Two Agents	0	None
	Binary Valuations	0	None
	Single-valued Valuations	1	1
EFX & PO	General	2ε	None
	Identical Single-valued Val.	1/2	1/2
EFX & PT	General	−	1/2

The second part explores the existence of EFX and PO allocations. By the observation that a PO allocation is also tight, without zero-value items ($v_i(g)$ is strictly larger than zero for any i and g). It leads to the negative result that there exists no α-EFX and PO allocation, where $\alpha > 2\varepsilon$. Then, for identical single-valued valuations, we show the existence of 1/2-EFX and PO allocations by using the leximin allocations, and 1/2 is also the upper bound of approximate EFX and PO.

In the last part, we relax the definition, tightness, to partial tightness (PT), where only request unenvied agents are tight and show the existence of 1/2-EFX and PT allocations for general valuations by presenting a pseudo-polynomial-time algorithm.

1.2   Related Work

In this paper, we focus on the existence of EFX allocations among indivisible items under budget constraints. The notion of EFX was initially proposed by [8] in the unconstrained setting, and the existence of EFX allocation is still a challenging open problem. Some research has answered on the side of limiting the number of agents, such as two or three agents [10,12,14]. On the other side of limiting agents' valuation functions, EFX allocations do exist for matroid-rank functions [1] and binary valuations [6]. There also are some additional results of EFX allocations [5,11,18]. Plaut and Roughgarden [17] showed the existence of 1/2-EFX allocations, which can be computed by a polynomial-time algorithm [9].

Besides the considerations of the fairness notion of EFX, we also consider the efficiency notion of PO. In the unconstrained setting, Barman et al. [4] devised a pseudo-polynomial time algorithm to compute EF1 and PO allocations. However, Plaut and Roughgarden [17] illustrated the negative result that no allocation was EFX and PO in general and provided some positive results for special cases, such as two agents.

Different from the above research, we study EFX allocations under budget constraints. Wu et al. [20] first studied fair allocations with budget constraints, they found that a budget-feasible allocation that maximized the Nash social welfare was 1/4-EF1 and PO. Later, several special cases were studied by Barman et al. [2] and Gan et al. [13]. Barman et al. [2] studied envy-freeness up to 2 items (EF2) allocations under budget constraints by proposing polynomial-time algorithms. Specially, they settled the existence of EF1 allocations for single-valued valuations where all items have the same value and the existence of EF1 allocations. However, we focus on EFX and tight (or PO) allocations for single-valued valuations. Until recently, Barman et al. [3] found that budget-feasible and EFX (with bounded charity) allocations existed by providing pseudo-polynomial-time algorithms. In this paper, we explore EFX and tight (or PO) allocations under budget constraints.

2   Preliminaries

Given an instance of allocation problem with an indivisible item (or good) set $M = \{g_1, \ldots, g_m\}$ and an agent set $N = \{1, \ldots, n\}$. In our budget settings, each item $g_j \in M$ has a size $s(g_j) \geq 0$, and each agent $i \in N$ has a budget $b_i \geq 0$. An item subset $T \subseteq M$ is budget-feasible for agent i if $s(T) = \sum_{g \in T} s(g) \leq b_i$. The goal is to allocate these m items to n agents under budget constraints, that is, every agent gets a budget-feasible bundle. $A = (A_1, \ldots, A_n)$ is a budget-feasible allocation if each bundle A_i is budget-feasible for agent i, $A_i \cap A_j = \emptyset$ for any $i \neq j$, and $\cup_{i=1}^n A_i \subseteq M$. When $\cup_{i=1}^n A_i = M$, the budget-feasible allocation A is a complete allocation. However, in our settings, there may exist some items unallocated, which we denote by $A_0 = M \setminus (\cup_{i=1}^n A_i)$ containing all unallocated items with respect to A.

Each agent $i \in N$ has an individual valuation function $v_i(\cdot) : 2^M \to \mathbb{R}_{\geq 0}$. Similar to majority of the literature on fair division, we assume that v_i is normalized ($v_i(\emptyset) = 0$), monotone ($v_i(S) \leq v_i(T)$ for all $S \subseteq T \subseteq M$), and additive ($v_i(T) = \sum_{g \in T} v_i(g)$ for any set $T \subseteq M$). In the remaining part of paper, we require every allocation to be feasible and write "allocation" instead of "budget-feasible allocation" unless otherwise specified.

Based on the above budget-feasible settings, our work is to find "fair" allocations, focusing on one central notion of fairness, EFX, under budget constraints.

Definition 1 (EFX). *Let $\alpha \in [0, 1]$ be any constant. An allocation $A = (A_1, \ldots, A_n)$ is α-approximate envy-free up to any item (α-EFX) if for any pair of agents i and j, and any subset $T \subseteq A_j$ with $s(T) \leq b_i$, we have that*

$$v_i(A_i) \geq \alpha \cdot v_i(T \setminus \{g\}), \quad \forall g \in T.$$

If the above α is equal to 1, 1-EFX allocations are also called EFX allocations.

Notice that a trivial allocation $A = (\emptyset, \ldots, \emptyset)$ is budget feasible, and it is already EF and EFX. But, it is not efficient by any measurement of efficiency since the total value of A is zero. Next, we will introduce tightness, PO and IO to measure efficiency under budget constraints.

Definition 2 (Tightness). *An allocation $A = (A_1, \ldots, A_n)$ is a tight allocation if for any agent $i \in N$, her bundle A_i is tight, that is,*

$$s(A_i \cup \{g\}) > b_i, \quad \forall g \in M \setminus (\cup_{i=1}^n A_i).$$

It means that, in a tight allocation, no agent can receive any additional unallocated item within her budget. In this paper, for the purpose of design and analysis, we add an "extra" item g_{m+1} which suffices that $s(g_{m+1}) = n \cdot \max_{j \in N} b_j$ and $v_i(\{g_{m+1}\}) = 0$ for any agent $i \in N$, in every instance of fair allocation problem.

Definition 3 (Pareto Optimality). *An allocation $A = (A_1, \ldots, A_n)$ is Pareto optimal (PO) if there exists no allocation X such that for some agent $i \in N$, and any other agent $k \in N \setminus \{i\}$,*

$$v_i(X_i) > v_i(A_i), \quad and \ v_k(X_k) \geq v_k(A_k), \quad \forall k \neq i.$$

2.1   Relationships of Tightness and PO

In this section, we explore the relationships of tightness and PO by two observations.

Observation 1. *Given an instance of allocation problem, every PO allocation can be extended into a tight allocation and maintain the values of agents' bundles.*

In fact, if a PO allocation $A = (A_1, \ldots, A_n)$ is not tight, there must exists some agent i and some unallocated good g, such that $v_i(A_i \cup \{g\}) = v_i(A_i)$, $s(A_i \cup \{g\}) \leq b_i$. Update the bundle $A_i = A_i \cup \{g\}$ until each $i \in N$ is tight, we can get a tight $A' = (A'_1, \ldots, A'_n)$ where $A'_i \supseteq A_i$ and $v_i(A'_i) = v_i(A_i)$ for any $i \in N$. The specific process can be seen in Example 1.

Example 1. Consider an instance with an item set $M = \{g_1, g_2, g_3, g_4, g_5\}$ and an agent set $N = \{1, 2\}$. All agents have identical budgets ($b_1 = b_2 = 1$). For any item $j \in M$, the size $s(g_j)$ and the value $v_i(g_j)$ are given in Table 2.

Let $A = (A_1, A_2)$ with $A_1 = \{g_1, g_2\}, A_2 = \{g_3\}$. We can verify that A is a PO allocation, but it is not tight. By setting $A'_1 = A_1, A'_2 = A_2 \cup \{g_4\}$, the PO allocation A can be extended to a tight A'.

Table 2. An instance that leads to a PO allocation but it is not tight.

g_j	g_1	g_2	g_3	g_4	g_5
$s(g_j)$	ε	ε	ε	$1-\varepsilon$	$1-\varepsilon$
$v_1(g_j)$	1	1	0	0	0
$v_2(g_j)$	1	0	1	0	0

Observation 2. *Given an instance $\mathcal{I}$ of allocation problem. Denote by $\mathcal{A}(PO) = \{A \mid A \text{ is } PO\}$, and $\mathcal{A}(Tight) = \{A \mid A \text{ is tight}\}$ as two allocation sets in $\mathcal{I}$. If zero-value items are not allowed, that is, $v_i(g) > 0$ for all $i \in N$, we have that*

$$\mathcal{A}(PO) \subset \mathcal{A}(Tight).$$

Let $A = (A_1, \ldots, A_n) \in \mathcal{A}(PO)$ be an arbitrary PO allocation. If A is not tight, there must exist some agent i and an unallocated item $g \in A_0$, such that $s(A_i \cup \{g\}) \le b_i$ and $v_i(g) > 0$. Then we can construct a budget-feasible allocation $B = (B_1, \ldots, B_n)$, where $B_i = A_i \cup \{g\}$ and $B_k = A_k$ for any $k \ne i$. It is clear that $v_i(B_i) > v_i(A_i)$, and $v_k(B_k) = v_k(A_k)$ for any $k \ne i$. It contradicts with $A \in \mathcal{A}(PO)$. Thus, the PO allocation A is tight.

3   EFX and Tight Allocations

In this section, we focus on EFX and tight allocations under budget constraints. Generally, any desired approximate EFX and tight allocations do not always exist. As we endeavoured to study single-valued valuations, we were surprised to find two results: Firstly, we can compute an EFX and tight allocation by a sequence-picking algorithm. Secondly, we find that EFX and PO are still incompatible, but a 1/2-EFX and PO allocation exists.

Theorem 3. *An α-EFX and tight allocation does not always exist under budget constraints, where $\alpha > 0$.*

Proof. Consider a binary instance with an item set $M = \{g_1, g_2, g_3, g_4, g_5\}$ and an agent set $N = \{1, 2\}$. Both agents have identical budgets ($b_1 = b_2 = 1$) and identical binary valuations $v(\cdot)$. For any item $j \in M$, the size $s(g_j)$ and the value $v(g_j)$ are given in Table 3.

Let $A = (A_1, A_2)$ be any budget-feasible and tight allocation. Denote by $\alpha > 0$ any positive constant. Due to tightness of A, we get that $g_1 \in A_1 \cup A_2$. Assume w.l.o.g. that $g_1 \in A_1$, we have that $|A_1| \ge 2$. Thus, there exists another item $g_k \in A_1$ with $v(g_k) = 0$. Thus, $v(A_2) = 0 < \alpha \cdot v(A_1 \setminus \{g_k\})$, which implies that the tight allocation A is not α-EFX.

Furthermore, this worse instance indicates that there exists no α-EFX and tight allocation even for two identical agents with binary valuations.   □

Table 3. An instance that exists no α-EFX and tight allocation.

g_j	g_1	g_2	g_3	g_4	g_5
$s(g_j)$	ε	ε	ε	ε	$1 - \varepsilon$
$v(g_j)$	1	0	0	0	0

3.1   EFX and Tight Allocations

This section introduces an efficient algorithm for computing EFX and tight allocations for single-valued valuations, which is a special class of bi-valued valuations.

Definition 4 (Single-valued Valuation). *A valuation* $v_i(\cdot) : 2^M \to \mathbb{R}_{\geq 0}$ *is a single-valued additive valuation if for any subset* $S \subseteq M$, $v_i(S) = \sum_{g \in S} c_i = c_i \cdot |S|$, *where* $c_i > 0$ *is a constant.*

For single-valued valuations, we design a polynomial-time Algorithm to compute an exact EFX and tight allocation.

Theorem 4. *For single-valued valuations, Algorithm 1 returns an EFX and tight allocation in polynomial time.*

Algorithm 1 Sequential-picking Algorithm

Input: $M = \{g_1, \ldots, g_m\}$ with sizes $s(g_e), e = 1, \ldots, m$, g_{m+1} with $s(g_{m+1}) = n \cdot \max_{j \in N} b_j$; $N = \{1, \ldots, n\}$ with budgets b_i and single-valued valuations $v_i(\cdot)$, $i = 1, \ldots, n$.
Output: An allocation A.
  1: Ordering all items by non-decreasing sizes, i.e.,
$$s(g_1) \leq \cdots \leq s(g_{m+1}).$$
  2: Initialize: $(A_1, \ldots, A_n) = (\emptyset, \ldots, \emptyset)$, $N_0 = \emptyset$, $i = 1$, $e = 1$.
  3: **while** $|N_0| < n$ **do**
  4:     Ordering agents in $N \setminus N_0$ by non-decreasing budgets, $\delta = \{\delta_1, \cdots, \delta_{|N \setminus N_0|}\}$;
  5:     $i = 1$;
  6:     **for** $i = 1$ to $|N \setminus N_0|$ **do**
  7:         **if** $s(A_{\delta_i} \cup g_e) \leq b_{\delta_i}$ **then**
  8:             $A_{\delta_i} = A_{\delta_i} \cup \{g_e\}$;
  9:             $M = M \setminus \{g_e\}$;
 10:             $e = e + 1$;
 11:         **else**
 12:             $N_0 = N_0 \cup \{\delta_i\}$;
 13:         **end if**
 14:         $i = i + 1$;
 15:     **end for**
 16: **end while**
 17: **return** $A = (A_1, \ldots, A_n)$

Proof. Let $A = (A_1, \ldots, A_n)$ be the output of Algorithm 1. Based on the If-Condition in line 7, it is certain that the allocation A is tight. Next, we will show that $A = (A_1, \ldots, A_n)$ is EFX, that is, for any two agents $i, j \in N$,

$$v_i(A_i) \geq v_i(X \setminus \{g\}), \quad \forall g \in X, \quad \forall X \subseteq A_j, \quad s(X) \leq b_i.$$

Let the for-loop (lines 6–15) be a round of the algorithm. Suppose that each agent $i \in N$ gets her last item at the ℓ_i-th round, that is, $|A_i| = \ell_i$. It implies that agent i must obtain one item in each ℓ-th round with $\ell < \ell_i$. Ordering all agents as π by non-decreasing budgets, i.e., $b_1 \leq \cdots \leq b_n$. The following will unfold by the sequence of agents i and j.

If agent i is arranged before j in π, then $b_i \leq b_j$. Assume that there exists a subset $T \subseteq A_j$, which suffices that $s(T) \leq b_i$ and $v_i(T) > v_i(A_i)$. Combined with single-valued valuations, we have that $|T| \geq |A_i| + 1$. Let $g^* \in T$ be the item with maximum size in the subset T, that is, $s(g^*) = \max\{s(g) \mid g \in T\}$. We assume that g^* is allocated to agent j at the ℓ_α-th round with $\ell_\alpha \geq \ell_i + 1$. Since agent i must obtain one item before agent j in each ℓ-th round with $\ell \leq \ell_i$ and all items are ordered by non-decreasing sizes, we can get that $s(g^*) \geq \max\{s(g) \mid g \in A_i\}$ and $s(T \setminus \{g^*\}) \geq s(A_i)$. Since item g^* is not arranged to agent i at the ℓ_α-th round, it is clear that $s(g^*) + s(A_i) > b_i$. Thus, $s(T) = s(T \setminus \{g^*\}) + s(g^*) \geq s(A_i) + s(g^*) > b_i$, which contradicts with $s(T) < b_i$.

If agent i is arranged after j in π, then $b_i \geq b_j$. It is clear that agent i does not envy agents when $l_i > l_j$. When $l_i < l_j$, we can get that $v_i(A_i) = v_i(A_j \setminus \{g\})$ for any $g \in A_j$ if $l_j = l_i + 1$. Thus, we will show that if $l_i < l_j$, $l_j = l_i + 1$. Denote by $g_j^{(\ell)}$ and $A_j^{(\ell)}$ the item allocated to agent j and the current item set of agent j at the end of ℓ-th round, respectively. For each $1 \leq \ell \leq \ell_i$, $s(g_j^{(\ell)}) \leq s(g_i^{(\ell)}) \leq s(g_i^{(\ell_i)})$. Assume that $\ell_j \geq \ell_i + 2$, we will show a contradiction. Consider the $(\ell_i + 1)$-th round. Let $g_s \in M$ be the item which will be allocated to agent i in the $(\ell_i + 1)$-th round, we have that $s(g_s) \geq s(g_j^{(\ell_i+1)})$ and $s(g_s) + s(A_i^{(\ell_i)}) > b_i$. Combining with the item ordering, we get that

$$s(A_j^{(\ell_i+1)}) \geq \sum_{k=2}^{\ell_i+1} s(g_j^{(k)}) \geq \sum_{k=1}^{\ell_i} s(g_i^{(k)}) = s(A_i^{(\ell_i)}) = s(A_i).$$

In the $(\ell_i + 2)$-th round,

$$s(A_j^{(\ell_i+2)}) = s(A_j^{(\ell_i+1)}) + s(g_j^{(\ell_i+2)}) \geq s(A_i^{(\ell_i)}) + s(g_s) > b_i \geq b_j,$$

where the first inequality holds since that $s(g_j^{(\ell_i+2)}) \geq s(g_s)$. It contradicts with $s(A_j^{(\ell_i+2)}) \leq b_j$. Thus, $|A_j| \leq |A_i| + 1$, and for any $X \subseteq A_j$ with $s(X) \leq b_i$ and any $g \in X$,

$$v_i(A_i) = |A_i| \geq |A_j| - 1 \geq v_i(X \setminus \{g\}).$$

$\square$

3.2   1/2-EFX and PO Allocations

This section firstly introduces the general negative result that any approximate EFX and PO allocations do not exist, even for two identical agents without zero-value items. Then we focus on the special setting of identical single-valued valuations and show the existence of 1/2-EFX and PO allocations.

Theorem 5. *Let $\varepsilon > 0$ be arbitrarily close to 0. An α-EFX and PO allocation does not always exist under budget constraints, where $\alpha > 2\varepsilon$.*

Proof. Based on Observation 2, every PO allocation is tight when zero-value items are not allowed. Thus, we prove the negative result by showing it holds for approximate EFX and tight.

Consider an instance with $M = \{g_1, g_2, g_3, g_4, g_5\}$ and $N = \{1, 2\}$. All agents have identical budgets ($b_1 = b_2 = 1$) and identical valuations ($v_1(\cdot) = v_2(\cdot) = v(\cdot)$). For any item $g_j \in M$, $s(g_j)$ and $v(g_j)$ are given in Table 4.

Table 4. A worse instance that leads to α-EFX and tight allocation.

g_j	g_1	g_2	g_3	g_4	g_5
$s(g_j)$	ε	$1 - \varepsilon$	ε	$1 - 2\varepsilon$	1
$v(g_j)$	ε	ε	$1 + 2\varepsilon$	ε	ε^2

Suppose that $A = (A_1, A_2)$ is an α-EFX and tight allocation in the instance, where $\alpha > 2\varepsilon$. Let $A_0 = M \setminus (A_1 \cup A_2)$ be unallocated items, we have that

$$s(A_i \cup \{g\}) > 1, \quad \forall g \in A_0, \quad \forall i \in N; \qquad \text{(Tightness)}$$
$$v_i(A_i) \geq \alpha \cdot v_i(A_j \setminus \{g\}), \quad \forall g \in A_j, \quad \forall i \neq j. \qquad (\alpha\text{-EFX})$$

Since $s(\{g_1, g_2, g_3, g_4, g_5\}) = \sum_{i=1}^{m} s(g_i) = 3 - \varepsilon > 2 = b_1 + b_2, A_0 \neq \emptyset$. If $g_3 \in A_0$, due to the tightness of A, we have that $s(A_i) > 1 - s(g_3) = 1 - \varepsilon$ for any agent $i \in \{1, 2\}$. Therefore, there must be that one agent gets $\{g_1, g_2\}$ and the other gets $\{g_5\}$. Assume w.l.o.g. that $A_1 = \{g_1, g_2\}$, $A_2 = \{g_5\}$. Then, we can get that $v_2(A_2) = \varepsilon \cdot v_1(A_1 \setminus \{g_1\})$, which is contradicted with $\alpha > 2\varepsilon$. If $g_3 \notin A_0$, we can assume w.l.o.g that $g_3 \in A_1$. Based on the tightness of A, there exists at least one another item $g' \neq g_3$, $g' \in A_1$, and $A_2 \neq \emptyset$. If $A_2 = \{g_5\}$, we have that $v_2(A_2) \leq \varepsilon^2 / (1 + 2\varepsilon) \cdot v_1(A_1 \setminus \{g'\}) < \varepsilon \cdot v_1(A_1 \setminus \{g'\})$, which is contradicted with $\alpha > 2\varepsilon$. Otherwise if $A_2 \neq \{g_5\}$, $A_2 \subset \{g_1, g_2, g_4\}$ by the tightness of A. Therefore, $v_2(A_2) \leq 2\varepsilon \leq 2\varepsilon \cdot v_1(A_1 \setminus \{g'\})$, which is contradicted with $\alpha > 2\varepsilon$. □

Theorem 6. *For identical single-valued valuations, an α-EFX and PO allocation does not always exist, where $\alpha > 1/2$.*

Proof. Consider an instance with $M = \{g_1, g_2, g_3, g_4, g_5\}$ and $N = \{1, 2\}$. All agents have identical budgets ($b = 1$) and identical valuations ($v(\cdot)$). For any item $g_j \in M$, the size $s(g_j)$ and the value $v(g_j)$ are given in Table 5. Notice that for any budget-feasible allocation $X = (X_1, X_2)$, $\sum_{i=1}^{2} v_i(X_i) \leq 4$.

Table 5. A worse instance for identical single-valued valuations.

g_j	g_1	g_2	g_3	g_4	g_5
$s(g_j)$	1	$1/3$	$1/3$	$1/3$	1
$v(g_j)$	1	1	1	1	1

Let $\alpha > 1/2$ be a constant. Assume that $A = (A_1, A_2)$ is an α-EFX and PO allocation. It is easy to verify that the PO allocations can only be the structure that one agent gets $\{g_1\}$ or $\{g_5\}$ and the other gets $\{g_2, g_3, g_4\}$. Without loss of generality, we assume that $A_1 = \{g_1\}$ and $A_2 = \{g_2, g_3, g_4\}$. However, $v_1(A_1) = 1/2 \cdot v_2(A_2 \setminus \{g_2\})$, which contradicts with α-EFX allocation. Thus, there exists no allocation is both PO and α-EFX with $\alpha > 1/2$. □

On the other side, for identical single-valued valuations, we find a 1/2-EFX and PO allocation by leximin ordering.

Comparison Operator $\prec$.   Let $A = (A_1, \ldots, A_n)$ and $B = (B_1, \ldots, B_n)$ be two feasible allocations.

(1) Ordering all agents by non-decreasing valuations of two allocations A and B, as $\pi^A = \{\pi_1^A, \cdots, \pi_n^A\}$ and $\pi^B = \{\pi_1^B, \cdots, \pi_n^B\}$ respectively.
(2) If there exists $i > 1$ which suffices that $v_{\pi_k^A}(A_{\pi_k^A}) = v_{\pi_k^B}(B_{\pi_k^B})$ for any $k < i$, and $v_{\pi_k^A}(A_{\pi_k^A}) < v_{\pi_k^B}(B_{\pi_k^B})$ for $k = i$, then $A \prec B$. Otherwise, $B \preceq A$.

Definition 5. *An allocation $A = (A_1, \ldots, A_n)$ is a leximin allocation if for any allocation B, $B \preceq A$.*

Based on leximin allocations, we find a best 1/2-EFX and PO allocation for single-valued valuations.

Theorem 7. *For identical single-valued valuations, there exists a 1/2-EFX and PO allocation.*

Proof. Consider an instance with $M = \{g_1, \ldots, g_m\}$ and $N = \{1, \ldots, n\}$. Let $v(\cdot)$ be the single-valued valuation for all agents. Let $A = (A_1, \ldots, A_n)$ be a leximin allocation, which means that A is the global maximum by the comparison operator. Thus, there exists no allocation X such that $v(X_i) > v(A_i)$ for some agent $i \in N$, and $v_k(X_k) \geq v_k(A_k)$ for all $k \neq i$. It implies that A is PO. The following will show that the leximin A is 1/2-EFX.

Suppose that there exists $i, j \in N$ such that for some $Q_{ij} \subseteq A_j$ and some item $g^* \in Q_{ij}$,
$$s(Q_{ij}) \leq b_i, \quad v(A_i) < 1/2 \cdot v(Q_{ij} \setminus \{g^*\}).$$

Notice that $|Q_{ij}| \geq 2$, otherwise $v(Q_{ij} \setminus g^*) = 0$ which is contradicted with the assumption. Let $\pi^A = \{\pi_1^A, \cdots, \pi_n^A\}$ be the order of all agents by non-decreasing valuations w.r.t. A. Without loss of generality, we assume that agent i lies in the last position among the agents with $\{k : v(A_k) \leq v(A_i)\}$ in π^A. Denote $t = |\{k : v(A_k) \leq v(A_i)\}|$. Construct an allocation $B = \{B_1, \ldots, B_n\}$ where $B_i \subseteq Q_{ij}$ with $|B_i| = \lceil \frac{|Q_{ij}|-1}{2} \rceil$, $B_j = A_j \setminus B_i$, and $B_k = A_k$ for any $k \neq i, j$. It is clear that B is a budget-feasible allocation and

$$v(B_i) \geq 1/2 \cdot v(Q_{ij} \setminus \{g^*\}) > v(A_i), \quad v(B_j) \geq v(B_i).$$

Similarly, let $\pi^B = \{\pi_1^B, \cdots, \pi_n^B\}$ be the order of all agents by non-decreasing valuations w.r.t B. If agent i lies in the position t of π^B, $v(\pi_t^B) = v(B_i) > v(A_i) = v(\pi_t^A)$. Otherwise, there exists an agent $r \in N \setminus \{k : v(A_k) \leq v(A_i)\}$ occurring in the position t of π^B, $v(\pi_t^B) = v(B_r) > v(A_i) = v(\pi_t^A)$. This indicates that the allocation $A \prec B$, which is contradicted by that A, is a leximin allocation.

Hence, the leximin allocation A is 1/2-EFX and PO.   □

4   EFX and Partial Tight Allocations

In this section, to further our efforts to find efficient and fair allocations, we focus on computing approximate EFX and partial tight (which is a relaxation notion of tightness) allocations.

Definition 6 (Partial Tightness, PT). *An allocation $A = (A_1, \ldots, A_n)$ is partial tight if for any $i \in N[unenvied]$,*

$$s(A_i \cup \{g\}) > b_i, \quad \forall g \in (M \setminus \cup_{i=1}^{n} A_i),$$

where $N[unenvied] = \{i \in N : v_u(T) \leq v_u(A_u), \forall T \subseteq A_i \text{ s.t. } s(T) \leq b_u, \forall u \in N\}$.

In addition, if an agent $i \in N[unenvied]$, she is referred to as an unenvied agent with respect to A.

Notice that every tight allocation is also PT. If all agents possess identical budgets, $N[unenvied]$ can be rewritten as $N[unenvied] = \{i \in N : v_u(A_i) \leq v_u(A_u), \forall u \in N\}$.

To better illustrate an efficient algorithms, we firstly introduce the envy graph under budget constraints.

Envy Graph: Given an allocation $A = (A_1, \ldots, A_n)$, the corresponding envy graph is defined as $G = (V, E)$, where the vertex set $V = N$, a directed edge $(i, j) \in E$ iff agent i envies agent j, i.e. $v_i(A_i) \leq v_i(T)$ for some $T \subseteq A_j$ with $s(T) \leq b_i$. Additionally, the directed circles on an envy graph are called **envy circles**.

Based on the procedure of Algorithm 2, we can obtain the following two lemmas.

Lemma 1. *In each iteration of the inner while-loop in lines 13–18 of Algorithm 2, let $\tilde{A} = (\tilde{A}_1, \ldots, \tilde{A}_n)$ and $A' = (A'_1, \ldots, A'_n)$ be the starting and the ending allocations respectively. If $\tilde{A}$ is a 1/2-EFX allocation, A' is still 1/2-EFX.*

Proof. Consider any iteration of the inner while-loop 13–18 in Algorithm 2. Let $\tilde{A} = (\tilde{A}_1, \ldots, \tilde{A}_n)$ and $A' = (A'_1, \ldots, A'_n)$ be the starting and the ending allocations respectively. Let G be the envy graph determined by allocation $\tilde{A} = (\tilde{A}_1, \ldots, \tilde{A}_n)$, and $C = (1, \ldots, d)$ be an envy cycle in the envy graph G. We have that $v_i(\tilde{A}_i) < v_i(\tilde{A}_{(i \mod d)+1})$ for any agent $i \in C$ by line 16. In the ending allocation A', for any agent $i \in C$, $A'_i = \arg\max\{v_i(Q) \mid Q \subseteq \tilde{A}_{(i \mod d)+1}, s(Q) \leq b_i\}$, and $A'_i = \tilde{A}_i$ for any agent $i \in N \setminus C$. Consider any two agents $i, j \in N$, we are sufficient to show that

$$v_i(A'_i) \geq \frac{1}{2} \cdot v_i(Q \setminus \{g\}), \quad \forall g \in Q, \ \forall Q \subseteq A'_j, s(Q) \leq b_i. \tag{1}$$

If $j \notin C$, it means that $A'_j = \tilde{A}_j$. Since A is 1/2-EFX, we have that $v_i(\tilde{A}_i) \geq 1/2 \cdot v_i(Q \setminus \{g\})$ for any $g \in Q$, and any $Q \subseteq \tilde{A}_j$ with $s(Q) \leq b_i$. Thus, the

Inequality (1) holds by $v_i(A_i') \geq v_i(\tilde{A}_i)$ and $v_j(A_j') = v_j(\tilde{A}_j)$. Otherwise, if $j \in C$. Based on the swapping rule of line 14, there exists an agent $t \in C$ ordered behind j in cycle C such that $A_j' = Q_t$, where $Q_t \subseteq \tilde{A}_t$ with $s(Q_t) \leq b_i$. Since $\tilde{A}$ is 1/2-EFX, we obtain that $v_i(\tilde{A}_i) \geq 1/2 \cdot v_i(Q \setminus \{g\})$ for any $g \in Q$, and any $Q \subseteq \tilde{A}_t$ with $s(Q) \leq b_i$. Since $v_i(A_i') \geq v_i(\tilde{A}_i)$ and $A_j' \subseteq \tilde{A}_t$, the Inequality (1) holds.

□

Algorithm 2 Greedy-packing Algorithm

Input: $M = \{g_1, \ldots, g_m\}$ with sizes $s(g_e), e = 1, \ldots, m$, g_{m+1} with size $s(g_{m+1})$, $N = \{1, \ldots, n\}$ with budgets $b_i, i = 1, \ldots, n$.
Output: An allocation.

1: Initialize: $A_i = \emptyset, i = 1, 2, \ldots, n$, $A_0 = M$;
2: **while** there exists $i \in N[unenvied]$ is not tight **do**
3:     Let i be the agent with the least number of items that is unenvied and not tight.
4:     Let $g^* \in \arg\max\{v_i(g) \mid g \in A_0, s(A_i \cup g) \leq b_i\}$;
5:     Let $g^h \in \arg\max\{v_i(g) \mid g \in A_0, s(g) \leq b_i\}$;
6:     **if** $v_i(A_i \cup \{g^*\}) \geq v_i(g^h)$ **then**
7:         $A_i = A_i \cup \{g^*\}$;
8:         $A_0 = A_0 \setminus \{g^*\}$;
9:     **else**
10:        $A_i = \{g^h\}$;
11:        $A_0 = A_0 \cup (A_i \setminus \{g^h\})$;
12:     **end if**
13:     **while** $N[unenvied] = \emptyset$ & there exists envy cycles in graph G **do**
14:        Find an envy-cycle $C = (1, \ldots, d)$;
15:        Let $A_i^C = \arg\max\{v_i(Q) \mid Q \subseteq A_{(i \bmod d)+1}, s(Q) \leq b_i\}$ for all $i \in C$;
16:        $A_0 = A_0 \cup (\cup_{i \in C}(A_i \setminus A_i^C))$;
17:        $A_i = A_i^C$ for all $i \in C$;
18:     **end while**
19: **end while**
20: **return** $A = (A_1, \ldots, A_n)$

Lemma 2. *Given an allocation $A = (A_1, \ldots, A_n)$, It can be determined if an agent $i \in N$ is unenvied with respect to A or not in $O(mn \cdot b_i)$ time.*

Proof. Consider an allocation A and any agent $i \in N$. Based on Definition 6, agent i is unenvied iff for any agent $k \in N$, $\max\{v_i(T) \mid T \subseteq A_k, s(T) \leq b_i\} \leq v_i(A_i)$. For a fixed A_k, compute $T^* = \arg\max\{v_i(T) \mid T \subseteq A_k, s(T) \leq b_i\}$ is equivalent to the weighted Knapsack problem. The above T^* can be computed in $O(m \cdot b_i)$ [15], and whether agent i is unenvied or not can be determined in $O(nm \cdot b_i)$.

□

Theorem 8. *Algorithm 2 computes a 1/2-EFX and PT allocation in pseudo-polynomial time.*

Proof. Let $A = (A_1, \ldots, A_n)$ be the output of Algorithm 2. We first prove that A is 1/2-EFX and PT, then we will analyze the running time of Algorithm 2.

(a) **The output A is PT.** Based on the While-loop condition in line 2, the output A is partial tight.

(b) **The output A is 1/2-EFX.** Consider any r-th round of outer while-loop in Algorithm 2. Let $\bar{A} = (\bar{A}_1, \ldots, \bar{A}_n)$ and $\hat{A} = (\hat{A}_1, \ldots, \hat{A}_n)$ be the starting and ending allocations in this round. To illustrate the conclusion, it is sufficient to prove that if $\bar{A}$ is 1/2-EFX, then $\hat{A}$ is 1/2-EFX. Let $A' = (A'_1, \ldots, A'_n)$ be the beginning allocation of the inner while-loop (line 13) in this round. If A' is 1/2-EFX, we have that $\hat{A}$ is a 1/2-EFX allocation based on Lemma 1. The following will show that A' is 1/2-EFX, i.e., for any $i, j \in N$,

$$v_i(A'_i) \geq 1/2 \cdot v_i(T \setminus \{g\}), \ \forall g \in T, \ \forall T \subseteq A'_j, s(T) \leq b_i. \tag{2}$$

If $|A'_j| = 1$, the Inequality (2) holds since for any subset $T \subseteq A'_j$, $v_i(T \setminus \{g\}) = 0$. If $A'_j = \bar{A}_j$, we have that the Inequality (2) holds since $\bar{A}$ is 1/2-EFX and $v_i(A'_i) \geq v_i(A_i)$. Otherwise if $A'_j \neq \bar{A}_j$, we obtain that $A'_j = \bar{A}_j \cup \{g^*\}$, where $g^* = \arg\max\{v_j(g) \mid g \in A'_0, s(\bar{A}'_j \cup \{g\}) \leq b_j\}$. To show the Inequality (2), it is sufficient to show that

$$v_i(A'_i) \geq v_i(Q \setminus \{g^*\}), \ \forall Q \subseteq A'_j, s(Q) \leq b_i; \ v_i(A'_i) \geq v_i(\{g^*\}). \tag{3}$$

Note that agent $j \in N[unenvied]$ is an unenvied agent at this moment. It means that $v_i(\bar{A}_i) \geq v_i(Q')$, for any $Q' \subseteq \bar{A}_j$ with $s(Q') \leq b_i$. Combined with $v_i(A'_i) = v_i(\bar{A}_i)$, we can get the first inequality in the Inequalities (3). Then, we will show that $v_i(\hat{A}_i) \geq v_i(g^*)$.

Since agent j is unenvied, we have that $|\bar{A}_i| \geq 1$, and $v_i(\bar{A}_i) > 0$. If g^* has never been allocated to any agent, then $v_i(A'_i) \geq v_i(g^*)$. If g^* has been allocated to agent i before the r-th round, g^* must be allocated to agent i, then we also have $v_i(A'_i) \geq v_i(\{g^*\})$. Otherwise, we assume that g^* has been allocated to agent t at the k-th round, where $k < r$. Suppose that $v_i(A'_i) < v_i(\{g^*\})$, we are going to construct a contradiction. Although agent t is an unenvied agent in the k-th round, based on $v_i(A'_i) < v_i(g^*)$, agent t will not belongs to $N[unenvied]$ in any k'-th round, where $r > k' > k$. It means that after the k-th round, agent t can only update her bundle in the inner while-loop, that is, agents swap bundles along the envy circle. Therefore, we can assume that agent h updates her bundle by choosing agent t's subset Q_h with $s(Q_h) \leq b_h$ and $g^* \notin Q_h$. It is contradicted with that agent t is an unenvied agent in the k-th round. Thus, $v_i(A'_i) \geq v_i(g^*)$. By Inequalities (3), it is certain that

$$v_i(A'_i) \geq 1/2 \cdot (v_i(Q \setminus \{g^*\}) + v_i(\{g^*\}) = 1/2 \cdot v_i(Q))$$
$$\geq 1/2 \cdot v_i(Q \setminus \{g\}), \ \forall g \in Q.$$

(c) **The running time is $O(n^5 m \cdot b_{\max} \cdot v_{\max})$.** Let $b_{\max} = \{b_i \mid i \in N\}$ and $v_{\max} = \{v_i(M) \mid i \in N\}$ be the maximum budget and the maximum total value respectively. Based on Lemma 2, we need to spend at most $O(n^2 m \cdot b_{\max})$ time to determine if it executes two conditions of the outer and inner while-loops respectively. In each inner while-loop, we need at most $O(m \cdot b_{\max})$ time to compute A_i^C. Thus, we need at most $O(n^4 m \cdot b_{\max})$ time to jump out of the inner while-loop. Observe that the total value of the ending allocation in each round will strictly increase while the round number increases. Thus, there are at most $n \cdot v_{\max}$ rounds of outer while-loop and the total running time is $O(n^5 m \cdot b_{\max} \cdot v_{\max})$. $\qquad$ $\square$

5 Conclusions

In this paper, we studied the problem of allocating goods where both fairness and efficiency requirements are met. We introduced two significant notions: Tightness and Partial Tightness (PT). It is worth mentioning that we analyzed several negative results that can not be unattainable for EFX and Tightness (or PO) allocations. On the other hand, when agents have single-valued (or identical) valuations, we showed that EFX and Tight (or approximate EFX and PO) allocations exist by algorithmic results.

There are several interesting directions for future work that follow our research agenda. Firstly, it would be interesting to explore the existence of EFX and tight (or PO) allocations for other special cases. Secondly, we proved that 1/2-EFX with PT allocations exists, but the upper bound of approximate EFX and PT allocation remains open. Finally, it is meaningful to explore EFX and tight (or PT) allocations under other constraints.

Acknowledgments. We are grateful to the anonymous reviewers for their invaluable comments, which helped us improve the quality of this manuscript.

References

1. Babaioff, M., Ezra, T., Feige, U.: Fair and truthful mechanisms for dichotomous valuations. In: Proceedings of the 35th AAAI Conference on Artificial Intelligence, vol. 35, no. 6, pp. 5119–5126 (2021)
2. Barman, S., Khan, A., Shyam, S., Sreenivas, K.V.N.: Finding fair allocations under budget constraints. In: Proceedings of the 37th AAAI Conference on Artificial Intelligence, vol. 37, no. 5, pp. 5481–5489 (2021)
3. Barman, S., Khan, A., Shyam, S., Sreenivas, K.V.N.: Guaranteeing envy-freeness under generalized assignment constraints. In: Proceedings of the 24th ACM Conference on Economics and Computation (EC'23), pp. 242–269 (2023)
4. Barman, S., Krishnamurthy, S.K., Vaish, R.: Finding fair and efficient allocations. In: Proceedings of the 19th ACM Conference on Economics and Computation (EC'18), pp. 557–574 (2018)
5. Berger, B., Cohen, A., Feldman, M., Fiat, A.: Almost full efx exists for four agents. In: Proceedings of the 36th AAAI Conference on Artificial Intelligence, vol. 36, no. 5, pp. 4826–4833 (2022)
6. Bu, X., Song, J., Yu., Z.: Efx allocations exist for binary valuations. In: Proceedings of International Workshop on Frontiers of Algorithmics (IJTCS-FAW'23), pp. 252–262 (2023)
7. Budish, E., Cantillon, E.: The multi-unit assignment problem: theory and evidence from course allocation at Harvard. Am. Econ. Rev. **102**(5), 2237–2271 (2012)

8. Caragiannis, I., Kurokawa, D., Moulin, H., Procaccia, A.D., Shah, N., Wang, J.: The unreasonable fairness of maximum nash welfare. In: Proceedings of the 17th ACM Conference on Economics and Computation (EC'16), pp. 305–322 (2016)
9. Chan, H., Chen, J., Li, B., Wu, X.: Maximin-aware allocations of indivisible goods. In: Proceedings of the 28th International Joint Conference on Artificial Intelligence (IJCAI'19), pp. 137–143 (2019)
10. Chaudhury, B.R., Garg, J., Mehlhorn, K.: Efx exists for three agents. In: Proceedings of the 21st ACM Conference on Economics and Computation (EC'20), pp. 1–19 (2020)
11. Chaudhury, B.R., Kavitha, T., Mehlhorn, K., Sgouritsa, A.: A little charity guarantees almost envy-freeness. SIAM J. Comput. **50**(4), 1336–1358 (2021)
12. Feldman, M., Mauras, S., Ponitka, T.: On optimal tradeoffs between efx and nash welfare (2023)
13. Gan, J., Li, B., Wu, X.: Approximation algorithm for computing budget-feasible ef1 allocations. In: Proceedings of the 21st International Conference on Autonomous Agents and Multiagent Systems (AAMAS'23), pp. 170–178 (2023)
14. Ghosal, P.V.V.P.H., Nimbhorkar, P., Varma, N.: Efx exists for four agents with three types of valuations (2023)
15. Kellerer, H., Pferschy, U., Pisinger, D.: Knapsack Problems, 1st edn. Springer-Verlag, Berlin, Heidelberg, GmbH, New York, NY (2004). https://doi.org/10.1007/978-1-4613-0303-9_5
16. Lipton, R.J., Markakis, E., Mossel, E., Saberi, A.: On approximately fair allocations of indivisible goods. In: Proceedings of the 5th ACM Conference on Electronic Commerce (EC'04), pp. 125–131 (2004)
17. Plaut, B., Roughgarden, T.: Almost envy-freeness with general valuations. SIAM J. Discret. Math. **34**(2), 1039–1068 (2020)
18. Ryoga, M.: Extension of additive valuations to general valuations on the existence of efx. In: Proceedings of the 29th Annual European Symposium on Algorithms (ESA'21), pp. 66:1–66:15 (2021)
19. Steinhaus, H.: The problem of fair division. Econometrica **16**, 101–104 (1948)
20. Wu, X., Li, B., Gan, J.: Budget-feasible maximum nash social welfare is almost envy-free. In: Proceedings of the 30th International Joint Conference on Artificial Intelligence (IJCAI'21), pp. 465–471 (2021)

Computations and Complexities of Tarski's Fixed Points and Supermodular Games

Chuangyin Dang[1], Qi Qi[2($\boxtimes$)], and Yinyu Ye[3]

[1] Department of Systems Engineering, City University of Hong Kong,
Hong Kong SAR, Kowloon Tong, China
sehead@cityu.edu.hk
[2] Gaoling School of Artificial Intelligence,
Renmin University of China, Beijing, China
qi.qi@ruc.edu.cn
[3] Department of Management Science and Engineering,
Stanford University, Stanford, CA 94305-4026, USA
yinyu-ye@stanford.edu

Abstract. We consider two models of computation for Tarski's order preserving function f related to fixed points in a complete lattice: the oracle function model and the polynomial function model. In both models, we find the first polynomial time algorithm for finding a Tarski's fixed point. In addition, we provide a matching oracle bound for determining the uniqueness in the oracle function model and prove it is Co-NP hard in the polynomial function model. The existence of the pure Nash equilibrium in supermodular games is proved by Tarski's fixed point theorem. Exploring the difference between supermodular games and Tarski's fixed point, we also develop the computational results for finding one pure Nash equilibrium and determining the uniqueness of the equilibrium in supermodular games.

Keywords: Fixed Point Theorem · Equilibrium Computation · Supermodular Game · Order Preserving Mapping · Co-NP Hardness

1 Introduction

Supermodular games, also known as the games of strategic complements, are formalized by Topkis in 1979 [22] and have been extensively studied in the literature, such as Bernstein and Federgruen [1,2], Cachon [4], Cachon and Lariviere [5], Fudenberg and Tirole [11], Lippman and McCardle [15], Milgrom and Roberts [16,17], Milgrom and Shannon [18], Topkis [23], and Vives [24,25]. In supermodular games, the utility function of every player has increasing differences. Then the best response of a player is a nondecreasing function of other

---

Some results in this paper are based on our working paper (2011) Computational Models and Complexities of Tarski's Fixed Points (https://web.stanford.edu/~yyye/unitarski1.pdf).

players' strategies. For example, if firm A's competing firm B starts spending more money on research it becomes more advisable for firm A to do the same.

Supermodular games arise in many applied models. They cover most static market models. For example, the investment games, Bertrand oligopoly, Cournot oligopoly all can be modeled as supermodular games. Many models in operations research have also been analyzed as supermodular games. For example, supply chain analysis, revenue management games, price and service competition, inventory competition etc. Recently, the problem of power control in cellular CDMA wireless network is also modeled as a supermodular game.

The existence of a pure Nash equilibrium in any supermodular game is proved by Tarski's fixed point theorem [20]. The well-known Tarski's fixed point theorem (Tarski) asserts that, if $(L, \preceq)$ is a complete lattice and f is order-preserving from L into itself, then there exists some $\mathbf{x}^* \in L$ such that $f(\mathbf{x}^*) = \mathbf{x}^*$.

This theorem plays a crucial role in the study of supermodular games for economic analysis and has other important applications. To compute a Nash equilibrium of a supermodular game, a generic approach is to convert it into the computation of a fixed point of an order preserving mapping. Recently, an algorithm has been proposed in Echenique [10] to find all pure strategy Nash equilibria of a supermodular game, which motivated the study in this paper.

An efficient computational algorithm for finding a Nash equilibrium has been a recognized important technical advantage in applications. Further, it is sometimes desirable to know if an already-found equilibrium for such applications is unique or not, for the decision whether additional resources should be spent to improve the already found solution. There were some interesting complexity results in algorithmic game theory research along this line, on determining whether or not a game has a unique equilibrium point. For the bimatrix game, Gilboa and Zemel [12] showed that it is NP-hard to determine whether or not there is a second Nash equilibrium. For this problem, computing even one equilibrium (which is known to exist), is already difficult and no polynomial time algorithms are known: Nash equilibrium for the bimatrix game is known to be PPAD-complete [9]. Similar cases are known for other problems such as the market equilibrium computation (Codenotti et al.) [3].

In this work, we first consider the fixed point computation of order preserving functions over a complete lattice, both for finding a solution and for determining the uniqueness of an already-found solution. Then we study the computational problems for finding one pure Nash equilibrium and determining the uniqueness of the equilibrium in supermodular games. We are interested in both the oracle function model and the polynomial function model. For both the fixed point problem and supermodular games, the domain space can be huge. Most interesting discussions consider a succinct representation (see Sect. 2.2) of the lattice $(L, \preceq)$ such that the input size is related to $\log |L|$. It is enough for the representation of a variable in a lattice of size $|L|$. Both the oracle function model and the polynomial time function model return the function value $f(x)$ on a lattice node x where x is of size $\log |L|$. They differ in the ways the functions are computed. The polynomial time function model computes $f(x)$ by an explicitly given algorithm, in time polynomial of $\log|L|$. The oracle model, on the other hand, always returns the value in one oracle step. More details comparing those two models can be found in Sect. 2.3.

1.1   Main Results and Related Work

A partially order set L is defined with $\preceq$ as a binary relation on the set L such that $\preceq$ is reflexive, transitive, and anti-symmetric. A lattice is a partially ordered set $(L, \preceq)$, in which any two elements $\mathbf{x}$ and $\mathbf{y}$ have a least upper bound (supremum), $\sup_L(\mathbf{x}, \mathbf{y}) = \inf\{\mathbf{z} \in L \mid \mathbf{x} \preceq \mathbf{z} \text{and} \mathbf{y} \preceq \mathbf{z}\}$, and a greatest lower bound (infimum), $\inf_L(\mathbf{x}, \mathbf{y}) = \sup\{\mathbf{z} \in L \mid \mathbf{z} \preceq \mathbf{x} \text{and} \mathbf{z} \preceq \mathbf{y}\}$, in the set. A lattice $(L, \preceq)$ is complete if every nonempty subset of L has a supremum and an infimum in L. Let f be a mapping from L to itself. f is order-preserving if $f(\mathbf{x}) \preceq f(\mathbf{y})$ for any $\mathbf{x}$ and $\mathbf{y}$ of L with $\mathbf{x} \preceq \mathbf{y}$.

We focus on the componentwise ordering and lexicographic ordering finite lattices. Let $L_d = \{\mathbf{x} \in Z^d \mid \mathbf{a} \leq \mathbf{x} \leq \mathbf{b}\}$, where $\mathbf{a}$ and $\mathbf{b}$ are two finite vectors of Z^d with $\mathbf{a} < \mathbf{b}$. We denote the componentwise ordering and the lexicographic ordering as $\leq_c$ and $\leq_l$ respectively. Clearly, $(L_d, \leq_c)$ is a finite lattice with componentwise ordering and $(L_d, \leq_l)$ is a finite lattice with lexicographic ordering.

Let f_c and f_l be an order preserving mapping from L_d into itself under the componentwise ordering and the lexicographic ordering respectively.

Tarski's Fixed Points: Oracle Function Model. When $f_l(\cdot)$ and $f_c(\cdot)$ are given as oracle functions and order preserving, we develop a complete understanding for finding a Tarski's fixed point as well as determining uniqueness of the Tarski's fixed point in both the lexicographic ordering and the componentwise ordering lattices.

We develop an algorithm of time complexity $O(\log^d |L|)$ to find a Tarski's fixed point on the componentwise ordering lattice $(L, \leq_c)$, for any constant dimension d. This algorithm is based on the binary search method. We first present the algorithm when $d = 2$. Follows the similar principle, this algorithm can be generalized to any constant dimension. This is the first known polynomial time algorithm for finding the Tarski's fixed point in terms of the componentwise ordering. In literature, we only have a polynomial time algorithm for the total order lattices (Chang et al.) [6].

On the other hand, given a componentwise ordering lattice $(L, \preceq_c)$ with one already known fixed point, we derive a $\Theta(N_1 + N_2 + \cdots + N_d)$ matching bound for determining the uniqueness of the fixed point, where $L = \{\mathbf{x} \in Z^d \mid \mathbf{a} \leq \mathbf{x} \leq \mathbf{b}\}$ and $N_i = b_i - a_i$.

A lexicographic ordering lattice can be viewed as a componentwise ordering lattice with dimension one by an appropriate polynomial time transformation to change the oracle function for the d-dimension space to an oracle function on the 1-dimension space. All the above results can be transplanted onto the lexicographic ordering lattice with a set of related parameters.

In literature, a polynomial time algorithm for finding a Tarski's fixed point is known only for the total order lattices.

Tarski's Fixed Points: Polynomial Function Model. Under the polynomial time function model, our polynomial time algorithm applies when the dimension is any finite constant. When the dimension is used as a part of the input size in unary, we first present a polynomial-time reduction of a 3-SAT problem to an order preserving mapping f from a componentwise ordering lattice L into itself. As a result of this reduction, we obtain that, given f as a polynomial time function, determining whether f has a unique fixed point in L is a Co-NP hard problem. Furthermore, even when the dimension is one, we show that determining the uniqueness of Tarski's fixed point in a lexicographic lattice is Co-NP hard, though there exists a polynomial-time algorithm for computing a Tarski's fixed point in a lexicographic lattice in any dimension.

Our main results for Tarki's fixed point computation are summarized in Table 1 and Table 2.

Supermodular Games. For supermodular games, we develop an algorithm to find a pure Nash equilbirum in polynomial time $O(\log N_1 \cdots \log N_{d-1})$ in the oracle function model, where d is the total number of players, N_i is the number of strategies of player i and $N_1 \leq N_2 \cdots \leq N_d$. It is the first polynomial time algorithm when d is a constant. Thus a pure Nash equilibirum can be found in time $O(poly(\log |L|) \cdot (\log N_1 \cdots \log N_{d-1})$ in the polynomial function model, where $|L| = N_1 \times N_2 \cdots \times N_d$. In the polynomial function model, we prove determining the uniqueness is Co-NP-hard.

In literature, Robinson(1951) [19] introduces the iterative method to solve a game and Topkis (1979) [22] uses this method to find a pure Nash equilibrium in a supermodular game which takes time $O(N_1 + N_2 + \cdots + N_d)$. Echenique in 2007 [10] proposes the first algorithm for finding all pure Nash equilibria in supermodular games. However, this algorithm takes expenontial time $O(N_1 \times N_2 \times \cdots \times N_d)$ to find the first pure equilibrium in the worst case.

Table 1. Main Results for Finding one Tarski's Fixed Point

	Polynomial Function	Oracle Function
Componentwise	$O(poly(\log |L|) \cdot (\log N_1 \cdots \log N_d))$	$O((\log N_1 \cdots \log N_d)$
Lexicographic	$O(poly(\log |L|) \cdot \log |L|)$	$O(\log |L|)$

Table 2. Main Results for Determining the Uniqueness of Tarski's Fixed Points

	Polynomial Function	Oracle Function
Componentwise	Co-NP-Complete	$\Theta(N_1 + N_2 + \cdots + N_d)$
Lexicographic	Co-NP-Complete	$\Theta(|L|)$

1.2   Organization

The rest of the paper is organized as follows. First, in Sect. 2, we present definitions as well as the difference of the polynomial function model and the oracle function model. We develop polynomial time algorithms in oracle function model for componentwise ordering and lexicographic ordering in Sect. 3. In Sect. 4, we derive the matching bound for determining the uniqueness of Tarski's fixed point under the oracle function model. We prove co-NP hardness for determining the uniqueness of Tarski's fixed point under the polynomial function model in Sect. 5. In Sect. 6, we develop the computational results for finding one pure Nash equilibrium and determining the uniqueness of the equilibrium in supermdular games. We conclude with discussion and remarks on our results and open problems in Sect. 7.

2   Preliminaries

In this section, we first introduce the formal definitions of the related concepts as well as the Tarski's fixed point theorem. We next compare the difference between the oracle function model and the polynomial function model.

2.1   The Lattice and Tarski's Fixed Point Theorem

Definition 1. *(Partial Order vs. Total Order) A relationship $\preceq$ on a set L is a partial order if it satisfies reflexivity ($\forall \mathbf{a} \in L : \mathbf{a} \preceq \mathbf{a}$); antisymmetry ($\mathbf{a} \preceq \mathbf{b}$ and $\mathbf{b} \preceq \mathbf{a}$ implies $\mathbf{a} = \mathbf{b}$); transitivity ($\mathbf{a} \preceq \mathbf{b}$ and $\mathbf{b} \preceq \mathbf{c}$ implies $\mathbf{a} \preceq \mathbf{c}$). It is a total order if $\forall \mathbf{a}, \mathbf{b} \in L$: either $\mathbf{a} \preceq \mathbf{b}$ or $\mathbf{b} \preceq \mathbf{a}$.*

Definition 2. *(Lattice) $(L, \preceq)$ is a lattice if*

1. L is a partial ordered set;
2. There are two operations: meet $\wedge$ and join $\vee$ on any pair of elements $\mathbf{a}, \mathbf{b}$ of L such that $\mathbf{a}, \mathbf{b} \preceq \mathbf{a} \vee \mathbf{b}$ and $\mathbf{a} \wedge \mathbf{b} \preceq \mathbf{a}, \mathbf{b}$

The lattice is complete lattice if for any subset $A = \{\mathbf{a}_1, \mathbf{a}_2, \cdots, \mathbf{a}_k\} \subseteq L$, there is a unique meet and a unique join: $\bigwedge A = (\mathbf{a}_1 \wedge \mathbf{a}_2 \wedge \cdots \wedge \mathbf{a}_k)$ and $\bigvee A = (\mathbf{a}_1 \vee \mathbf{a}_2 \vee \cdots \vee \mathbf{a}_k)$.

For simplicity, we use L for a lattice when no ambiguity exists on $\preceq$. We should specify $\preceq$ whenever it is necessary.

Definition 3. *(Order Preserving Function) A function f on a lattice $(L, \preceq)$ is order preserving if $\mathbf{a} \preceq \mathbf{b}$ implies $f(\mathbf{a}) \preceq f(\mathbf{b})$.*

Theorem 1. *(Tarski's Fixed Point Theorem) [20]. If L is a complete lattice and f an increasing function from L to itself, there exists some $\mathbf{x}^* \in L$ such that $f(\mathbf{x}^*) = \mathbf{x}^*$, which is a fixed point of f.*

This theorem guarantees the existence of fixed points of any order-preserving function $f : L \to L$ on any nonempty complete lattice.

Definition 4. *(Lexicographic Ordering Function). Given a set of points on a d-dimensional space R^d, the lexicographic ordering function $\leq_l$ is defined as:*

$\forall \mathbf{x}, \mathbf{y} \in R^d$, $\mathbf{x} \leq_l \mathbf{y}$ *if either* $\mathbf{x} = \mathbf{y}$ *or* $x_i = y_i$, *for* $i = 1, 2, \ldots, k-1$, *and* $x_k < y_k$ *for some* $k \leq d$.

Definition 5. *(Componentwise Ordering Function). Given a set of points on a d-dimensional space R^d, the componentwise ordering function $\leq_c$ is defined as:*

$\forall \mathbf{x}, \mathbf{y} \in R^d$, $\mathbf{x} \leq_c \mathbf{y}$ *if* $\forall i \in \{1, 2, \cdots, d\} : x_i \leq y_i$.

2.2   Succinct Representation

For the problems we consider in this work, there are usually 2^{d*n} nodes where d is a constant and n is an input parameter. Therefore, the input size is exponential in the input parameter n. We need to represent such input data succinctly. As an example, for the set $N = \{0, 1, 2, \cdots, 2^n - 1\}$, the input can be described as all the integers i: $0 \leq i \leq 2^n - 1$. Each such integer i can be written by up to n bits.

2.3   The Oracle Function Model Versus the Polynomial Time Function Model

The two succinctly represented function models are the oracle function model and the polynomial function model.

For the oracle model, we treat the function as a black box that outputs the function value for every domain variable once a request is sent in to the oracle. The output of the oracle is arbitrary on the first query but it cannot change a function value after a query is already made to the oracle on the same variable. For example, let $N = \{0, 1\}$. Let $f : N \to N$ be an oracle function. When we ask for $f(0)$, the oracle could answer anything, either 0 or 1. Suppose the oracle answers $f(0) = 1$ in the first query. Later, if we need to use $f(0)$ again, it must be the same 1. Equivalently, we may assume that the function values are stored in the harddisk. After a query, it is saved in the memory cache. Later uses of the same query will be the value in the memory cache and there is no need to check with the harddisk again. It is important to note that, the oracle funciton model contains all the functions $f : N \to M$ where N is its domain and M is its range. This is very different from the polynomial function we are going to introduce next.

For the polynomial function model, the input function is an algorithm that gives the answer for the function value on the input data. The algorithm returns the answer in time polynomial in the input parameter n. Alternatively, the polynomial time algorithm can be replaced by a polynomial size logical circuits consisting of gates $\{AND, NOT, OR\}$ of Boolean variables.

Clearly oracle function admits much more functions than those computable in polynomial time. Therefore a problem is usually much harder under the oracle function model than under the polynomial time function model.

3   Polynomial Time Algorithm Under Oracle Function Model

In this section, we consider the complexity of finding a Tarski's fixed point in any constant dimension d with the function value f given by an oracle. Chang et al. [6] proved that a fixed point can be found in time polynomial when the given lattice is total order.

Define $L = \{\mathbf{x} \in Z^d \mid \mathbf{a} \le \mathbf{x} \le \mathbf{b}\}$, where $\mathbf{a}$ and $\mathbf{b}$ are two finite vectors of Z^d with $\mathbf{a} < \mathbf{b}$.

Theorem 2 *(Chang et al.) [6]. When $(L, \preceq)$ is given as an input and the order preserving function f is given as an oracle, a Tarski's fixed point can be found in time $O(\log |L|)$ on a finite lattice when $\preceq$ is a total order on L.*

Since any two vectors in the lexicographic ordering is comparable, the lexicographic ordering is a total order. We have

Corollary 1. *When $(L, \preceq)$ is given as an input and the order preserving function f is given as an oracle, a Tarski's fixed point can be found in time $O(\log |L|)$ on a finite lattice when $\preceq$ is a lexicographic ordering in L.*

The proof is rather standard utilizing the total order property of the lexicographic ordering. As the componentwise ordering lattice cannot be modelled as a total order, it leaves open the oracle complexity of finding a fixed point in a componentwise ordering lattice. Here we show that this problem is also polynomial time solvable, by designing a polynomial algorithm to find a fixed point of f in time $O((\log |L|)^d)$ given a componentwise ordering lattice L.

The algorithm exploits the order properties of the componentwise lattice and applying the binary search method with a dimension reduction technique. To illustrate the main ideas, we first consider the 2D case before moving on to the general case.

Without loss of generality, we assume L is a $N \times N$ square centred at point $(0,0)$. The componentwise ordering is denoted as $\le_c$.

Algorithm 1 *Point_check() (A polynomial algorithm for 2D lattice)*

- *Input:*
    A 2-dimensional lattice $(L, \leq_c)$, $|L| = N^2$ (Input size to the oracle is $2 \log N$
    since the input size for both dimensions to the oracle is $\log N$.)
    An oracle function f. f is a order preserving function. $\forall \mathbf{x} \in L, f(\mathbf{x}) \in L$
    and $f(\mathbf{x}) \leq_c f(\mathbf{y})$ if $\mathbf{x} \leq_c \mathbf{y}, \forall \mathbf{x}, \mathbf{y} \in L$
- *Point_check(L, f)*
    Let $\mathbf{x}^0$ be the center point in L. Let $\mathbf{x}^L$ be the left most point in L such that
    $\mathbf{x}_2^L = \mathbf{x}_2^0$. Let $\mathbf{x}^R$ be the right most point in L such that $\mathbf{x}_2^R = \mathbf{x}_2^0$.
        1. If $f(\mathbf{x}^0) = \mathbf{x}^0$, return($\mathbf{x}^0$); end;
        2. If $f(\mathbf{x}^0) \geq_c \mathbf{x}^0$, $L' = \{\mathbf{x} | \mathbf{x} \geq_c \mathbf{x}^0, \mathbf{x} \in L\}$. Point_check(L', f);
        3. If $f(\mathbf{x}^0) \leq_c \mathbf{x}^0$, $L'' = \{\mathbf{x} | \mathbf{x} \leq_c \mathbf{x}^0, \mathbf{x} \in L\}$. Point_check(L'', f);
        4. If $f(\mathbf{x}^0)_1 < x_1^0$ and $f(\mathbf{x}^0)_2 > x_2^0$, Binary_Search($\mathbf{x}^L, \mathbf{x}^0$);
        5. If $f(\mathbf{x}^0)_1 > x_1^0$ and $f(\mathbf{x}^0)_2 < x_2^0$, Binary_Search($\mathbf{x}^0, \mathbf{x}^R$);
- *Binary_Search(x, y)*
    Let $\mathbf{x}^m = \lfloor 1/2(\mathbf{x} + \mathbf{y}) \rfloor$
        1. If $f(\mathbf{x}^m) = \mathbf{x}^m$, return($\mathbf{x}^m$); end;
        2. If $f(\mathbf{x}^m) \geq_c \mathbf{x}^m$, $L' = \{\mathbf{x} | \mathbf{x} \geq_c \mathbf{x}^m, \mathbf{x} \in L\}$. Point_check(L', f);
        3. If $f(\mathbf{x}^m) \leq_c \mathbf{x}^m$, $L'' = \{\mathbf{x} | \mathbf{x} \leq_c \mathbf{x}^m, \mathbf{x} \in L\}$. Point_check(L'', f);
        4. If $f(\mathbf{x}^m)_1 < x_1^m$ and $f(\mathbf{x}^m)_2 > x_2^m$, Binary_Search($\mathbf{x}, \mathbf{x}^m$);
        5. If $f(\mathbf{x}^m)_1 > x_1^m$ and $f(\mathbf{x}^m)_2 < x_2^m$, Binary_Search($\mathbf{x}^m, \mathbf{y}$);

Theorem 3. *When the order preserving function f is given as an oracle, a Tarski's fixed point can be found in time $O(\log^2 N)$ on a finite 2D lattice formed by integer points of a box with side length N by using Algorithm 1 Point_check.*

Proof. Start from a lattice of size $|L|$, we first prove that in at most $O(\log N)$ steps the above algorithm either finds the fixed point or reduces the input lattice to size $|L|/2$ (Fig. 1).

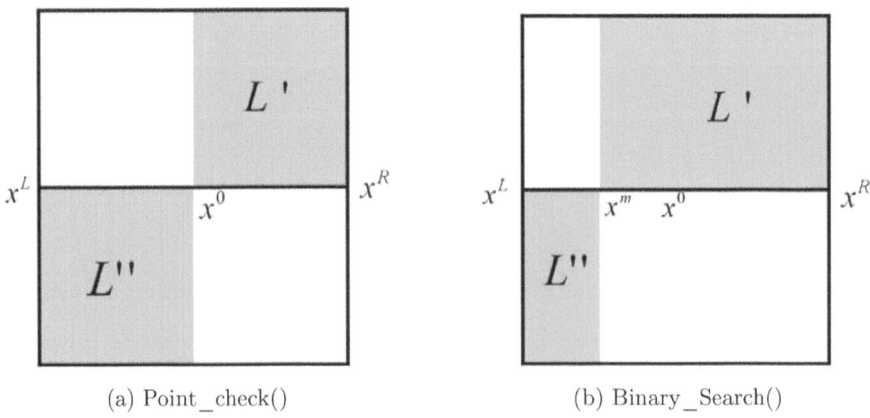

(a) Point_check()   (b) Binary_Search()

Fig. 1. A polynomial algorithm for 2D Lattice

Consider the algorithm Point_check(L, f).

1. Case I: If $f(\mathbf{x}^0) = \mathbf{x}^0$, $\mathbf{x}^0$ is the fixed point which is found in 1 step.
2. Case II: If $f(\mathbf{x}^0) \geq_c \mathbf{x}^0$, since f is a order preserving function, $\forall \mathbf{y} \geq_c \mathbf{x}^0$, we have $f(\mathbf{y}) \geq_c f(\mathbf{x}^0) \geq_c \mathbf{x}^0$. Let $L' = \{\mathbf{x}|\mathbf{x} \geq_c \mathbf{x}^0, \mathbf{x} \in L\}$. Define $f'(\mathbf{x}) = f(\mathbf{x}), \forall \mathbf{x} \in L'$. Then $f' : L' \rightarrow L'$ is a order preserving function on the complete lattice L'. By Tarski's fixed point theorem, there must exist a fixed point in L'. Next we only need to check L' which is only $1/4$ size of $|L|$.
3. Case III: If $f(\mathbf{x}^0) \leq_c \mathbf{x}^0$, similar to the analysis in Case II, we only need to consider $L'' = \{\mathbf{x}|\mathbf{x} \leq_c \mathbf{x}^0, \mathbf{x} \in L\}$ which is only $1/4$ size of $|L|$ in the next step.
4. Case IV: If $f(\mathbf{x}^0)_1 < x_1^0$ and $f(\mathbf{x}^0)_2 > x_2^0$, we prove that Binary_Search($\mathbf{x}^L, \mathbf{x}^0$) finds a fixed point or reduce the size of the lattice by half in $\log \frac{N}{2}$ steps. Since f is a order preserving function, for all adjacent points $\mathbf{u} \leq_c \mathbf{v} \in L$, it is impossible that $f(\mathbf{u})_1 > u_1$ and $f(\mathbf{v})_1 < v_1$. Thus, on a line segment $[\mathbf{x}, \mathbf{y}]$ where $x_2 = y_2$, if $f(\mathbf{x})_1 \geq x_1$ and $f(\mathbf{y})_1 < y_1$, there must exist a point $\mathbf{z}$ such that $f(\mathbf{z})_1 = z_1$. On the other hand, we have $f(\mathbf{x}^0)_1 < x_1^0$ and by the boundary condition $f(\mathbf{x}^L)_1 \geq x_1^L$, therefore, there must exist a point $\mathbf{x}' \in [\mathbf{x}^l, \mathbf{x}^0]$ such that $f(\mathbf{x}')_1 = x_1'$. This point $\mathbf{x}'$ can be found in time $\log \frac{N}{2}$ by using binary search. If $f(\mathbf{x}')_2 > x_2'$, similar to the analysis in Case II, we only need to consider $L' = \{\mathbf{x}|\mathbf{x} \geq_c \mathbf{x}', \mathbf{x} \in L\}$ which is at most $1/2$ size of $|L|$ in the next step. If $f(\mathbf{x}')_2 < x_2'$, we only need to consider $L'' = \{\mathbf{x}|\mathbf{x} \leq_c \mathbf{x}', \mathbf{x} \in L\}$ which is at most $1/4$ size of $|L|$ in the next step. If $f(\mathbf{x}')_2 = x_2'$, then $\mathbf{x}'$ is the fixed point.
5. Case V: If $f(\mathbf{x}^0)_1 > x_1^0$ and $f(\mathbf{x}^0)_2 < x_2^0$, similarly, we can prove that Binary_Search($\mathbf{x}^0, \mathbf{x}^R$) finds a fixed point or reduce the size of the lattice by half in $\log \frac{N}{2}$ steps.

The size of the lattice is reduced by half in every $O(\log N)$ steps. Therefore, the algorithm finds a fixed point in at most $O(\log N \times \log L) = O(\log^2 N)$ steps. $\qquad\square$

The above algorithm can be generalized to any constant dimensional lattice with $L = \{\mathbf{x} \in Z^d \mid \mathbf{a} \leq \mathbf{x} \leq \mathbf{b}\}$, where $\mathbf{a}$ and $\mathbf{b}$ are two finite vectors of Z^d with $\mathbf{a} < \mathbf{b}$. We reduce a $(d+1)$-dimension problem to a d-dimension one. Assume we have an algorithm for a d-dimensional problem with time complexity $O(\log^d |L|)$. Let the algorithm be $A_d(L, f)$.

Consider a $d + 1$-dimensional lattice $(L, \leq_c)$. Choose the central point in L, and denote it by $\mathbf{O} = (O_1, O_2, \cdots, O_{d+1})^T$. Take the section of L by a hyperplane parallel to $x_{d+1} = 0$ passing through $\mathbf{O}$. Denote it as L^d. Clearly, it is a d-dimensional lattice. We define a new oracle function f_d on L^d, based on the oracle function f on L. Define $f_d(x_1, x_2, \cdots, x_d) = (y_1, y_2, \cdots, y_d)$, if $f(x_1, x_2, \cdots, x_d, O_{d+1}) = (y_1, y_2, \cdots, y_d, y_{d+1})$. We apply the algorithm $A_d(L^d, f_d)$ to obtain a Tarski's fixed point in time $(\log |L|)^d$. Let the fixed point be denoted by $\mathbf{x}^*$. Therefore, $f(\mathbf{x}^*) = (\mathbf{x}^*, O_{d+1}) + a\mathbf{e}_{d+1}$ or $f(\mathbf{x}^*) = (\mathbf{x}^*, O_{d+1}) - a\mathbf{e}_{d+1}$, where a is some constant, $\mathbf{e}_{d+1}$ is a $d + 1$ dimensional unit vector with 1 on its $d + 1$th position.

In either case, we obtain a new box $\mathbf{B}$ with size no more than half of the original box defined by $[\mathbf{a}, \mathbf{b}]$, such that $f(\cdot)$ maps all points in $\mathbf{B}$ into $\mathbf{B}$ and is order preserving. We can apply the algorithm recursively on $\mathbf{B}$. The base case can be handled easily. Therefore the total time is

$$T(|L|^{d+1}) \leq T(\frac{|L|^{d+1}}{2}) + O(\log^d |L|).$$

It follows that $T(|L|^{d+1}) = O(\log^d |L|)$.

Formally, the polynomial time algorithm for finding a Tarski's fixed point in a d-dimensional componentwise ordering lattice is described as follows.

Algorithm 2 *Fixed_point() (A polynomial algorithm for any constant dimensional lattice)*

- *Input:*
   A d dimensional lattice L^d, WLOG, $|L^d| = N^d$ (Input size to the oracle is $d \log N$ since the input size for both dimensions to the oracle is $\log N$.). An oracle function f^d. f^d is a order preserving function. $\forall \mathbf{x} \in L^d, f^d(\mathbf{x}) \in L^d$ and $f^d(\mathbf{x}) \leq_c f^d(\mathbf{y})$ if $\mathbf{x} \leq_c \mathbf{y}, \forall \mathbf{x}, \mathbf{y} \in L^d$.
- *Fixed_point(L^d)*
   1. *If $d > 1$*
      (a) *Let $\mathbf{x}^0$ be the center point in L^d.*
      (b) *Let $L^{d-1} = \{\mathbf{x} = (x_1, x_2, \cdots, x_{d-1}) | (\mathbf{x}, x_d^0) \in L^d\}$.*
      (c) *Let $f^{d-1}(\mathbf{x}) = (f^d(\mathbf{x}, x_d^0)_1, f^d(\mathbf{x}, x_d^0)_2, \cdots, f^d(\mathbf{x}, x_d^0)_{d-1})$.*
      (d) *$\mathbf{x}^* =$Fixed_point(L^{d-1}).*
      (e) *If $f^d(\mathbf{x}^*, x_d^0)_d > x_d^0$, $L^d = \{\mathbf{x} | \mathbf{x} \geq (\mathbf{x}^*, x_d^0)\}$; Fixed_point($L^d$);*
      (f) *If $f^d(\mathbf{x}^*, x_d^0)_d < x_d^0$, $L^d = \{\mathbf{x} | \mathbf{x} \leq (\mathbf{x}^*, x_d^0)\}$; Fixed_point($L^d$);*
      (g) *If $f^d(\mathbf{x}^*, x_d^0)_d = x_d^0$, return $(\mathbf{x}^*, x_d^0)$; end;*
   2. *If $d = 1$, let $\mathbf{x}^L$ be the left end point and $\mathbf{x}^R$ be the right end point. binary_search($\mathbf{x}^L, \mathbf{x}^R, f^d$).*
- *binary_search($\mathbf{x}, \mathbf{y}, f$)*
   1. *If $f(\mathbf{x}^L) = 0$, output $\mathbf{x}^L$;*
   2. *else if $f(\mathbf{x}^R) = 0$, output $\mathbf{x}^R$;*
   3. *else*
      (a) *If $f(\lfloor 1/2(\mathbf{x}^L + \mathbf{x}^R) \rfloor) < \lfloor 1/2(\mathbf{x}^L + \mathbf{x}^R) \rfloor$, binary_search($\mathbf{x}^L, \lfloor 1/2(\mathbf{x}^L + \mathbf{x}^R) \rfloor, f$);*
      (b) *If $f(\lfloor 1/2(\mathbf{x}^L + \mathbf{x}^R) \rfloor) > \lfloor 1/2(\mathbf{x}^L + \mathbf{x}^R) \rfloor$, binary_search($\lfloor 1/2(\mathbf{x}^L + \mathbf{x}^R) \rfloor, \mathbf{x}^R, f$);*
      (c) *else output $\mathbf{x}^*$.*

Theorem 4. *When the order preserving function f is given as an oracle, a Tarski's fixed point can be found in time $O(\log^d |L|)$ on a componentwise ordering lattice $(L, \leq_c)$.*

4   Determining Uniqueness Under Oracle Function Model

It has been a natural question to check whether there is another fixed point after finding the first one, such as in the applications for finding all Nash equilibria

(Echenique) [10]. For the componentwise ordering lattice, we derive a $\Theta(N_1 + N_2 + \cdots + N_d)$ matching bounds for determining the uniqueness of the fixed point even for randomized algorithms. The technique builds on and further reveals crucial properties of mathematical structures for fixed points.

Theorem 5. *Given the componentwise lattice $L = N_1 \times N_2 \times \cdots \times N_d$ of d dimensions, an order preserving function f and a fixed point $\mathbf{x}^0$, the deterministic oracle complexity is $\Theta(N_1 + N_2 + \cdots + N_d)$ to decide whether there is a unique fixed point.*

Proof. For dimension $d \geq 2$, let $L = \{\mathbf{x} \in Z : \mathbf{0} \leq_c \mathbf{x} \leq_c (N_1, N_2, \cdots, N_d)\}$. For $\mathbf{x} = (x_1, x_2, \cdots, x_d)$, let $maxindex(\mathbf{x}) = \max\{i : x_i > 0\}$ for any nonzero vector $\mathbf{x}$. Define $auxi(\mathbf{x}) = -\mathbf{e}_{maxindex(\mathbf{x})}$ where $\mathbf{e}_i$ is a unit vector in i-th coordinate. Therefore, $auxi(\cdot)$ is well defined on nonzero vectors in the lattice L. One example of two dimension case is demonstrated in Fig. 2. The fixed point is denoted by the red color. The direction of all the other points are defined by the function $auxi(\cdot)$.

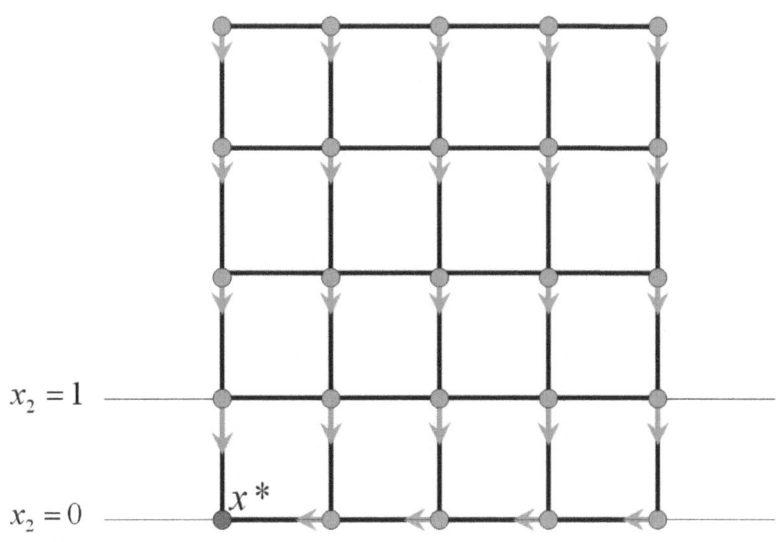

Fig. 2. $auxi(\mathbf{x})$

The adversary will set $g(\mathbf{x}) = f(\mathbf{x}) - \mathbf{x}$ to be $auxi(\mathbf{x})$ except at certain points (to be decided according to the algorithm) where it may hide a zero point.

1. Proof of the Lower Bound:
   First consider $\mathbf{x}$ such that $x_d = 0$. It constitutes a solution of $d-1$ dimension. By the inductive hypothesis, it requires time $N_1 + N_2 + \cdots + N_{d-1}$ to decide whether or not there is one zero point at $x_d = 0$.

Second, when there is no such zero point, we need to decide if there is a zero point at $\mathbf{x}$ with $x_d > 0$. Fixing any $i > 0$, we will set, for all $\mathbf{x}$ with $x_d = i$, $g(\mathbf{x}) = 0$ whenever none of such $\mathbf{x}$ is queried, and set $g(\mathbf{x}) = -e_d$ otherwise. This will take N_d queries.

One may note that the adversary always answers a non-zero value. In fact, for any pair $i = maxindex(\mathbf{x})$ and $j = x_i$ not query, the adversary can make $g(\mathbf{x}) = 0$ without violating the order preserving property.

2. Proof of the Upper Bound:

We design an algorithm which always queries the componentwise maximum point of the lattice $\mathbf{x}^{max} = (N_1, N_2, \cdots, N_d)$. We should have $g(\mathbf{x}^{max}) \leq_c \mathbf{0}$. We are done if it is zero. Otherwise, there must exist some i, such that $g(\mathbf{x}^{max})_i < 0$. The problem is reduced to a smaller lattice $L' = \{\mathbf{x} \in Z : \mathbf{0} \leq_c \mathbf{x} \leq_c (N_1, N_2, \cdots, N_{i-1}, N_i - 1, N_{i+1}, \cdots, N_d)\}$ which has a total sum of side lengths at most $N_1 + N_2 + \cdots + N_d - 1$. The claim follows.

$\square$

5   Determining Uniqueness Under Polynomial Function Model

In this section, we consider the dimension as a part of the input size in unary and develop a hardness proof for the polynomial function model for determining the uniqueness of a given fixed point. We start with a polynomial-time reduction from the 3-SAT problem which is NP-complete to one of finding a second Tarski's fixed point, by deriving an order preserving mapping f from a componentwise ordering lattice L into itself, with a given fixed point. Therefore, given f as a polynomial time function with a known fixed point, determining whether f has another fixed point in L is NP-hard. In other words, determining the uniqueness of a Tarski's fixed point is co-NP-hard.

Furthermore, even for the case when the dimension is one, the uniqueness problem is still co-NP-hard. This can be proved by designing a polynomial-time reduction from the 3-SAT problem to the uniqueness of Tarski's fixed point in a lexicographic lattice. As the lexicographic order defines a total order, it can be reduced to a one dimensional problem by finding a polynomial time algorithm for the order function calculation. It then follows that determining the uniqueness of Tarski's fixed point in a lexicographic lattice is Co-NP hard though there exists a polynomial-time algorithm for finding one Tarski's fixed point in a lexicographic lattice in any dimension.

We start with one of the NP-complete problems, 3-SAT, defined as follows.

Definition 6 (*3CNF-formula*). *A literal is a boolean variable. A clause is several literals connected with $\vee$'s. A boolean formula is in conjuctive normal form (CNF) if it is made of clauses connected with $\wedge$'s. If every clause has exactly 3 literals, the CNF-formula is called 3CNF-formula.*

Definition 7 *(3-SAT Problem).*

Input: n boolean variables $x_1, x_2, \cdots, x_n$ m clauses $C_1, C_2, \cdots, C_m$, each consisting of three literals from the set $\{x_1, \bar{x}_1, x_2, \bar{x}_2, \cdots, x_n, \bar{x}_n\}$.

Output: An assignment of $\{0, 1\}$ to the boolean variables $x_1, x_2, \cdots, x_n$, such that the 3CNF-formula $F : C_1 \wedge C_2 \cdots \wedge C_m = true$, i.e., there is at least one true literal for every clause.

Theorem 6 *[13]. 3-SAT is NP-complete*

For both lexicographic ordering and componentwise ordering, the Co-NP-hardness results can be derived from a reduction from 3-SAT problem.

Corollary 2. *Given lattice $(L, \leq_l)$ and an order preserving mapping f as a polynomial function, determining that f has a unique fixed point in L is a Co-NP hard problem.*

Corollary 3. *Given lattice $(L, \leq_c)$ and an order preserving mapping f as a polynomial function, determining that f has a unique fixed point in L is a Co-NP hard problem.*

6 Finding Equilibria in Supermodular Games

In previous sections, we solved the computational problems of the Tarski's fixed point. We are still interested in how to find a pure Nash equilibrium and how to determine the uniqueness of the pure Nash equilibrium in a supermodular game. Because of the strong connection between the equilibrium of supermodular games and the Tarski's fixed point, the question is whether the previous results hold for supermodular games.

In this section, we develop a polynomial time algorithm to find a pure Nash equilibirum which is more efficient than the algorithm we design for Tarski's fixed point before. But the Co-NP-hardness result still holds for supermodular games.

6.1 Supermodular Games and Tarski's Fixed Points

We will start with the formal definition of supermodular games.

Definition 8 *(Supermodular Games). Let $\Gamma = \{(S_i, u_i) : i = 1, \cdots, d\}$ be a finite supermodular game with d players:*

- *S_i is a finite subset of $\mathbf{R}$;*
- *u_i has increasing differences in (s_i, s_{-i}), where s_{-i} is the strategy set of all other players except player i. I.e.,*

$$u(s_i', s_{-i}') - u(s_i, s_{-i}') \geq u(s_i', s_{-i}) - u(s_i, s_{-i}), \forall s_i' \geq s_i, s_{-i}' \geq s_{-i}$$

In the following discussion, without loss of generality, we assume $S_i = \{0, 1, \cdots, N_i - 1\}$. The model can be viewed as a discretized version of a game with continuous strategy spaces, where each S_i is an interval. Then $S = \times_{i=1}^{d} S_i$ is a componentwise lattice.

Let B_i denote i's best-response function in Γ.

$$B_i(s_{-i}) := arg \max_{s_i \in S_i} u_i(s_i, s_{-i}).$$

Denote $\overline{B}_i(s_{-i})$ as the greatest element and $\underline{B}_i(s_{-i})$ as the least element in $B_i(s_{-i})$.

In supermodular games, by Topkis' theorem [21], we have,

$$\overline{B}_i(s_{-i}) \geq \overline{B}_i(s'_{-i}) \text{and} \underline{B}_i(s_{-i}) \geq \underline{B}_i(s'_{-i}) \text{if} s_{-i} \geq s'_{-i}.$$

Let $\underline{B}(s) = \{\underline{B}_i(s_{-i}) : i = 1, \cdots, d\}$ be the least best-response function of the game, then $\underline{B} : S \to S$ is order preserving.

Tarski's fixed point theorem guarantees the existence of fixed points of any order preserving function $f : L \to L$ on any nonempty complete lattice. A supermodular game $(S, \preceq)$ is a complete lattice and the least best-response function $\underline{B}$ is order-preserving from S to itself. Therefore, there exists an equilibrium point $x^* \in S$ such that $\underline{B}(x^*) = x^*$.

6.2   Equilibrium Computation in Supermodular Games

Recall $\underline{B}(s) = \{\underline{B}_i(s_{-i}) : i = 1, \cdots, d\}$, where $s \in S$ and $S = \times_{i=1}^{d} S_i$ is the best-response function of the supermodular game. We assume the strategy set for each player i is $S_i = \{0, 1, \cdots, N_i - 1\}$. In Tarski's fixed point theorem, the only requirement for function f is order-preserving. In a supermodular game, $\underline{B}$ is not only order-preserving but also needs to be consistent, since for the same s_{-i}, the value of $\underline{B}_i(s_{-i})$ should be the same. Therefore. if there exists an algorithm A that finds a Tarski's fixed point for any componentwise lattice with order-preserving function f in time T, A can find an equilibrium in supermodular game in time T. However, not vice versa.

Without loss of generality, we assume $N_1 \leq N_2 \leq \cdots \leq N_d$.

Theorem 7. *When the best response function $\underline{B}$ is given as an oracle, a pure Nash equilibrium can be found in time $O(\log N_1 \log N_2 \cdots \log N_{d-1})$ in a supermodular game Γ.*

Proof. The algorithm is similar to the proof of Theorem 4 for finding one Tarski's fixed point. The only difference here is for the 2D case. On a $N_1 \times N_2$ box, we start with the node $(\lfloor \frac{N_1}{2} \rfloor, y)$, where $1 \leq y \leq N_2$ can be any integer. We query the value of $\underline{B}(\lfloor \frac{N_1}{2} \rfloor, y)$. Assume the value is (x', y'). Next we query the value of $\underline{B}(\lfloor \frac{N_1}{2} \rfloor, y')$. Because in the previous query we have already known that $\underline{B}_2(\lfloor \frac{N_1}{2} \rfloor) = y'$, we must have $\underline{B}(\lfloor \frac{N_1}{2} \rfloor, y') = (x'', y')$.

1. If $x'' = \lfloor \frac{N_1}{2} \rfloor$, $(\lfloor \frac{N_1}{2} \rfloor, y')$ is a pure Nash equilibrium.
2. If $x'' < \lfloor \frac{N_1}{2} \rfloor$, all nodes smaller than (x'', y') form a complete lattice and the size is less than half of the original lattice.
3. If $x'' > \lfloor \frac{N_1}{2} \rfloor$, all nodes greater than (x'', y') form a complete lattice and the size is also less than half of the original lattice.

Therefore, by using the property of the best response function, a pure Nash can be found in time $2 \log N_1$ for 2D case. Recall that finding a Tarski's fixed point takes time $O(\log N_1 \log N_2)$ for 2D case. The generalization for the higher dimensional cases is similar to what we do for finding one Tarski's fixed point. Thus we can find a pure Nash in time $O(\log N_1 \log N_2 \cdots \log N_{d-1})$. $\square$

Again, by a reduction from the 3-SAT problem, we obtain

Theorem 8. *Given a d-player supermodular game Γ with strategy set $S = \times_{i=1}^{d} S_i : \forall i, S_i = \{-1, 0, \cdots, 2^n - 1\}$ and a best response function B as a polynomial function determining that Γ has a unique Nash equilibrium is a Co-NP hard problem.*

7   Conclusion and Open Problems

Results on the Tarski's fixed points contrast with past results for the general fixed point computation in several ways. First in the oracle function model, several fixed point computational problems are known to require an exponential number of queries for constant dimensions, including the two dimensional case (Chen and Deng; Hirsch et al.) [8,14]. Our results prove the problem of computing a Tarski's fixed point to be polynomial in the oracle model. It also follows that it is so for the polynomial function model, which is also different for those fixed point computational problems which are known to be PPAD-complete for constant dimensions, including the two dimensional case (Chen and Deng) [7].

In the polynomial function model, we prove that determining the uniqueness is co-NP-complete. In comparison, the uniqueness of Nash equilibrium is known to be co-NP-complete but its existence is in PPAD.

The above comparisons with previous work leave the following outstanding open problem: Is it PPAD-complete to find a Tarski's fixed point in the variable dimension n for the polynomial function model? This problem is known to be true for finding a Sperner simplex in dimension n when n is a variable. We conjecture that this is also true for finding a Tarski's fixed point.

References

1. Bernstein, F., Federgruen, A.: Decentralized supply chains with competing retailers under demand uncertainty. Manag. Sci. **51**, 18–29 (2005)
2. Bernstein, F., Federgruen, A.: A general equilibrium model for industries with price and service competition. Oper. Res. **52**, 868–886 (2004)

3. Codenotti, B., Saberi, A., Varadarajan, K.R., Ye, Y.: The complexity of equilibria: hardness results for economies via a correspondence with games. Theor. Comput. Sci. **408**(2–3), 188–198 (2008)
4. Cachon, G.P.: Stcok wars: inventory competition in a two echelon supply chain. Oper. Res. **49**, 658–674 (2001)
5. Cachon, G.P., Lariviere, M.A.: Capacity choice and allocation: strategic behavior and supply chain performance. Manag. Sci. **45**, 1091–1108 (1999)
6. Chang, C.L., Lyuu, Y.D., Ti, Y.W.: The complexity of Tarski's fixed point theorem. Theor. Comput. Sci. **401**, 228–235 (2008)
7. Chen, X., Deng, X.: On the complexity of 2D discrete fixed point problem. Theor. Comput. Sci. **410**(44), 4448–4456 (2009)
8. Chen, X., Deng, X.: Matching algorithmic bounds for finding a Brouwer fixed point. J. ACM **55**(3), 13:1–13:26 (2008)
9. Chen, X., Deng, X., Teng, S.: Settling the computational complexity of two player Nash equilibrium. J. ACM (JACM) **56**(3) (2008)
10. Echenique, F.: Finding all equilibria in games of strategic complements. J. Econ. Theory **135**, 514–532 (2007)
11. Fudenberg, D., Tirole, J.: Game Theory. MIT Press, Cambridge (1991)
12. Gilboa, I., Zemmel, E.: Nash and correlated equilibria: some complexity considerations. Games Econ. Behav. **1**, 80–93 (1989)
13. Karp, R.M.: Reducibility Among Combinatorial Problems. In: Miller, R.E., Thatcher, J.W., Bohlinger, J.D. (eds.) Complexity of Computer Computations, pp. 85–103. Plenum, New York (1972)
14. Hirsch, M.D., Papadimitriou, C.H., Vavasis, S.: Exponential lower bounds for finding brouwer fixed points. J. Complex. **5**, 379–416 (1989)
15. Lippman, S.A., McCardle, K.F.: The competitive newsboy. Oper. Res. **45**, 54–65 (1997)
16. Milgrom, P., Roberts, J.: Rationalizability, learning, and equilibrium in games with strategic complementarities. Econometrica **58**, 155–1277 (1990)
17. Milgrom, P., Roberts, J.: Comparing equilibria. Am. Econ. Rev. **84**, 441–459 (1994)
18. Milgrom, P., Shannon, C.: Monotone comparative statics. Econometrica **62**, 157–180 (1994)
19. Robinson, J.: An iterative method of solving a game. Ann. Math. **54**(2), 296–301 (1951)
20. Tarski, A.: A lattice-theoretical fixpoint theorem and its applications. Pac. J. Math. **5**, 285–308 (1955)
21. Topkis, D.M.: Ordered Optimal Decisions. Ph.D. Dissertation, Stanford University (1968)
22. Topkis, D.M.: Equilibrium points in nonzero-sum n-person submodular games. SIAM J. Control Optim. **17**, 773–787 (1979)
23. Topkis, D.M.: Supermodularity and Complementarity. Princeton University Press, Princeton (1998)
24. Vives, X.: Nash equilibrium with strategic complementarities. J. Math. Econ. **19**, 305–321 (1990)
25. Vives, X.: Complemetarities and games: new developments. J. Econ. Lit. **XLIII**, 437–479 (2005)

Parity-Constrained k-Supplier Problem

Xinlan Xia[1], Lu Han[1(✉)], and Lili Mei[2]

[1] School of Science, Beijing University of Posts and Telecommunications, Beijing,
People's Republic of China
{xiaxinlan,hl}@bupt.edu.cn
[2] School of Cyberspace, Hangzhou Dianzi University, Hangzhou,
People's Republic of China
meilili@zju.edu.cn

Abstract. In this paper, we propose and study the parity-constrained k-supplier (PAR k-supplier) problem, generalizing the classical (unconstrained) k-supplier problem. In the PAR k-supplier problem, we are given a set of facilities and a set of clients in a metric space with distances. An integer k is also given. Each facility is associated with an odd or even parity requirement. The goal is to open at most k facilities and assign each client to an opened facility, such that the number of clients assigned to a facility meets its parity requirement, and the maximum distance of any client to its assigned facility is minimized.

As our main contribution, we provide a constant-factor approximation algorithm for the PAR k-supplier problem with a ratio of 9. Our algorithm proceeds by first ignoring all parity requirements and constructing an (unconstrained) k-supplier instance. Next, we design an algorithm to solve the constructed instance and obtain a solution that may have some invalid facilities. An invalid facility is an opened facility whose parity requirement is violated. Last, to obtain a feasible solution, we match up all the invalid facilities and make reassignments according to each invalid pair.

Keywords: approximation algorithms · parity constraints · k-supplier

1 Introduction

The k-supplier problem is one of the most well-known NP-hard clustering problems in the literature of operations research and combinatorial optimization [3,8,10,14–16,20]. In this problem, we are given a set of facilities and a set of clients in a metric space with distances, along with an integer k. The objective is to open at most k facilities and assign each client to an opened facility, such that the maximum distance between a client and its assigned facility is minimized. Many researchers focus on developing approximation algorithms for

Lu Han is supported by Beijing Municipal Natural Science Foundation (No. Z220004), National Natural Science Foundation of China (No. 12371321). Lili Mei is supported by National Natural Science Foundation of China (No. 12201594).

the k-supplier problem and its generalizations. For a minimization problem, an α-approximation algorithm is a polynomial time algorithm that for any instance of the problem can output a solution whose objective value is within a factor of α of the value of an optimal solution. Note that when the set of facilities and the set of clients are equal, the k-supplier problem simplifies to the classical k-center problem. Gonzalez [5] and Hochbaum and Shmoys [9,10] proposed the first 2-approximation algorithm for the k-center problem and 3-approximation algorithm for the k-supplier problem, respectively. Hochbaum and Shmoys [10] also demonstrated that there exist no approximation algorithms for the k-center problem and k-supplier problem with ratios strictly better than 2 and 3, unless P = NP.

A great deal of the generalizations of the k-supplier problem are stimulated by the corresponding version of the k-center problem. Recently, for the k-center problem, Kim et al. [11] proposed a meaningful generalization called the parity-constrained k-center (PAR k-center) problem and designed its 6-approximation algorithm. Parity has applications in the fields of matroid optimization [19], graph theory [4], submodular optimization [6,7], quantum optimization [12], etc. Compared with the k-center problem, in the PAR k-center problem, each facility has an odd or even parity requirement. The goal is to minimize the maximum distance of any client to its assigned facility, guaranteeing that the number of clients assigned to a facility meets its parity requirement. Parity is common in many activities, for example, we need an odd number of people when voting, and we need an even number of people when organizing a competitive game. The work of Kim et al. [11] encourages us to study the parity-constrained k-supplier (PAR k-supplier) problem. Besides the PAR k-supplier problem, other generalizations of the k-supplier problem have also been studied, including the k-supplier problem with outliers [1], the priority k-supplier problem [18], etc. [2, 13,17].

Our Contribution. As our main contribution, we propose the PAR k-supplier problem and design a constant-factor approximation algorithm with a ratio of 9. Our algorithm is inspired by the previous works of Kim et al. [11] on the PAR k-center problem. For any instance of the PAR k-supplier problem, we first get rid of the parity requirements of all the given facilities to construct an instance of the (unconstrained) k-supplier problem. Then, by designing an algorithm for the k-supplier problem, we solve the constructed instance and obtain a solution. Note that in the obtained solution, there could exist some opened facilities whose parity requirements are not satisfied and we call these facilities invalid facilities. It can be demonstrated that the number of invalid facilities in the solution is even, if not, we can fix this situation via a careful modification of the solution. Therefore, it is possible to match all these invalid facilities in pairs. For each pair of invalid facilities, we attempt to find a path between them whose internal points are all opened facilities. Finally, to obtain a feasible solution, we reassign clients along the path so that the number of clients connected to the two invalid facilities, as endpoints of the paths, can be successfully reversed in parity.

Organization of Our Paper. The remainder of the paper is structured as follows. Section 2 gives the formal descriptions of the PAR k-supplier problem. Section 3 presents our 9-approximation algorithm along with the analysis.

2 Preliminaries

Here is a formal description of the PAR k-supplier problem. In a PAR k-supplier instance $\mathcal{I}$, we are given a set of facilities $\mathcal{F}$ and a set of clients $\mathcal{D}$ in a metric space a metric space (V, d) with a distance function $d : V \times V \to \mathbb{R}_{\geq 0}$, where $V = \mathcal{F} \cup \mathcal{D}$. Assume that the distances are metric, that means the following three assumptions hold for the distances:

- Non-negative: $\forall i, j \in V$, $d(i, j) \geq 0$.
- Symmetric: $\forall i, j \in V$, $d(i, j) = d(j, i)$.
- Satisfy the triangle inequality: $\forall i, j, k \in V$, $d(i, j) + d(j, k) \geq d(i, k)$.

An integer k is also given. Each facility is associated with an odd or even parity requirement represented by a parity constraint function $\pi : \mathcal{F} \to \{\text{odd}, \text{even}\}$. The goal is to open at most k facilities and assign each client to an opened facility, such that the number of clients assigned to a facility meets its parity requirement, and the maximum distance of any client to its assigned facility is minimized.

Denote by (S, σ) a solution of the PAR k-supplier instance $\mathcal{I}$, where $S \subseteq \mathcal{F}$ is the set of opened facilities, and $\sigma : \mathcal{D} \to S$ is a mapping from each client in $\mathcal{D}$ to its assigned facility in S. For a solution (S, σ), if $|S| \leq k$ and each facility $i \in S$ satisfies that $|\sigma^{-1}(i)| = \pi(i)$, we say that the solution is a feasible solution. Note that for any PAR k-supplier and k-supplier instances, if their inputs of $\mathcal{F}$, $\mathcal{D}$, and k are the same, it is apparent that any feasible solution to the PAR k-supplier instance is also a feasible solution to the k-supplier instance.

Denote by (S^*, σ^*) an optimal solution of the PAR k-supplier instance $\mathcal{I}$, and let τ^* be the objective value of the optimal solution, i.e.,

$$\tau^* = \max_{j \in \mathcal{D}} d(j, \sigma^*(j)).$$

It is worth noting that there are polynomial numbers of possible distances between clients and facilities, and the value of the optimal solution is one of the distances. Therefore, when designing an algorithm for the PAR k-supplier problem, we could suppose that τ^* is known.

For complete graph $G = (V, E)$ with distances of all the vertex pairs (i.e., the distances of all edges) and a parameter $\tau \in \mathbb{R}_{\geq 0}$, denote by $G_{\leq \tau} = (V', E')$ a subgraph of G, in which $V' = V$ and for each $u, v \in V$, we have that $(u, v) \in E'$ if and only if $d(u, v) \leq \tau$ under the graph G. For a non-negative integer L, define $[L] = \{1, 2, ..., L\}$.

3   Approximation Algorithm and Analysis

In this section, we present the first constant-factor approximation algorithm for the PAR k-supplier problem. Note that in the PAR k-supplier problem, two kinds of constraints need to be satisfied, the first kind is the cardinality constraint (i.e., the number of opened facilities is no more than k), and the second kind is the parity constraints (the number of clients assigned to each opened facility meets its parity). The main idea of our algorithm is very natural, we first try to deal with the cardinality constraint, and then deal with the parity constraints. Therefore, our algorithm includes two major phases, named the cardinality-constrained satisfied phase and the parity-constrained satisfied phase. Due to space limitations, the proofs of all the essential lemmas are removed but will appear in a full version of this paper.

3.1   The Cardinality-Constrained Satisfied Phase

In the cardinality-constrained satisfied phase, we first get rid of all the parity requirements of the given PAR k-supplier instance $\mathcal{I} = (\mathcal{F}, \mathcal{D}, k, d, \pi)$ to obtain an (unconstrained) k-supplier instance $\mathcal{I}_{\mathrm{cs}} = (\mathcal{F}, \mathcal{D}, k, d)$. Then, we design an algorithm to solve $\mathcal{I}_{\mathrm{cs}}$ and obtain a solution satisfying the cardinality constraint. Here is a brief description of how to solve $\mathcal{I}_{\mathrm{cs}}$. We focus on each facility-client pair whose distance is no more than the optimal value of τ^* of $\mathcal{I}$. Mainly by connecting edges between these pairs, we construct a graph G. Note that graph G may have several connected components. For each connected component, we select some facilities, that have reasonable distances from each other, for opening. Combining all the opened facilities in each connected component finally yields a solution satisfying the cardinality constraint. The formal description of this phase is given in the Algorithm 1.

Note that for any given PAR k-supplier instance, before we run the Algorithm 1 to obtain a solution satisfying the cardinality constraint, it is important to first check whether a feasible solution exists for the instance. More specifically, if the total number of given clients is odd and the parity requirement of each facility is even, it is not hard to see that there exists no feasible solution. It is also worth noting that based on the facilities chosen in each connected component, we construct a group of trees in the Algorithm 1, and these trees will play a crucial role in dealing with the parity constraints.

Algorithm 1 provides a solution $(S_{\mathrm{cs}}, \sigma_{\mathrm{cs}})$, where S_{cs} is the set of opened facilities, and σ_{cs} is a mapping from each client to some facility in S_{cs}. The following theorem shows that the solution $(S_{\mathrm{cs}}, \sigma_{\mathrm{cs}})$ satisfies the cardinality constraint and its objective value has an upper bound related to the optimal value of τ^* of $\mathcal{I}$.

Lemma 1. *For the constructed k-supplier instance $\mathcal{I}_{\mathrm{cs}} = (\mathcal{F}, \mathcal{D}, k, d)$, the Algorithm 1 outputs a feasible solution $(S_{\mathrm{cs}}, \sigma_{\mathrm{cs}})$, i.e., we have that*

$$|S_{\mathrm{cs}}| \leq k,$$

Algorithm 1: The cardinality-constrained satisfied phase.

Input: A PAR k-supplier instance $\mathcal{I}$ and its optimal value of τ^*.

Output: A solution $(S_{\mathrm{cs}}, \sigma_{\mathrm{cs}})$ and a group of solutions and trees.

Step 1 Construct several connected components.

For the given PAR k-supplier instance $\mathcal{I} = (\mathcal{F}, \mathcal{D}, k, d, \pi)$, we get rid of the parity requirements of all the facilities to obtain an (unconstrained) k-supplier instance $\mathcal{I}_{\mathrm{cs}} = (\mathcal{F}, \mathcal{D}, k, d)$. Construct a graph $G = (\hat{\mathcal{F}} \cup \mathcal{D}, E)$, where

$$\hat{\mathcal{F}} := \{ i \in \mathcal{F} : \text{there exists } j \in \mathcal{D} \text{ such that } d(i, j) \leq \tau^* \},$$

$$E := \{(i, j) : \text{for } i \in \mathcal{F} \text{ and } j \in \mathcal{D}, \text{ we have that } d(i, j) \leq \tau^* \}.$$

Note that the graph $G = (\hat{\mathcal{F}} \cup \mathcal{D}, E)$ may consist of several connected components. Denote by L the total number of connected components, and $G_l = (\hat{\mathcal{F}}_l \cup \mathcal{D}_l, E_l)$ the lth components, where $l \in [L]$, and $\hat{\mathcal{F}}_l$ and $\mathcal{D}_l$ are the sets of facilities and clients in this component, and $\hat{\mathcal{F}}_l \cup \mathcal{D}_l$ and E_l are its vertex set and edge set. For any two vertices i, j in some connected component G_l, define $d_{G_l}(i, j)$ as the number of edges used on the shortest path from i to j in the component. For any vertex i and a vertex set S_l in connected component G_l, define $d_{G_l}(i, S_l)$ as the least number of edges used on the path from i to some vertex in S_l.

Step 2 For each $G_l = (\hat{\mathcal{F}}_l \cup \mathcal{D}_l, E_l)$, we do the following process.

Step 2.1 Initialization

Set $S_l := \emptyset$, and set $\sigma_l(j) := \emptyset$ for each $j \in \mathcal{D}_l$. Define $\mathcal{F}_l^{\mathrm{can}} := \hat{\mathcal{F}}_l$, and $\mathcal{D}_l^{\mathrm{unc}} := \mathcal{D}_l$. Let T_l be a tree in the connected component of G_l, and set its vertex set as $V(T_l) := \emptyset$ and its edge set as $E(T_l) := \emptyset$. For each facility $i \in \hat{\mathcal{F}}_l$, define

$$\mathcal{D}_l^{\mathrm{nea}}(i) := \{ j \in \mathcal{D}_l^{\mathrm{unc}} : d_{G_l}(i, j) \leq 3 \},$$

$$\mathcal{F}_l^{\mathrm{nea}}(i) := \{ i' \in \mathcal{F}_l^{\mathrm{can}} : d_{G_l}(i, i') \leq 2 \}.$$

Step 2.2 Choose the first opened facility

Arbitrarily choose a facility $i \in \mathcal{F}_l^{\mathrm{can}}$, and update $S_l := S_l \cup \{i\}$ and $V(T_l) := V(T_l) \cup \{i\}$. For each client $j \in \mathcal{D}_l^{\mathrm{nea}}(i)$, assign it to facility i, and update $\sigma_l(j) := i$. Update $\mathcal{F}_l^{\mathrm{can}} := \mathcal{F}_l^{\mathrm{can}} \backslash \mathcal{F}_l^{\mathrm{nea}}(i)$, and $\mathcal{D}_l^{\mathrm{unc}} := \mathcal{D}_l^{\mathrm{unc}} \backslash \mathcal{D}_l^{\mathrm{nea}}(i)$.

Step 2.3 Choose the remaining opened facilities

while $\mathcal{D}_l^{\mathrm{unc}} \neq \emptyset$ **do**

Arbitrarily choose a facility i from $\mathcal{F}_l^{\mathrm{can}}$ satisfying $d_{G_l}(i, S_l) = 4$ and is within a distance τ^* to some unassigned client. Update $S_l := S_l \cup \{i\}$, $V(T_l) := V(T_l) \cup \{i\}$. Add i to tree T_l by letting it become a child of a node i' in T_l such that $d_{G_l}(i, i') = 4$. Update $E(T_l) := E(T_l) \cup \{(i, i')\}$. For each client $j \in \mathcal{D}_l^{\mathrm{nea}}(i)$, assign it to facility i, and update $\sigma_l(j) := i$. Update $\mathcal{F}_l^{\mathrm{can}} := \mathcal{F}_l^{\mathrm{can}} \backslash \mathcal{F}_l^{\mathrm{nea}}(i)$ and $\mathcal{D}_l^{\mathrm{unc}} := \mathcal{D}_l^{\mathrm{unc}} \backslash \mathcal{D}_l^{\mathrm{nea}}(i)$.

Step 3 Construct a solution satisfying the candinality constraint.

Set $S_{\mathrm{cs}} := \bigcup_{l \in [L]} S_l$ and $\sigma_{\mathrm{cs}}(j) := \sigma_l(j)$ for any $j \in \mathcal{D}$, where $l \in [L]$.

Output the solution $(S_{\mathrm{cs}}, \sigma_{\mathrm{cs}})$, and a group of solutions (S_l, σ_l) and trees T_l, where $l \in [L]$.

---

and the objective value of the solution (S_{cs}, σ_{cs}) *can be bounded by a factor of the optimal value of* τ^* *of the given PAR k-supplier instance* $\mathcal{I}$, *which is,*

$$\max_{j \in \mathcal{D}} d(j, \sigma_{cs}(j)) \leq 3\tau^*.$$

3.2   The Parity-Constrained Satisfied Phase

In the parity-constrained satisfied phase, we modify the solution of (S_{cs}, σ_{cs}) to obtain a solution (S, σ) that also satisfies the parity constraint. The modification process focuses on each solution of (S_l, σ_l) obtained from the Algorithm 1, where $l \in [L]$, separately and splits into two cases, a simple case when $|S_l| = 1$, and a more complicated case when $|S_l| \geq 2$. The main structure of the parity-constrained satisfied phase is given in the Algorithm 2, in which we will call two sub-algorithms, namely the Algorithm 3 and Algorithm 4, to modify the simple and complicated cases respectively.

Algorithm 2: The parity-constrained satisfied phase.

Input: The group of solutions (S_l, σ_l), where $l \in [L]$.
Output: A solution (S, σ) for the PAR k-supplier instance $\mathcal{I}$.
Step 1 Modify each solution (S_l, σ_l), **where** $l \in [L]$.
    for each solution (S_l, σ_l) with $l \in [l]$ **do**
        if $|S_l| = 1$ **then**
            Call Algorithm 3 to obtain a modified solution (S_l', σ_l').
        else $|S_l| \geq 2$
            Call Algorithm 4 to obtain a modified solution (S_l', σ_l').
Step 2 Construct a solution satisfying the parity constraint.
    Set $S := \bigcup_{l \in [L]} S_l'$ and $\sigma(j) := \sigma_l'(j)$ for any $j \in \mathcal{D}$, where $l \in [L]$.
    Output the solution (S, σ).

Here is the formal description of the Algorithm 3, which deals with each constructed connected component $G_l = (\hat{\mathcal{F}}_l \cup \mathcal{D}_l, E_l)$, where $l \in [L]$, with a solution (S_l, σ_l) satisfying $|S_l| = 1$.

Lemma 2. *For each connected component* $G_l = (\hat{\mathcal{F}}_l \cup \mathcal{D}_l, E_l)$, *when its related solution* (S_l, σ_l) *has that* $|S_l| = 1$, *the Algorithm 3 outputs a modified solution* (S_l', σ_l') *satisfying the parity constraint of each facility* $i \in S_l'$, *i.e., we have that*

$$|\{j \in \mathcal{D}_l : \sigma_l'(j) = i\}| = \pi(i),$$

for each $i \in S_l'$, *and the distance of each client* $j \in \mathcal{D}_l$ *can be bounded by a factor of the optimal value of* τ^* *of the given PAR k-supplier instance* $\mathcal{I}$, *which is,*

$$d(j, \sigma_l'(j)) \leq 7\tau^*.$$

Algorithm 3: Modification of the simple case.

Input: A connected component $G_l = (\hat{\mathcal{F}}_l \cup \mathcal{D}_l, E_l)$, the solution (S_l, σ_l) on G_l, the parity requirement $\pi(i)$ for each facility $i \in \hat{\mathcal{F}}_l$.

Output: A solution (S'_l, σ'_l).

if there exists a facility $i \in \hat{\mathcal{F}}_l$ such that $\pi(i)$ and $|\mathcal{D}_l|$ have the same parity **then**

   Set $S'_l := \{i\}$ and $\sigma'_l(j) := i$ for every $j \in \mathcal{D}_l$.

else

   Arbitrarily choose a facility i from $\hat{\mathcal{F}}_l \backslash S_l$, and update $S'_l := S_l \cup \{i\}$. Let i be assigned by one of the clients j adjacent to i in G_l and update $\sigma'_l(j) := i$. For each client $j' \in \mathcal{D}_l \backslash \{j\}$, update $\sigma'_l(j') := \sigma_l(j')$.

Output the solution (S'_l, σ'_l).

Here is the formal description of the Algorithm 4, which deals with each constructed connected component $G_l = (\hat{\mathcal{F}}_l \cup \mathcal{D}_l, E_l)$, where $l \in [L]$, with a solution (S_l, σ_l) satisfying $|S_l| \geq 2$. For a better understanding of Step 2 of the Algorithm 4, Fig. 1 shows a simple example of the reassigning paths and exposed paths.

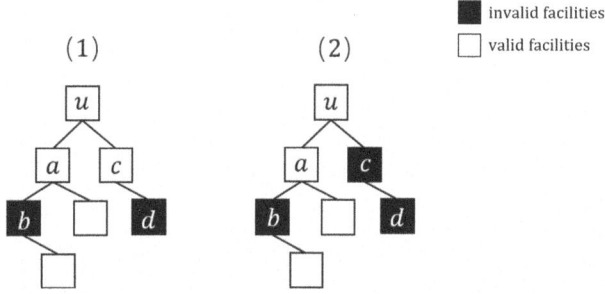

Fig. 1. Two examples of searching for node-disjoint reassigning paths within the subtree $T_l(u)$. On the left, there are exposed paths $p_1 = (b, a)$, $p_2 = (d, c)$; pairing them up forms the node-disjoint reassigning path $p = (b, a, c, d)$. On the right, $p = (b, a)$ is the remaining exposed path. Connecting it to the root u forms a new exposed path $p' = (b, a, u)$.

The feasibility of the Algorithm 4 can be seen from the following lemma.

Lemma 3. *For each connected component $G_l = (\hat{\mathcal{F}}_l \cup \mathcal{D}_l, E_l)$, when its related solution (S_l, σ_l) has that $|S_l| \geq 2$, the Algorithm 4 focuses on three cases in Step 1. For Cases 2 and 3, by opening and closing some facilities, we could guarantee that at the end of Step 1, the number of facilities of S_l^{inv} is even.*

Lemma 4. *For each connected component $G_l = (\hat{\mathcal{F}}_l \cup \mathcal{D}_l, E_l)$, when its related solution (S_l, σ_l) has that $|S_l| \geq 2$, the Algorithm 4 outputs a modified solution*

Algorithm 4: Modification of the complicated case.

Input: A connected component $G_l = (\hat{\mathcal{F}}_l \cup \mathcal{D}_l, E_l)$, the solution (S_l, σ_l) on G_l, and tree $\mathcal{T}_l$, where $l \in [L]$, the parity requirement $\pi(i)$ for each facility $i \in \hat{\mathcal{F}}_l$.

Output: A solution (S_l', σ_l').

Step 1 Let the total number of the invalid facilities become even.

  Denote by S_l^{inv} as a set of the invalid facilities from S_l, where an invalid facility is an opened facility whose parity requirement is violated. At this point, one of the following three cases may happen.

  Case 1. $|S_l^{\mathrm{inv}}|$ is even.

  Case 2. $|S_l^{\mathrm{inv}}|$ is odd, and there exists an opened facility $i \in S_l$ whose $\pi(i)$ is odd.

  Case 3. $|S_l^{\mathrm{inv}}|$ is odd, and there does not exist any opened facility $i \in S_l$ whose $\pi(i)$ is odd.

  if Case 1 **then**

    Let an arbitrary node $i \in V(\mathcal{T}_l)$ be the root of tree $\mathcal{T}_l$.

  if Case 2 **then**

    Arbitrarily choose one of the odd-constrained facilities $r \in V(\mathcal{T}_l)$ to be the root of tree $\mathcal{T}_l$. Arbitrarily choose one of r's children c. Update $\sigma_l(j) := c$ for every client assigned by facility r. Update $S_l := S_l \backslash \{r\}$.

  if Case 3 **then**

    Arbitrarily choose one of the odd-constrained facilities $i \in \hat{\mathcal{F}}_l \backslash S_l$ and update $S_l := S_l \cup \{i\}$. Add i to tree $\mathcal{T}_l$ by letting i become a child of a node $i' \in V(\mathcal{T}_l)$ such that $d_{G_l}(i, i') \leq 4$. Arbitrarily choose a facility r from $S_l^{\mathrm{inv}} \backslash \{i\}$ to be the root of tree $\mathcal{T}_l$. Arbitrarily choose one of r's child c. Update $\sigma_l(j) := c$ for every client assigned by facility r. Update $S_l := S_l \backslash \{r\}$.

Step 2 Find node-disjoint reassigning paths on tree $\mathcal{T}_l$.

  For the given tree $\mathcal{T}_l$ consists of a subset of facilities from $\hat{\mathcal{F}}_l$, denote $\mathcal{T}_l(u)$ as a subtree that rooted at $u \in V(\mathcal{T}_l)$, denote $\mathcal{T}_l^2$ as the 2-th power graph of $\mathcal{T}_l$, whose vertex set $V(\mathcal{T}_l^2) := V(\mathcal{T}_l)$, edge set $E(\mathcal{T}_l^2) := \{(i, i') : i, i' \in V(\mathcal{T}_l), d_{\mathcal{T}_l}(i, i') \leq 2\}$. Reassigning paths are node-disjoint paths on $\mathcal{T}_l^2$ between invalid facilities, and exposed paths are node-disjoint paths on $\mathcal{T}_l^2$ with one and only one of its endpoints invalid.

  for each subtree $\mathcal{T}_l(u)$ **do**

    if u is one of $\mathcal{T}_l$'s leaves **then**

      If u is an invalid facility, identify $p = (u)$ as an exposed path within $\mathcal{T}_l$.

    else

      Collect the exposed paths from each subtree $\mathcal{T}_l(c)$ of $\mathcal{T}_l(u)$, where c is the child of u. Pair each two of the exposed paths arbitrarily by connecting the valid endpoints. For the remaining exposed path(if any), extend it to u to form a new exposed path.

    end

    Update u to be its father node in $\mathcal{T}_l$.

Step 3 Modify the assignment through node-disjoint reassigning paths.

  for each node-disjoint reassigning path $p = (s_1, s_2, ..., s_n)$ **do**

    Arbitrarily choose one of the clients j such that $\sigma_l(j) = s_i$ and $d(j, s_i) \leq \tau^*$, reassign it to facility s_{i+1}, and update $\sigma_l(j) := s_{i+1}$ for each $i = 1, ..., n-1$.

  end

  Output the solution (S_l', σ_l'), where $S_l' := S_l$ and $\sigma_l'(j) := \sigma_l(j)$ for any $j \in \mathcal{D}_l$.

(S'_l, σ'_l) satisfying the parity constraint of each facility $i \in S'_l$, i.e., we have that

$$|\{j \in \mathcal{D}_l : \sigma'_l(j) = i\}| = \pi(i),$$

for each $i \in S'_l$, and the distance of each client $j \in \mathcal{D}_l$ can be bounded by a factor of the optimal value of τ^* of the given PAR k-supplier instance $\mathcal{I}$, which is,

$$d(j, \sigma'_l(j)) \leq 9\tau^*.$$

3.3   The Approximation Ratio

Combining Lemmas 2 and 4, we obtain the main results of the Algorithm 2.

Theorem 1. *For the given PAR k-supplier instance $\mathcal{I} = (\mathcal{F}, \mathcal{D}, k, d, \pi)$, Algorithm 2 outputs a feasible solution (S, σ), i.e., we have that*

$$|S| \leq k,$$

and that

$$|\{j \in \mathcal{D} : \sigma'(j) = i\}| = \pi(i),$$

for each facility $i \in S$. The objective value of the solution (S, σ) can be bounded by a factor of the optimal value of τ^* of the given PAR k-supplier instance $\mathcal{I}$, which is,

$$\max_{j \in \mathcal{D}} d(j, \sigma(j)) \leq 9\tau^*.$$

References

1. Chakrabarty, D., Goyal, P., Krishnaswamy, R.: The non-uniform k-center problem. ACM Trans. Algorithms **16**(4), 1–19 (2020)
2. Chen, X., Ji, S., Wu, C., Xu, Y., Yang, Y.: An approximation algorithm for diversity-aware fair k-supplier problem. Theor. Comput. Sci. 114305 (2023)
3. Chen, X., Xu, D., Xu, Y., Zhang, Y.: Parameterized approximation algorithms for sum of radii clustering and variants. In: Proceedings of the AAAI Conference on Artificial Intelligence, vol. 38, no. 18, pp. 20666–20673 (2024)
4. Frank, A., Király, Z.: Graph orientations with edge-connection and parity constraints. Combinatorica **22**, 47–70 (2002)
5. Gonzalez, T.F.: Clustering to minimize the maximum intercluster distance. Theoret. Comput. Sci. **38**, 293–306 (1985)
6. Grötschel, M., Lovász, L., Schrijver, A.: The ellipsoid method and its consequences in combinatorial optimization. Combinatorica **1**, 169–197 (1981)
7. Grötschel, M., Lovász, L., Schrijver, A.: Corrigendum to our paper "the ellipsoid method and its consequences in combinatorial optimization". Combinatorica **4**(4), 291–295 (1984)
8. Han, L., Xu, D., Xu, Y., Yang, P.: Approximation algorithms for the individually fair k-center with outliers. J. Global Optim. **87**(2), 603–618 (2023)
9. Hochbaum, D.S., Shmoys, D.B.: A best possible heuristic for the k-center problem. Math. Oper. Res. **10**(2), 180–184 (1985)

10. Hochbaum, D.S., Shmoys, D.B.: A unified approach to approximation algorithms for bottleneck problems. J. ACM **33**(3), 533–550 (1986)
11. Kim, K., Shin, Y., An, H.C.: Constant-factor approximation algorithms for parity-constrained facility location and k-center. Algorithmica **85**(7), 1883–1911 (2023)
12. Lanthaler, M., Lechner, W.: Minimal constraints in the parity formulation of optimization problems. New J. Phys. **23**(8), 083039 (2021)
13. Lee, E., Nagarajan, V., Wang, L.: On some variants of Euclidean k-supplier. Oper. Res. Lett. **50**(2), 115–121 (2022)
14. Li, S.: A 1.488 approximation algorithm for the uncapacitated facility location problem. Inf. Comput. **222**, 45–58 (2013)
15. Li, S., Svensson, O.: Approximating k-median via pseudo-approximation. In: Proceedings of the Annual ACM Symposium on Theory of Computing, pp. 901–910 (2013)
16. Luo, H., Han, L., Shuai, T., Wang, F.: Approximation and heuristic algorithms for the priority facility location problem with outliers. Tsinghua Sci. Technol. **29**(6), 1694–1702 (2024)
17. Nagarajan, V., Schieber, B., Shachnai, H.: The Euclidean k-supplier problem. Math. Oper. Res. **45**(1), 1–14 (2020)
18. Plesník, J.: A heuristic for the p-center problems in graphs. Discret. Appl. Math. **17**(3), 263–268 (1987)
19. Tong, P., Lawler, E.L., Vazirani, V.V.: Solving the weighted parity problem for gammoids by reduction to graphic matching. In: Progress in Combinatorial Optimization, pp. 363–374 (1984)
20. Xu, Y., Chau, V., Wu, C., Zhang, Y., Zissimopoulos, V., Zou, Y.: A semi brute-force search approach for (balanced) clustering. Algorithmica **86**(1), 130–146 (2024)

Approximating Principal-Agent Problem Under Bayesian

Qinqin Gong[1], Ling Gai[2,3], Yijing Wang[4], Dachuan Xu[1], and Ruiqi Yang[1(✉)]

[1] Institute of Operations Research and Information Engineering, Beijing University of Technology, Beijing 100124, People's Republic of China
gongqinqin@emails.bjut.edu.cn, {xudc,yangruiqi}@bjut.edu.cn
[2] Business School, University of Shanghai for Science and Technology, Shanghai 200093, People's Republic of China
lgai@usst.edu.cn
[3] School of Intelligent Emergency Management, University of Shanghai for Science and Technology, Shanghai 200093, People's Republic of China
[4] Consulting Center for Strategic Assessment, Academy of Military Science, Beijing 100091, People's Republic of China
yjwang@amss.ac.cn

Abstract. In the classical principal-agent model, the principal delegates a task to an agent who takes an action that is unobservable to the principal and that is costly. We examine this issue within a Bayesian framework where the agent's privacy type, representing the unit cost of effort, is single-dimensional and follows a publicly known probability distribution. Each action necessitates distinct levels of effort. This conception of effort and cost per unit underpins the model's objective to assess the approximation guarantee of linear contracts compared to optimal contracts within a Bayesian framework. Linear contracts encompass allocation rules mapping types to actions and payment rules associating types with random outcomes for a selected set of actions. We approach this problem through a computational lens. Our primary focus lies on cases where the agent determines an action set to fulfill a complex task delegated by the principal, with success being uncertain. We demonstrate that if the probability of success behaves as a submodular function of the action set, linear contracts approximate optimality under sufficient uncertainty in the principal-agent relationship.

Keywords: Contracts theory · Bayesian principal-agent model · Submodularity · Linear contracts

1 Introduction

The principal-agent problem has garnered increasing attention in both economic and computational spheres in recent years. This problem examines a scenario in which a principal delegates a task to an agent and seeks to incentivize the agent to exert effort [29]. However, the principal cannot directly observe the

B. Li et al. (Eds.): IJTCS-FAW 2024, LNCS 14752, pp. 185–198, 2025.
https://doi.org/10.1007/978-981-97-7752-5_15

agent's actions (referred to as "hidden action"), only the randomized outcome stemming from those actions. Each action incurs a cost borne by the agent (e.g., levels of effort), while the principal benefits from the resultant outcome. The principal's objective is to motivate the agent to undertake actions that benefit the principal, leading to a profitable outcome. To achieve this objective, the principal furnishes the agent with a *contract* specifying the amount of reward to be disbursed based on the outcomes achieved. The principal-agent problem finds diverse applications across various domains, including labor markets, insurance, and crowdsourcing platforms. Contract theory, with principal-agent problems at its core, holds significant relevance in the service market [15]. The evolution of contract theory has become an indispensable analytical tool for addressing these challenges.

The classic model fails to encompass scenarios involving adverse selection (or "hidden information"). In such cases, multiple agent types, following a known prior distribution, exist unbeknownst to the principal, constituting Bayesian settings. As recommended actions and subsequent outcomes may differ by type, it becomes imperative to examine the ramifications of type uncertainty on contracting. Additionally, the classical model overlooks situations where agents must undertake a series of actions to accomplish a complex task. Here, the principal's reward is contingent upon the simultaneous execution of these actions, adhering to the principle of diminishing marginal gain. Linear contracts, which allocate a portion of the principal's reward as the agent's payment, represent a significant category within contract theory. They have garnered considerable attention in the literature on principal-agent problems due to their favorable analytical characteristics and straightforward nature. Notably, Dütting et al. [18] demonstrate that linear contracts are optimal for binary outcomes (e.g., success or failure of a task) and robustly optimal for broader outcome spaces when the expected reward for each set of agents is known. Alon et al. [3] delineate structural properties of allocation rules for discrete and continuous types within a single-dimensional private type model utilizing linear programming techniques. Additionally, Alon et al. [4] illustrate the approximation of linear contracts through the adoption of approximation theory, contingent upon specific assumptions satisfied by the principal-agent problem. In this model, the agent has the option to undertake one of n costly actions, each action entailing a level of effort to yield outcomes governed by a stochastic distribution, thereby determining the principal's potential reward. The agent possesses a private type denoting the unit effort cost, with the cost of effort linearly correlated to their type and the level of effort requisite for the chosen action.

In contrast to Alon et al. [4], we propose an extended model that transitions from recommending a single action to recommending a subset of actions. We further assume binary outcomes, comprising success and failure, with the success probability function characterized as a monotonic submodular function defined over the action subset. While it may be feasible to identify the optimal action subset from the 2^n possibilities through enumeration, the computational expense associated with considering all action subsets for computing the optimal contract renders this approach impractical, especially when the size of actions is an input parameter. Consequently, we aim to explore the approximate optimality of linear contracts through the lens of approximation algorithms. Our study demonstrates that the action subset identified by the algorithms can achieve an approximation ratio nearly identical to that in the work by Alon et al. [4].

1.1 Our Contribution and Techniques

We investigate the computational complexity associated with identifying optimal contracts in Bayesian principal-single agent problems. Our study establishes the parameterized approximate optimality of linear contracts within the aforementioned extension of the principal-agent model. The complexity of the problem escalates when transitioning from recommending a single action to recommending a subset of actions. Even in scenarios where the reward function exhibits a well-structured format, such as a complement-free hierarchy [34] (e.g., submodular, XOS), the problem retains its NP-hard status.

Our model explores scenarios wherein the agent possesses a single-dimensional private type. The agent executes tasks by selecting combinations of action subsets, yielding binary random outcomes (both success and failure). We concentrate on instances where the probability of success follows a non-negative monotone submodular function. Each action corresponds to a specific effort requirement, and the total effort of each action subset equals the sum of efforts within that subset. Consequently, the principal's reward is contingent upon the successful outcome of the action subset. The agent's private type reflects their unit effort cost, which constitutes a linear function of both their private type and the effort amount. The principal's objective is to devise contracts that maximize its revenues.

Linear contracts lose their approximate optimality when the agent type distribution collapses into a point mass (c.f., [4, 18]). In such scenarios, there exists no ambiguity concerning the agent's type. We adopt the small-tail assumption, which delineates the knife-edge case of point-mass-like settings from conventional settings featuring ample uncertainty. Furthermore, by leveraging the slowly-increasing assumption regarding the distribution function of types, we derive parameterized approximation guarantees for linear contracts that align with optimal social welfare. The summary of our findings is presented below:

Main Result 1 (Formally, see Theorem 1). *Consider a principal-agent problem with constants $\omega, \beta, \eta \in (0, 1]$ and $\kappa \in [\underline{t}/\omega, \overline{t}]$. If the problem adheres to the (κ, η)-small-tail condition and the distribution of agent types satisfies the (ω, β, κ)-slowly-increasing criterion, then a linear contract with parameter ω yields the principal's expected utility with an approximation of $\frac{1}{\theta_2(1-\omega)\beta\eta}$ the optimal expected welfare.*

The steps outlining the key ingredients for formulating the parameterized approximation guarantee for linear contracts with a submodular success probability function are as follows:

As demonstrated in Dütting et al., [18], linear contracts exhibit arbitrarily poor performance in comparison to the principal's optimal utility (a.k.a. the second best) in the type-deterministic scenario. However, under sufficient uncertainty, approximation guarantees may exist when comparing to the optimal welfare (a.k.a. the first best). We denote this as: $\text{WF}_{[x,y]} = \int_x^y [R_{S^*(1,t)} - e(S^*(1,t)) \cdot t]g(t)dt$ the welfare contribution from types on interval $[x,y] \subseteq [\underline{t}, \overline{t}]$, where g is the density of type distribution function G.

For any true type $t \in T$ of the agent and the corresponding payment p^t, it is observed that the best response action set $S^*(p^t, t)$ is equivalent to the welfare maximizing action subset for the type $\frac{t}{\omega}$, denoted as $S^*\left(1, \frac{t}{\omega}\right)$. In other words, $S^*(p^t, t) = \omega \cdot \arg\max_{S \in 2^A} \left\{ f(S) - e(S) \cdot \frac{t}{\omega} \right\} = S^*\left(1, \frac{t}{\omega}\right)$.

We employ the concept of weak (θ_1, θ_2)–incentive compatibility, which guarantees that the agent truthfully reports its type while also agreeing with the principal to accept a recommended subset that offers competitive approximate utility compared to the optimum (see Definition 2 for details).

We denote the principal's expected utility of a linear contract with parameter ω as PEU. Utilizing the small-tail assumption (refer to Assumption 1 for details) and the (ω, β, κ)-slowly-increasing assumption (refer to Assumption 2 for details), we conclude:

$$\text{PEU} \geq (1 - \omega)\theta_2\beta \int_\kappa^{\overline{t}/\omega} e\left(S^*(1,t)\right) G(t)dt \geq (1 - \omega)\theta_2\beta\eta\text{WF}_{[\underline{t},\overline{t}]}.$$

Thus, we establish the claim.

Theorem 1 relies on the concept of a slowly-increasing distribution. Interestingly, quartiles can be utilized to establish sufficient conditions for this assumption, thereby negating the necessity of the slowly-increasing distribution assumption and yielding the subsequent theorem.

Main Result 2 (Formally, see Theorem 2). *Consider ω, η, and α all within the range of $(0, 1]$. For a principal-agent problem where $G(t_\alpha) = \alpha$, if the problem conforms to the $\left(\frac{t_\alpha}{\omega}, \eta\right)$-small-tail condition, then a linear contract with parameter ω yields the principal's expected utility with an approximation of $\frac{1}{\theta_2(1-\omega)\alpha\eta}$ of the optimal expected welfare.*

1.2   Related Work

Contract theory, a cornerstone of modern economics, has undergone rapid evolution in recent decades and finds extensive application in real-world contexts [1]. The seminal work by Grossman and Hart [24] introduced the fundamental principal-single agent model and showcased its efficient computation through LP solutions for each action. Carroll [9] and Dütting et al. [18] established that linear contracts are max-min optimal in certain scenarios. Diamond [14] demonstrated that linear contracts are no more costly than other effort-inducing contracts under conditions encompassing all fair gambles and hedges. Holmström and Milgrom [30] elucidated that linear contracts are optimal within a dynamic setting.

Bayesian Contracts. Our study is closely aligned with existing literature [13,23,36] focusing on the intersection of moral hazard and adverse selection. Guruganesh et al. [25] explored the computational approximation of linear contracts, both in single contract and contract menu settings. The problem of approximability for randomized linear contract menus was addressed by Guruganesh et al. [26]. The complexity of computing the optimal single or menu contracts was discussed in [11,12]. Alon et al. [3] illustrated implementable allocation rules under discrete and continuous types using LP duality theory. They also demonstrated, in [4], that linear contracts approach optimality under conditions of sufficient uncertainty. Recently, Castiglioni et al. [10] introduced an almost approximation-preserving polynomial-time reduction from multi-parameter Bayesian contract design to the single-parameter case.

Combinatorial Contracts. Babaioff et al. [7] examined a scenario where the principal incentivizes a subset of agents to exert costly effort to complete a task, with the outcome depending on various combinations of agents' efforts. Dütting et al. [15] investigated a similar scenario involving a principal and an agent, where the agent can select a subset of actions to accomplish a task. They demonstrated that linear contracts remain robustly optimal for higher-dimensional result spaces under first moment constraints. Additional references to combinatorial contracts can be found in [5,6,16,17,20,21,39]

The works mentioned above solely address moral hazard and do not account for the adverse selection setting, where the agent's private type is a factor. In our model, we approach the Bayesian principal-agent problem through the lens of combinatorial contracts. Agents with various types select subsets of actions within a binary outcome space. Our findings indicate that linear contracts offer approximate optimality under conditions of sufficient uncertainty.

Furthermore, extensive research examines contract design through a computational lens, as evidenced by works such as [19,28], and [32].

1.3 Organization

We outline our problem and establish the model in Sect. 2. Section 3 presents our main results, which entail approximate guarantees for linear contracts contingent upon specific key assumptions. Finally, Sect. 4 concludes our work.

2 Preliminaries

2.1 Hidden-Action Principal-Agent with Type Costs

In our principal-agent problem involving two players, a single principal assigns a task to a solitary agent whose type cost, denoted by t, remains private. The agent has at their disposal an action set $A = \{1, ..., n\}$, from which they can choose a subset $S \in 2^A$ of actions under any given type t. We operate within the framework of a binary outcome space $\Omega = \{0, 1\}$, where "1" signifies task success and "0" denotes task failure. The principal receives a reward of r_1 for outcome

"1" and a reward of r_0 for outcome "0". To maintain consistency with Dütting et al. [15], we standardize $r_1 = 1$ and $r_0 = 0$. For outcome "1", each strategy $S \in 2^A$ implemented by the agent yields a success probability denoted by $f(S) : 2^A \rightarrow [0,1]$. We stipulate that $f(\emptyset) = 0$ and that f is monotonically non-decreasing and submodular, as described by the condition $f(S') \leq f(S)$ for any $S' \subseteq S \in 2^A$ and the inequality $f(S) + f(T) \geq f(S \cup T) + f(S \cap T), \forall S, T \in 2^A$. The agent possesses a privately known type $t \in \mathbf{R}_{\geq 0}$, representing the agent's cost per unit of effort. We assume the type cost follows a publicly-known distribution G with density g, and it falls within the support $T = [\underline{t}, \overline{t}]$ where $0 \leq \underline{t} \leq \overline{t} < \infty$. Each action $i \in A$ necessitates $e_i \in \mathbf{R}_{\geq 0}$ units of effort from the agent. Consequently, when an agent with type t adopts the strategy S, the agent's cost is $e(S) \cdot t = \sum_{i \in S} e_i \cdot t$. Naturally, the cost of taking no action (i.e., $S = \emptyset$) is zero.

In summary, akin to the classic model in contract literature, we embrace the concept of hidden action, positing that the agent's actions remain concealed, and the actions undertaken by the agent are not directly observable by the principal. Instead, the principal only perceives the randomized outcome resulting from the selected action set. Additionally, the agent's true type t constitutes private information known solely to the agent.

2.2   Contracts

Our contract extends the standard model introduced by [3,4], comprising an allocation rule a and a payment rule p. In this contract, denoted as (a, p), the *allocation rule* $a : T \rightarrow 2^A$ maps the agent's reported type $t' \in T$ to a recommended action set $a(t') \in 2^A$, while the *payment rule* $p : T \rightarrow \mathbf{R}_{\geq 0}^2$ maps the agent's reported type t' to a vector of payments for the binary outcome. Specifically, given the reported type t', $p_0^{t'}$ and $p_1^{t'}$ represent the payments made by the principal to the agent for outcomes 0 and 1 respectively. We assume that payments are non-negative for all types t' and outcomes, a standard condition in the contract literature known as *limited liability*. Economists primarily focus on the concept of *moral hazard*, where the principal specifies payments based on the effort exerted by the agent, as the agent lacks motivation to undertake costly actions.

We assume that if outcome 0 is realized, the principal receives no reward and pays nothing, implying $p_0^t = 0$ for any $t \in T$ (as indicated in Dütting et al. [15]). The timeline of events for a contract (a, p) is summarized as follows:

- The principal offers a contract (a, p) based on the public distribution types of the agent.
- Upon observing the contract, the agent, who possesses type t, reports a type t'. Subsequently, the contract recommends a subset of actions $a(t') \in 2^A$ and provides a payment profile $p^{t'} = (p_0^{t'}, p_1^{t'})$.
- The agent privately selects a set of actions $S \in 2^A$, which may different from $a(t')$, incurring a cost of $\sum_{i \in S} e_i \cdot t$.
- Outcome 1 occurs with a success probability of $f(S)$. The principal receives the corresponding reward r_1 and pays the agent a price of $p_1^{t'}$.

We focus on linear contracts, which have been extensively studied in previous contract literature. For any $\omega > 0$, type t, and $S \in 2^A$, we denote by $p^t :\stackrel{\Delta}{=} p_1^t = \omega \cdot r_1$ the linear contract with a single parameter ω. Assuming t is the true type of the agent, t' is the reported type, and S is the set of actions chosen by the agent, we recall that $r_0 = 0$, $p_0^t = 0$, and we normalize $r_1 = 1$. Let $P_S^{t'} = \omega \cdot f(S)$ denote the expected payment from the principal to the agent, and let $R_S = f(S)$ denote the expected reward for the principal. Then the expected utility of the agent is given by:

$$P_S^{t'} - e(S) \cdot t = \omega \cdot f(S) - e(S) \cdot t \tag{1}$$

and the expected utility of the principal is given by: $R_S - P_S^{t'} = (1 - \omega) \cdot f(S)$. We calculate the expected welfare of type t with the chosen set S of actions as the sum of the expected utilities of the principal and the agent. $R_S - e(S) \cdot t = f(S) - e(S) \cdot t$. When considering the agent with true type t and reported type t', he will rationally choose the subset of actions that maximizes his utility as stated in (1), specifically $S^*(p^{t'}, t) \in \arg\max_{S \in 2^A} \left\{ P_S^{t'} - e(S) \cdot t \right\}$. We adopt the assumption from [9,25] that in case of multiple subsets of actions with the same maximum expected utility from the agent's perspective, the tie-breaking favors the principal. Consequently, $S^*(p^{t'}, t)$ is unambiguously defined. Then, the agent will choose t' to maximize his expected utility, considering his anticipated selected subset of actions, i.e., $P_{S^*(p^{t'}, t)}^{t'} - e(S^*(p^{t'}, t)) \cdot t$.

Definition 1 (IC). *A linear contract (a, ω) is considered incentive compatible for the agent with a true type t if the agent's utility is maximized by the following conditions:*

1. *Willingly accepting the recommended set of actions, i.e. $a(t) = S^*(p^t, t)$.*
2. *Truthfully reporting his type, i.e. $t \in \arg\max_{t' \in T} \left\{ P_{S^*(p^{t'}, t)}^{t'} - e(S^*(p^{t'}, t)) \cdot t \right\}$.*

The agent's expected utility is guaranteed to be at least zero if he selects the optimal set of actions that meet the incentive compatible constraints. Otherwise, choosing the null action set $\emptyset$ confirms this assertion. Therefore, incentive compatibility implies individual rationality (IR).

Contract design, especially in auction scenarios, presents an optimization challenge aimed at maximizing specific objectives within various constraints. A pivotal constraint in this context is Incentive Compatibility (IC), which guarantees that participants are motivated to honestly disclose their preferences and engage in favorable actions. The objective is to formulate contracts that yield efficient outcomes by motivating participants to act in desired ways while adhering to the IC constraint.

Cutting-edge optimal auction design often involves transitioning from strict Incentive Compatibility (IC) constraints to approximate Incentive Compatibility (ϵ-IC). This approach relaxes the requirement for participants to perfectly align their incentives with truthfully revealing their preferences, permitting minor deviations or manipulations within a specified tolerance level (ϵ). This relaxation widens the scope of feasible auction designs and can sometimes enhance outcomes in terms of revenue generation or efficiency. However, it's crucial to strike a balance between relaxation and maintaining sufficient incentive compatibility to prevent significant distortions or strategic behavior by participants. The relaxation of ϵ-IC constraints has been explored across various domains beyond auction design, including voting, matching, and competitive equilibrium contexts. Furthermore, within the framework of Nash equilibrium, the relaxation of ϵ-IC constraints has been examined in prior work [38]. These studies underscore the significance of exploring ϵ-IC relaxations across diverse economic contexts to attain favorable outcomes while addressing strategic behavior and incentives. Additionally, Dütting et al. [19] introduced and investigated another relaxation known as δ-IC.

We introduce the concept of weak (θ_1, θ_2)-incentive compatibility (or weak (θ_1, θ_2)-IC), wherein the agent truthfully reports his type and is willing to accept a recommended subset, compromising with the principal as long as the approximate utility is competitive compared to the optimum. Formally, we define weak (θ_1, θ_2)-incentive compatibility in detail in Definition 2.

Definition 2 *(weak (θ_1, θ_2)-IC).* *For any pair $\theta_1, \theta_2 > 0$, a linear contract (a, ω) is weak (θ_1, θ_2)-incentive compatible if the agent with type t satisfies:*

1. *Reporting his type truthfully, i.e. $t \in \arg\max_{t' \in T} \left\{ P^{t'}_{S^*(p^{t'}, t)} - e(S^*(p^{t'}, t)) \right.$*
   $\left. \cdot t \right\}$,
2. *there exists a subset $S' \in 2^A$ such that*

$$f(S') - e(S') \cdot \frac{t}{\omega} \geq \theta_1 \cdot f(S^*(1, \frac{t}{\omega})) - \theta_2 \cdot e(S^*(1, \frac{t}{\omega})) \cdot \frac{t}{\omega},$$

where $S^*(1, \frac{t}{\omega}) = \omega \cdot \arg\max_{S \in 2^A} \left\{ f(S) - e(S) \cdot \frac{t}{\omega} \right\} = S^*(\omega, t)$.

This concept is considered for several reasons. First, the agent "gifts" effort to the principal employing him (c.f., Akerlof [2]), and there exists an expanding literature work in economics addressing behavioral biases in contract theory under competitive markets (c.f., Koszegi et al. [33]). The second reason is for the tractability of the problem. We observe that finding the optimal recommended set of actions is NP-hard for both scenarios where the action size n is an input parameter and when the success function f is part of the complement-free hierarchy. This contrasts with the concept of recommended single action from Alon et al. [3,4], where the optimal recommended action can be found in polynomial time. From an optimization standpoint, maximizing submodular functions is an NP-hard problem (c.f., Feige [22]), where submodularity is a subclass of the complement-free hierarchy and can be efficiently approximated from (c.f., Nemhauser et al. [37]).

In the principal-agent model, the parameters θ_1 and θ_2 represent the level of approximation for the recommended subset after compromise. These parameters are typically selected as positive values greater than zero to ensure that the approximation accurately captures the intended subset of recommendations.

The aim of this study is to optimize the principal's revenue by implementing weak (θ_1, θ_2)-IC contracts following compromise. The expectation considers the agent's random type t drawn from G, and the resulting randomized outcome derived from the set of actions $a(t)$ recommended to the agent.

$$\text{OPT} = \max_{(a,\omega) \text{ is weak } (\theta_1,\theta_2)\text{-IC}} E_{t\sim G}[R_{a(t)} - P^t_{a(t)}].$$

3 Optimality of Linear Contracts

In this section, we elaborate on our principal finding by delineating the approximation guarantees for linear contracts predicated on fundamental assumptions. Within the framework outlined by Alon et al. [4], the notion of the "small-tail assumption" was introduced. This assumption serves to differentiate between two distinct scenarios: those resembling point-mass distributions and those characterized by significant uncertainty.

The small-tail assumption constitutes a pivotal condition delineating the level of uncertainty within the setting. In scenarios resembling point-mass distributions, uncertain factors are predominantly concentrated in a limited number of extreme outcomes, culminating in a "knife-edge" scenario. Under such circumstances, relaxing IC constraints without substantial compromise to desired outcomes becomes challenging. Conversely, in conventional settings characterized by ample uncertainty, the distribution of uncertain factors is more dispersed, affording greater flexibility in relaxing IC constraints while still attaining desirable outcomes.

Denote $\text{WF}_{[x,y]}$ as the welfare contribution from types on interval $[x, y] \subseteq [\underline{t}, \overline{t}]$. It is expressed as:

$$\text{WF}_{[x,y]} = \int_x^y [R_{S^*(1,t)} - e(S^*(1,t)) \cdot t]g(t)dt$$
$$= \int_x^y [f(S^*(1,t)) - e(S^*(1,t)) \cdot t]g(t)dt$$

For completeness, we restate the small-tail assumption as follows:

Assumption 1. *[4] An instance of a principal-agent problem whose types are supported on $[\underline{t}, \overline{t}]$ is defined as (κ, η)-small-tail, if for any parameters $\eta \in (0, 1]$ and $\kappa \in [\underline{t}, \overline{t}]$, it holds that $\text{WF}_{[\kappa,\overline{t}]} \geq \eta \cdot \text{WF}_{[\underline{t},\overline{t}]}$.*

The (κ, η)-small-tail condition quantifies the concentration of welfare among the lowest types. Consequently, as κ increases and η approaches 1, the tail diminishes, pushing the setting further from a point mass distribution.

We further adhere to the assumption of a *slowly-increasing distribution* as proposed by Alon et al. [4], which elucidates the progression of the private type cost distribution G. This assumption posits that if the type cost escalates from ωt to t at a rate of $1/\omega$, then the cumulative distribution function G experiences an expansion by a factor of up to $1/\beta$.

Assumption 2. *[4] A distribution G with a support range of $[\underline{t}, \overline{t}]$ is termed (ω, β, κ)-**slowly-increasing** if, for any parameters $\omega, \beta \in (0,1]$ and $\kappa \in [\underline{t}, \overline{t}]$, it satisfies the condition: $G(\omega t) \geq \beta \cdot G(t)$ for all $t \geq \kappa$.*

Building upon the aforementioned assumptions, we encapsulate our principal results in the following theorem.

Theorem 1. *For any constant $\omega, \beta, \eta \in (0,1]$, let us consider a principal-agent problem with $\kappa \in [\frac{t}{\omega}, \overline{t}]$. If the problem adheres to the (κ, η)-small-tail condition, and the distribution of the agent types conforms to the (ω, β, κ)-slowly-increasing criterion, then a linear contract with parameter ω yields the principal's expected utility with an approximation of $\frac{1}{\theta_2(1-\omega)\beta\eta}$ of the optimal expected welfare.*

Proof. Let's revisit the definition of $S^*(p^{t'}, t)$ as mentioned earlier. The substitution of $p^{t'}$ for 1 implies that the reward is paid in full to the agent. For an agent of type t, $S^*(1,t)$ represents the action set that maximizes welfare. The welfare from types supported on $[\kappa, \overline{t}]$ can be expressed as:

$$WF_{[\kappa,\overline{t}]} = \int_{\kappa}^{\overline{t}} [f(S^*(1,t)) - e(S^*(1,t)) \cdot t] g(t) dt$$

$$= -G(\kappa)[f(S^*(1,\kappa)) - e(S^*(1,\kappa)) \cdot \kappa] + \int_{\kappa}^{\overline{t}} e(S^*(1,t)) G(t) dt,$$

where the second equality follows from the assumption that $u(\overline{t}) = f(S^*(1,\overline{t})) - e(S^*(1,\overline{t})) \cdot \overline{t} = 0$. Note that:

$$S^*(p^t, t) = \arg\max_{S \in 2^A} \left\{ P_S^t - e(S) \cdot t \right\}$$

$$= \arg\max_{S \in 2^A} \left\{ \omega f(S) - e(S) \cdot t \right\}$$

$$= \omega \arg\max_{S \in 2^A} \left\{ f(S) - e(S) \cdot \frac{t}{\omega} \right\}$$

$$= S^* \left(1, \frac{t}{\omega} \right).$$

Assuming S^A is a recommended set after compromises satisfying weak (θ_1, θ_2)-IC constraints, which can be efficiently fond in polynomial time, we denote the principal's expected utility of a linear contract with parameter ω as PEU. Then, it follows:

$$\text{PEU} = \mathbf{E}_{t \sim G}[(1-\omega)f(S^A)]$$

$$= (1-\omega)\int_{\underline{t}}^{\overline{t}} f(S^A) g(t) dt$$

$$\geq (1-\omega)\int_{\underline{t}}^{\overline{t}} \left[f(S^A) - e(S^A) \cdot \frac{t}{\omega} \right] g(t) dt$$

$$\geq (1-\omega)\int_{\underline{t}}^{\overline{t}} \left[\theta_1 f\left(S^*\left(1, \frac{t}{\omega}\right)\right) - \theta_2 e\left(S^*\left(1, \frac{t}{\omega}\right)\right) \cdot \frac{t}{\omega} \right] g(t) dt,$$

where the first inequality follows from the non-negativity of $e(S^A) \cdot \frac{t}{\omega}$, the second inequality stems from the definition of S^A. Utilizing integration by parts:

$$\text{PEU} \geq \frac{(1-\omega)\theta_2}{\omega} \int_{\underline{t}}^{\overline{t}} e\left(S^*\left(1, \frac{t}{\omega}\right)\right) G(t)dt$$

$$= (1-\omega)\theta_2 \int_{\underline{t}/\omega}^{\overline{t}/\omega} e\left(S^*\left(1, t\right)\right) G(\omega t)dt$$

$$\geq (1-\omega)\theta_2\beta \int_{\kappa}^{\overline{t}/\omega} e\left(S^*\left(1, t\right)\right) G(t)dt,$$

where the first equality employs the method of integration by substitution, and the second inequality arises from $G(\omega t) \geq \beta \cdot G(t)$ for all $t \geq \kappa$. Since $\kappa \geq \frac{t}{\omega}$, it follows that:

$$\text{PEU} \geq (1-\omega)\theta_2\beta \int_{\kappa}^{\overline{t}/\omega} e\left(S^*\left(1, t\right)\right) G(t)dt$$

$$\geq (1-\omega)\theta_2\beta\text{WF}_{[\kappa,\overline{t}]}$$

$$\geq (1-\omega)\theta_2\beta\eta\text{WF}_{[\underline{t},\overline{t}]}$$

where the third inequality is a consequence of the small-tail assumption.

We proceed to establish Theorem 2 by employing the small-tail assumption (i.e., Assumption 1) in conjunction with Theorem 1.

Theorem 2. *Consider* $\omega, \eta, \alpha \in (0,1]$. *For a principal-agent problem with* $G(t_\alpha) = \alpha$, *if the problem adheres to the* $(\frac{t_\alpha}{\omega}, \eta)$-*small-tail assumption (as stated in Assumption 1), then a linear contract with parameter* ω *yields the principal's expected utility with an approximation of* $\frac{1}{\theta_2(1-\omega)\alpha\eta}$ *of the optimal expected welfare.*

Proof. To achieve $G(\omega\kappa) = \alpha$, we straightforwardly set $\kappa = \frac{t_\alpha}{\omega}$. Consequently, we ensure $\kappa \geq \frac{t}{\omega}$ since the agent's type $t_\alpha \geq \underline{t}$. Given that the cumulative distribution function G is monotonically increasing and $G(t) \in [0,1]$ for any t, it follows that for any cost $t \geq \kappa$ and $\omega \in (0,1]$, we have

$$G(\omega t) \geq G(\omega\kappa) \geq G(\omega\kappa) \cdot G(t).$$

The agent type distribution G conforms to the $(\omega, G(\omega\kappa), \kappa)$-slowly-increasing assumption. Furthermore, assuming the satisfaction of the (κ, η)-small-tail condition for type costs, according to Theorem 1, the approximation ratio is

$$\frac{1}{\theta_2(1-\omega)G(\omega\kappa)\eta} = \frac{1}{\theta_2(1-\omega)\alpha\eta}.$$

Remark. The benchmark for the approximation ratio is the optimal welfare, which may significantly surpass the optimal principal's utility. Thus, the actual approximation guarantee may be better when compared to the optimal principal's utility. To efficiently address the assumption of computing the com promise subset S^A, we draw motivation from the work of Harshaw et al. [27], which presents a distorted-greedy algorithm with an approximation ratio of $\theta_1 = 1-1/e$ and $\theta_2 = 1$ when the reward function f is monotone submodular. Additionally, Jin et al. [31] introduce an improved approximation with $\theta_1 = 1$ and $\theta_2 = 1 + \ln \frac{f(S^*(1,\frac{t}{\omega}))}{e(S^*(1,\frac{t}{\omega}))\cdot\frac{t}{\omega}}$.

Finally, to handle with the assumption of computing the compromise subset S^A efficiently, we motivate the work by Harshaw et al. [27], which provides a distorted-greedy with an approximation ratio $\theta_1 = 1 - 1/e$ and $\theta_2 = 1$ when reward function f is monotone submodular. Moreover, the work by Jin et al. [31] presents an improved approximation with $\theta_1 = 1$ and $\theta_2 = 1 + \ln \frac{f(S^*(1,\frac{t}{\omega}))}{e(S^*(1,\frac{t}{\omega}))\cdot\frac{t}{\omega}}$.

Furthermore, our analytical techniques are applicable to non-monotone settings. Previous research by Lu et al. [35] investigates scenarios where the submodular function is non-monotone, providing approximation guarantees with $\theta_1 = 1/e$ and $\theta_2 = 1$. Further related work can be found in [8].

4   Conclusion

In this study, we investigate the principal-agent model within a Bayesian framework, where the agent, possessing a privately known type, can select any subset of actions. The principal's objective is to devise a contract that maximizes her expected utility. We investigate the design of approximately optimal contracts when the expected reward functions fall within the complement-free hierarchy. Demonstrating the near-optimality of linear contracts using weak (θ_1, θ_2)-IC under conditions of adequate uncertainty, our inquiry raises several pertinent questions. We inquire whether linear contracts remain approximately optimal when the cost function is also a complement-free set function, such as XOS or subadditive. In our analysis, we assume the reward function to be monotonic. Consequently, a natural extension would be to address the question of whether there exists an approximate guarantee for linear contracts when the reward function is non-monotonic. Furthermore, exploring the realm of multi-agent contract problems, wherein each agent selects a subset of actions, presents an intriguing avenue for future research into designing near-optimal contracts in such scenarios.

Acknowledgements. The first and fourth authors are supported by Natural Science Foundation of China (No. 12131003). The second author is supported by Natural Science Foundation of China (No. 11201333). The third author is supported by Natural Science Foundation of China (No. 12201619). The fifth author is supported by Natural Science Foundation of China (No. 12101587) and China Postdoctoral Science Foundation (No. 2022M720329).

References

1. Scientific background on the 2016 Nobel price in economic sciences. Royal Swedish Academy of Sciences (2016)
2. Akerlof, G.: Labor contracts as partial gift-exchange. Q. J. Econ. **97**(4), 543–569 (1982)
3. Alon, T., Dütting, P., Talgam-Cohen, I.: Contracts with private cost per unit-of-effort. In: Proceedings of ACM EC, pp. 52–69 (2021)
4. Alon, T., Dütting, P., Li, Y., Talgam-Cohen, I.: Bayesian analysis of linear contracts. arXiv:2211.06850 (2022)
5. Babaioff, M., Feldman, M., Nisan, N.: Free-riding and free-labor in combinatorial agency. In: Proceedings of SAGT, pp. 109–121 (2009)
6. Babaioff, M., Feldman, M., Nisan, N.: Mixed strategies in combinatorial agency. J. Artif. Intell. Res. **38**, 339–369 (2010)
7. Babaioff, M., Feldman, M., Nisan, N.: Combinatorial agency. J. Econ. Theory **147**(3), 999–1034 (2012)
8. Bodek, K., Feldman, M.: Maximizing sums of non-monotone submodular and linear functions: understanding the unconstrained case. arXiv:2204.03412 (2022)
9. Carroll, G.: Robustness and linear contracts. Am. Econ. Rev. **105**(2), 536–563 (2015)
10. Castiglioni, M., Chen, J., Li, M., Xu, H., Zuo, S.: A reduction from multi-parameter to single-parameter Bayesian contract design. arXiv:2404.03476v1 (2024)
11. Castiglioni, M., Marchesi, A., Gatti, N.: Bayesian agency: linear versus tractable contracts. In: Proceedings of ACM EC, pp. 285–286 (2021)
12. Castiglioni, M., Marchesi, A., Gatti, N.: Designing menus of contracts efficiently: the power of randomization. In: Proceedings of ACM EC, pp. 705–735 (2022)
13. Castro-Pires, H., Chade, H., Swinkels, J.: Disentangling moral hazard and adverse selection. Am. Econ. Rev. **114**(1), 1–37 (2024)
14. Diamond, P.: Managerial incentives: on the near linearity of optimal compensation. J. Polit. Econ. **106**(5), 931–57 (1998)
15. Dütting, P., Ezra, T., Feldman, M., Kesselheim, T.: Combinatorial contracts. In: Proceedings of FOCS, pp. 815–826 (2022)
16. Dütting, P., Ezra, T., Feldman, M., Kesselheim, T.: Multi-agent contracts. In: Proceedings of STOC, pp. 1311–1324 (2023)
17. Dütting, P., Feldman, M., Gal Tzur, Y.: Combinatorial contracts beyond gross substitutes. In: Proceedings of SODA, pp. 92–108 (2024)
18. Dütting, P., Roughgarden, T., Talgam-Cohen, I.: Simple versus optimal contracts. In: Proceedings of ACM EC, pp. 369–387 (2019)
19. Dütting, P., Roughgarden, T., Talgam-Cohen, I.: The complexity of contracts. SIAM J. Comput. **50**(1), 211–254 (2021)
20. Emek, Y., Feldman, M.: Computing optimal contracts in combinatorial agencies. Theor. Comput. Sci. **452**, 56–74 (2012)
21. Ezra, T., Feldman, M., Schlesinger, M.: On the (in) approximability of combinatorial contracts. In: Proceedings of ITCS (2024)
22. Feige, U.: A threshold of $\ln n$ for approximating set cover. J. ACM **45**(4), 634–652 (1998)
23. Gottlieb, D., Moreira, H.: Simple contracts with adverse selection and moral hazard. Theor. Econ. **17**(3), 1357–1401 (2022)
24. Grossman, S., Hart, O.: An analysis of the principal-agent problem. Econometrica **51**(1), 7–45 (1983)

25. Guruganesh, G., Schneider, J., Wang, J.: Contracts under moral hazard and adverse selection. In: Proceedings of ACM EC, pp. 563–582 (2021)
26. Guruganesh, G., Schneider, J., Wang, J., Zhao, J.: The power of menus in contract design. In: Proceedings of ACM EC, pp. 818–848 (2023)
27. Harshaw, C., Feldman, M., Ward, J., Karbasi, A.: Submodular maximization beyond non-negativity: guarantees, fast algorithms, and applications. In: Proceedings of ICML, pp. 2634–2643 (2019)
28. Ho, C., Slivkins, A., Vaughan, J.: Adaptive contract design for crowdsourcing markets: bandit algorithms for repeated principal-agent problems. J. Artif. Intell. Res. **55**, 317–359 (2016)
29. Holmström, B.: Moral hazard and observability. Bell J. Econ. **10**(1), 74–91 (1979)
30. Holmström, B., Milgrom, P.: Aggregation and linearity in the provision of intertemporal incentives. Econometrica **55**(2), 303–328 (1987)
31. Jin, T., Yang, Y., Yang, R., Shi, J., Huang, K., Xiao, X.: Unconstrained submodular maximization with modular costs: tight approximation and application to profit maximization. Proc. VLDB Endow. **14**(10), 1756–1768 (2021)
32. Kleinberg, J., Raghavan, M.: How do classifiers induce agents to invest effort strategically? In: Proceedings of ACM EC, pp. 825–844 (2019)
33. Köszegi, B.: Behavioral contract theory. J. Econ. Lit. **52**(4), 1075–1118 (2014)
34. Lehmann, B., Lehmann, D., Nisan, N.: Combinatorial auctions with decreasing marginal utilities. Games Econom. Behav. **55**(2), 270–296 (2006)
35. Lu, C., Yang, W., Gao, S.: Regularized nonmonotone submodular maximization. Optimization 1–27 (2023)
36. Myerson, R.: Optimal coordination mechanisms in generalized principal-agent problems. J. Math. Econ. **10**(1), 67–81 (1982)
37. Nemhauser, G., Wolsey, L., Fisher, M.: An analysis of approximations for maximizing submodular set functions-I. Math. Program. **14**, 265–294 (1978)
38. Papadimitriou, C.: The complexity of finding Nash equilibria. Algorithmic Game Theory **2**, 30 (2007)
39. Vuong, R., Dughmi, S., Patel, N., Prasad, A.: On supermodular contracts and dense subgraphs. In: Proceedings of ACM-SIAM SODA, pp. 109–132 (2024)

Robust Facility Leasing Problem with Penalties

Baoyi Duan[1], Lu Han[1(✉)], Sai Ji[2], and Lili Mei[3]

[1] School of Science, Beijing University of Posts and Telecommunications, Beijing, People's Republic of China
{dby,hl}@bupt.edu.cn
[2] Institute of Mathematics, Hebei University of Technology, Tianjin, People's Republic of China
jisai@hebut.edu.cn
[3] School of Cyberspace, Hangzhou Dianzi University, Hangzhou, People's Republic of China
meilili@zju.edu.cn

Abstract. In this paper, we propose and study the robust facility leasing problem with penalties (RFLEP), generalizing several well-known problems including the facility leasing problem (FLE). In the RFLEP, we are given the locations of facilities and clients, along with an integer q. A facility can be leased to open at its location for a certain time interval it chooses and is available for connecting clients during its leasing time. Leasing a facility incurs a leasing cost. Each client arrives at a specific given time and could choose one of these three states when it arrives, which are to be connected to some leased facility and pay a connection cost, reject to be connected and pay a penalty cost, and decide to be an outlier. Each connection cost is related to the locations of the corresponding pair of facility and client, and assume that the connection costs are non-negative, symmetric, and satisfy the triangle inequality. The objective is to lease some facilities and decide the state of each client, such that the total number of outliers is at most q, and the total leasing, connection, and penalty cost is minimized.

As our main contribution, we design a primal-dual 3-approximation algorithm for the RFLEP. One of the main difficulties in solving the RFLEP is that the integrality gap of its natural linear program relaxation is unbounded. To overcome this difficulty, we design our algorithm based on a modified linear program relaxation, which guarantees that some very expensive leasing costs will not be incurred during the dual ascent process and leads to a constant-factor approximation ratio.

Keywords: Facility Leasing · Outliers · Penalty · Approximation Algorithm · Primal-Dual

Lu Han is supported by Beijing Municipal Natural Science Foundation (No. Z220004), National Natural Science Foundation of China (No. 12371321). Sai Ji is supported by Science and Technology Project of Hebei Education Department (No. BJK2023076) and National Natural Science Foundation of China (No. 12101594). Lili Mei is supported by National Natural Science Foundation of China (No. 12201594).

B. Li et al. (Eds.): IJTCS-FAW 2024, LNCS 14752, pp. 199–210, 2025.
https://doi.org/10.1007/978-981-97-7752-5_16

1 Introduction

The uncapacitated facility location problem (UFLP) was introduced and studied as early as the 1960s [12], and it is one of the classical clustering problems in the literature of combinatorial optimization [4,5,8,9,13,14,16,22]. The inputs of the UFLP are sets of facility and client locations, opening cost of each facility, and connection costs between each facility-client pair. Assume that the connection costs are non-negative, symmetric, and satisfy the triangle inequality. The aim is to open some facilities and connect each client to an opened facility, minimizing the incurred total opening and connection costs. The NP-hardness of this problem was proven, and the main research involves designing approximation algorithms across different fields [2,7,10,11]. Moreover, a great deal of the variants of the UFLP had been studied due to different practical backgrounds [6,16,20,21].

The performance of an approximation algorithm is typically measured by its approximation ratio, which characterizes the gap between the objective value of an optimal solution and the solution obtained from the algorithm. More specifically, a polynomial time algorithm for a minimization problem is a ρ-approximation algorithm, if the algorithm can be applied to solve all the instances of the problem, and the objective value of the solution obtained from the algorithm is no more than ρ times the objective value of an optimal solution. Among all existing approximation algorithms for the UFLP, Shmoys et al. [18] proposed the first constant-factor 3.16-approximation algorithm, based on the technique of LP-rounding, and Li [13] gave the currently best approximation ratio of 1.488. The well-known unapproximable lower bound of the UFLP is 1.463 [7].

In the follow-up research on the UFLP, considering whether all clients need to be connected by facilities, Charikar et al. [3] introduced the concept of outliers and formally proposed the robust facility location problem (RFLP) and the facility location problem with penalties (FLPP). In both the RFLP and FLPP, a client could reject to be connected. The main difference between these two problems is that in the RFLP, a certain number of clients could be chosen as outliers and reject to be connected, and in the FLPP if a client rejects to be connected, it would pay a penalty cost. Charikar et al. [3] gave primal-dual 3-approximation algorithms for both the RFLP and FLPP. For the FLPP, Li et al. [15] introduced an LP-rounding method after the results of Xu et al. [20,21], and obtained the currently best approximation ratio of 1.514. Wang et al. [19] proposed the robust facility location problem with penalties (RFLPP), and used the technique of primal-dual [11] to present a 3-approximation algorithm. By using greedy augmentation, they further improved the approximation ratio to 2.

The facility leasing problem (FLE) is another interesting variant of the UFLP, which is introduced by Gupta et al. [1]. Compared with the UFLP, in the FLE, a facility cannot be permanently opened, but can be leased at its location for a certain time interval it chooses. Each client will arrive at a specific given time and can be connected to a facility only if the facility is being leased at that time.

Nagarajan et al. [17] designed a primal-dual 3-approximation algorithm for the FLE.

In this paper, inspired by the works of Wang et al. [19] on the RFLPP and Nagarajan et al. [17] on the FLE, we propose our robust facility leasing problem with penalties (RFLEP) and design a primal-dual based approximation algorithm. Compared with the FLE, in the RFLEP, some clients could reject to be connected, including those who are chosen as outliers, and those who pay their penalty costs. It is worth mentioning that the natural linear program relaxation of the RFLEP has an unbounded integrality gap. Our approximation algorithm is designed based on a modified linear program relaxation, which leads to a constant-factor approximation ratio of 3.

The main structure of this paper is as follows. Section 2 introduces formal descriptions of the RFLEP. Sections 3 and 4 present a 3-approximation algorithm and its analysis for the RFLEP, respectively. Some further discussions are given in Sect. 5.

2   Preliminaries

In a RFLEP instance $\mathcal{I}$, we are given a set of client locations N_c, several time period from 1 to T, and an integer q. At each time period, some client locations will have arriving clients. Assume that the locations of the arrival clients at each time period are a partition of N_c. We define a client as a binary (j, t), where $j \in N_c$ is its location and $t \in \{1, 2, ..., T\}$ is its arrival time. Each client (j, t) has a penalty cost of p_j. We are also given a set of facility locations N_f and a set of lease types K. Each lease type $k \in K$ is associated with a lease duration l_k, and assume that l_k is an integer. Each facility location could be leased in a certain type at some time period. We define a facility as a triple (i, k, s), where $i \in N_f$ is its location, $k \in K$ is its lease type, and $s \in \{1, 2, ..., T\}$ is its start time of leasing. For each facility (i, k, s), it has a leasing cost of f_i^k and will be opened during the time interval of $[s, s + l_k)$. Each pair of facility (i, k, s) and client (j, t) has a connection cost of c_{ij}. Assume that the connection costs are non-negative, symmetric, and satisfy the triangle inequality. A client (j, t) can be connected to some facility (i, k, s) only if its arrival time t is within the opening interval $[s, s + l_k)$ of the facility. We also could reject to connect a client by penalizing it (i.e., paying its penalty cost) or selecting it as an outlier. The RFLEP aims to lease some facilities, decide to connect each client, or penalize it, or select it as an outlier, such that the total number of outliers is at most q, and the total leasing, connection, and penalty cost is minimized.

Denote by $\mathcal{F}$ the set of all facility triples, and $\mathcal{D}$ the set of all client binaries. Denote by $(\hat{\mathcal{F}}, \hat{\mathcal{C}}, \hat{\mathcal{P}}, \hat{\mathcal{O}}, \sigma)$ a solution of the RFLEP, in which $\hat{\mathcal{F}}$ is the set of leased facilities, and $\hat{\mathcal{C}}$ is the set of clients being connected, and $\hat{\mathcal{P}}$ is the sets of clients being penalized, and $\hat{\mathcal{O}}$ is the set of clients selected as outliers, and $\sigma : \hat{\mathcal{C}} \to \hat{\mathcal{F}}$ is a mapping reflecting the connection between clients in $\hat{\mathcal{C}}$ and the leased facilities in $\hat{\mathcal{F}}$. For a solution $(\hat{\mathcal{F}}, \hat{\mathcal{C}}, \hat{\mathcal{P}}, \hat{\mathcal{O}}, \sigma)$, if $|\hat{\mathcal{O}}| \leq q$, and the arrival time t of each client $(j, t) \in \hat{\mathcal{C}}$ is within the opening interval $[s, s + l_k)$ of some leased facility

$(i, k, s) \in \hat{\mathcal{F}}$ satisfying that $\sigma(j, t) = (i, k, s)$, we say that the solution is feasible. Denote by I_s^k the time interval of $[s, s + l_k)$.

We introduce the following variables to give the integer program of the RFLEP.

- For a pair of facility $(i, k, s) \in \mathcal{F}$ and client $(j, t) \in \mathcal{D}$, use variable $x_{iks,jt}$ to indicate whether the client (j, t) is connected to the facility (i, k, s).
- For a facility $(i, k, s) \in \mathcal{F}$, use y_{iks} to indicate whether the facility location i will be leased in type k starting from the time period of s.
- For a client $(j, t) \in \mathcal{D}$, use z_{jt} to indicate whether the client (j, t) is being penalized.
- For a client $(j, t) \in \mathcal{D}$, use r_{jt} to indicate whether the client (j, t) is being selected as an outlier.

The RFLEP can be formulated as the following integer program:

$$\min \sum_{(i,k,s)\in\mathcal{F}} f_i^k y_{iks} + \sum_{(i,k,s)\in\mathcal{F}} \sum_{(j,t)\in\mathcal{D}} c_{ij} x_{iks,jt} + \sum_{(j,t)\in\mathcal{D}} p_j z_{jt} \tag{1}$$

$$\text{s.t.} \sum_{(i,k,s)\in\mathcal{F}:t\in I_s^k} x_{iks,jt} + z_{jt} + r_{jt} \geq 1, \qquad \forall (j, t) \in \mathcal{D},$$

$$x_{iks,jt} \leq y_{iks}, \qquad \forall (i, k, s) \in \mathcal{F}, (j, t) \in \mathcal{D},$$

$$\sum_{(j,t)\in\mathcal{D}} r_{jt} \leq q,$$

$$x_{iks,jt}, y_{iks}, z_{jt}, r_{jt} \in \{0, 1\}, \qquad \forall (i, k, s) \in \mathcal{F}, (j, t) \in \mathcal{D}.$$

The objective function is the sum of leasing, connection, and penalty costs. The first constraints ensure that any client must be connected, or penalized, or selected as an outlier. The second constraints guarantee that if a client is connected to a facility, the facility must be leased. The third constraint ensures that the number of outliers is no more than q.

Relaxing all the variables in the program (1) yields the following linear program relaxation:

$$\min \sum_{(i,k,s)\in\mathcal{F}} f_i^k y_{iks} + \sum_{(i,k,s)\in\mathcal{F}} \sum_{(j,t)\in\mathcal{D}} c_{ij} x_{iks,jt} + \sum_{(j,t)\in\mathcal{D}} p_j z_{jt} \tag{2}$$

$$\text{s.t.} \sum_{(i,k,s)\in\mathcal{F}:t\in I_s^k} x_{iks,jt} + z_{jt} + r_{jt} \geq 1, \qquad \forall (j, t) \in \mathcal{D},$$

$$x_{iks,jt} \leq y_{iks}, \qquad \forall (i, k, s) \in \mathcal{F}, (j, t) \in \mathcal{D},$$

$$\sum_{(j,t)\in\mathcal{D}} r_{jt} \leq q,$$

$$x_{iks,jt}, y_{iks}, z_{jt}, r_{jt} \geq 0, \qquad \forall (i, k, s) \in \mathcal{F}, (j, t) \in \mathcal{D}.$$

By introducing the corresponding dual variables, we obtain the following dual program of the program (2):

$$\max \sum_{(j,t)\in\mathcal{D}} v_{jt} - q\theta \tag{3}$$

$$
\begin{aligned}
\text{s.t.} \quad & v_{jt} - w_{iks,jt} \leq c_{ij}, && \forall (i,k,s) \in \mathcal{F}, (j,t) \in \mathcal{D}, t \in I_s^k \\
& \sum_{(j,t)\in\mathcal{D}} w_{iks,jt} \leq f_i^k, && \forall (i,k,s) \in \mathcal{F}, \\
& v_{jt} \leq p_j, && \forall (j,t) \in \mathcal{D}, \\
& v_{jt} \leq \theta, && \forall (j,t) \in \mathcal{D}, \\
& v_{jt}, w_{iks,jt}, \theta \geq 0. && \forall (i,k,s) \in \mathcal{F}, (j,t) \in \mathcal{D}.
\end{aligned}
$$

For the first and the second constraints in the above dual program, the variable v_{jt} of each client (j,t) can be viewed as a budget that (j,t) is willing to pay for being connected to a facility. The third constraints imply that the budget of any client (j,t) should be less than its penalty cost. The fourth constraints suggest that we could define a positive upper bound θ for the budget of each client.

Due to space limitations, the proof of all the lemmas in this paper are removed but will appear in a full version.

3   A Primal-Dual Approximation Algorithm

Our main difficulty in solving the RFLEP is that its natural linear program relaxation has an unbounded integrality gap. To overcome this difficulty, we focus on constructing a modified linear program relaxation with a bounded integrality gap at the very beginning. For any given RFLEP instance $\mathcal{I}$, to obtain the modified linear program relaxation, we need to construct a modified RFLEP instance $\mathcal{I}^{(1)}$. The instance $\mathcal{I}^{(1)}$ is constructed from guessing the facility with the most expensive leasing cost in an optimal solution and updating the leasing costs of some related facilities. The construction process of instance $\mathcal{I}^{(1)}$ is formally given in the Algorithm 1.

For the modified RFLEP instance $\mathcal{I}^{(1)}$, we try to construct a dual feasible solution of its corresponding dual program. The construction process of the dual solution is formally given in the Algorithm 2, which also outputs the sets of temporarily leased facilities, connected clients, penalized clients, and selected outliers represented by the symbols of $\widetilde{\mathcal{F}}$, $\widetilde{\mathcal{C}}$, $\widetilde{\mathcal{P}}$, and $\widetilde{\mathcal{O}}$, respectively. Note that all the temporary sets do not necessarily lead to a feasible solution of the given RFLEP instance $\mathcal{I}$, but it is possible to construct a feasible one based on them. A brief description of the Algorithm 2 is that we start by setting all dual variables to 0 and uniformly raising the values of some dual variables in each iteration of the dual construction process, and we terminate the algorithm when all the dual variables cannot be raised anymore. In the Algorithm 2, we dummy a timing variable t_{dum} and for each facility (i,k,s), denote by t_{iks} its temporarily leased time. For each client (j,t), denote by $\text{wit}(j,t)$ the facility that causes the dual

---

Algorithm 1: Construct a modified RFLEP instance $\mathcal{I}^{(1)}$.

Input: A given RFLEP instance $\mathcal{I}$.

Output: A modified RFLEP instance $\mathcal{I}^{(1)}$.

    Step 1 For the given RFLEP instance $\mathcal{I}$, guess the facility $(i_{\max}, k_{\max}, s_{\max})$, which has the most expensive leasing cost $f_{i_{\max}}^{k_{\max}}$ in an optimal solution.

    Step 2 For the facility $(i_{\max}, k_{\max}, s_{\max})$, update its leasing cost to 0. For each facility $(i, k, s) \in \mathcal{F} \setminus (i_{\max}, k_{\max}, s_{\max})$, if its leasing cost is greater than $f_{i_{\max}}^{k_{\max}}$, update the leasing cost to ∞.

    Step 3 Output the modified RFLEP instance $\mathcal{I}^{(1)}$ which has the same inputs of the given instance $\mathcal{I}$ except for the updated leasing costs.

---

variable of v_{jt} to stop raising, and we call wit(j, t) the connecting witness of the client (j, t).

Based on the constructed dual feasible solution and the sets of $\widetilde{\mathcal{F}}$, $\widetilde{\mathcal{C}}$, $\widetilde{\mathcal{P}}$, and $\widetilde{\mathcal{O}}$, it is possible to construct a feasible solution of the given RFLEP instance $\mathcal{I}$. The construction process of the feasible solution of $\mathcal{I}$ is formally given in the Algorithm 3. In the Algorithm 3, let sets $\hat{\mathcal{F}}$, $\hat{\mathcal{C}}$, $\hat{\mathcal{P}}$, and $\hat{\mathcal{O}}$ respectively denote the set of finally leased facilities, clients being connected, clients being penalized, clients selected as outliers and denote by $\sigma : \hat{\mathcal{C}} \to \hat{\mathcal{F}}$ the connection between clients in $\hat{\mathcal{C}}$ and the leased facilities in $\hat{\mathcal{F}}$.

The following lemma prove that step 5 in the Algorithm 3 can be successfully performed, and makes it clear that the Algorithm 3 can provide a feasible solution.

Lemma 1. *For each client $(j, t) \in \hat{\mathcal{C}}$, there exists some facility $(i, k, s) \in \hat{\mathcal{F}}$ satisfying $t \in I_s^k$.*

4 The Analysis of the Approximation Ratio

Denote by OPT and OPT$^{(1)}$ the value of the optimal solutions of the given and modified RFLEP instances $\mathcal{I}$ and $\mathcal{I}^{(1)}$, respectively. Recall that the facility $(i_{\max}, k_{\max}, s_{\max})$ is the facility with the most expensive leasing cost $f_{i_{\max}}^{k_{\max}}$ in an optimal solution. According to the construction process of the instance $\mathcal{I}^{(1)}$, we have that

$$f_{i_{\max}}^{k_{\max}} + \text{OPT}^{(1)} \le \text{OPT}.$$

Respectively denote LC, CC, and PC as the total leasing, connection, and penalty cost of the feasible solution $(\hat{\mathcal{F}}, \hat{\mathcal{C}}, \hat{\mathcal{P}}, \hat{\mathcal{O}}, \sigma)$ obtained from the Algorithm 3. We divide the set of clients $\hat{\mathcal{C}}$ into two sets $\hat{\mathcal{C}}_1$ and $\hat{\mathcal{C}}_2$, where for each client $(j, t) \in \hat{\mathcal{C}}_1$, there exists a facility $(i, k, s) \in \mathcal{F}_{\text{ind}}$ satisfying $w_{iks,jt} > 0$, and for each client $(j, t) \in \hat{\mathcal{C}}_2$, there exists no facility $(i, k, s) \in \mathcal{F}_{\text{ind}}$ satisfying $w_{iks,jt} > 0$. For each client $(j, t) \in \hat{\mathcal{C}}_1$, define

$$v_{jt}^{\text{LC}} := w_{iks,jt} \text{ and } v_{jt}^{\text{CC}} := c_{ij},$$

Algorithm 2: Construct a dual feasible solution related to $\mathcal{I}^{(1)}$.

Input: The modified RFLEP instance $\mathcal{I}^{(1)}$.

Output: A dual feasible solution related to $\mathcal{I}^{(1)}$ and sets $\widetilde{\mathcal{F}}$, $\widetilde{\mathcal{C}}$, $\widetilde{\mathcal{P}}$, and $\widetilde{\mathcal{O}}$.

Step 1 Initialization.

For each client $(j,t) \in \mathcal{D}$ and facility $(i,k,s) \in \mathcal{F}$, set dual variables $v_{jt} := 0$ and $w_{iks,jt} := 0$. Set $\theta := 0$. For each facility $(i,k,s) \in \mathcal{F}$, set $t_{iks} := \infty$. Set $t_{\text{dum}} := 0$. Set $\widetilde{\mathcal{F}} := \emptyset$, $\widetilde{\mathcal{C}} := \emptyset$, $\widetilde{\mathcal{P}} := \emptyset$, $\widetilde{\mathcal{O}} := \mathcal{D}$. For each facility $(i,k,s) \in \mathcal{F}$, set $N_{iks}^{\text{wit}} := \emptyset$. For each client $(j,t) \in \mathcal{D}$, set wit$(j,t) := \emptyset$.

Step 2 Dual ascent process.

While $|\widetilde{\mathcal{O}}| > q$ **do**

Raise t_{dum} along with each dual variable $v_{jt} \in \widetilde{\mathcal{O}}$ uniformly until one of the following four events occurs. If more than one events happen at the same time, we arbitrarily break ties.

Event 1. There exist some facility $(i,k,s) \in \mathcal{F} \setminus \widetilde{\mathcal{F}}$, client $(j,t) \in \widetilde{\mathcal{O}}$, such that

$$v_{jt} = c_{ij} \text{ and } t \in I_s^k.$$

If Event 1 happens, we update $w_{iks,jt} := v_{jt} - c_{ij}$, and start to increase $w_{iks,jt}$ uniformly with the increasing of v_{jt}.

Event 2. There exist some facility $(i,k,s) \in \widetilde{\mathcal{F}}$, client $(j,t) \in \widetilde{\mathcal{O}}$, such that

$$v_{jt} = c_{ij} \text{ and } t \in I_s^k.$$

If Event 2 happens, we stop to increase v_{jt}. Update $N_{iks}^{\text{wit}} := N_{iks}^{\text{wit}} \cup \{(j,t)\}$, $\widetilde{\mathcal{C}} := \widetilde{\mathcal{C}} \cup \{(j,t)\}$, $\widetilde{\mathcal{O}} := \widetilde{\mathcal{O}} \setminus \{(j,t)\}$, and wit$(j,t) := (i,k,s)$.

Event 3. There exist some facility $(i,k,s) \in \mathcal{F} \setminus \widetilde{\mathcal{F}}$, such that

$$\sum_{(j,t) \in \mathcal{D}} w_{iks,jt} = f_i^k.$$

If Event 3 happens, we stop to increase v_{jt} for each client $(j,t) \in \widetilde{\mathcal{O}}$ satisfying $w_{iks,jt} > 0$. Define

$$N_{iks}^{\text{wit}} := \{(j,t) \in \widetilde{\mathcal{O}} : w_{iks,jt} > 0\}.$$

Update $\widetilde{\mathcal{F}} := \widetilde{\mathcal{F}} \cup \{(i,k,s)\}$, $\widetilde{\mathcal{C}} := \widetilde{\mathcal{C}} \cup N_{iks}^{\text{wit}}$, $\widetilde{\mathcal{O}} := \widetilde{\mathcal{O}} \setminus N_{iks}^{\text{wit}}$. Update $t_{iks} := t_{\text{dum}}$ and for each client $(j,t) \in N_{iks}^{\text{wit}}$, update wit$(j,t) := (i,k,s)$.

Event 4. There exist some client $(j,t) \in \widetilde{\mathcal{O}}$, such that

$$v_{jt} = p_j.$$

If Event 4 happens, we stop to increase v_{jt}, and update $\widetilde{\mathcal{P}} := \widetilde{\mathcal{P}} \cup \{(j,t)\}$, and $\widetilde{\mathcal{O}} := \widetilde{\mathcal{O}} \setminus \{(j,t)\}$.

Stop raise t_{dum} along with each dual variable v_{jt} satisfying $(j,t) \in \widetilde{\mathcal{O}}$, and update $\theta := t_{\text{dum}}$.

Output Dual variables $\{v_{jt}\}_{(j,t) \in \mathcal{D}}$, $\{w_{isk,jt}\}_{(i,k,s) \in \mathcal{F},(j,t) \in \mathcal{D}}$ and θ. Sets $\widetilde{\mathcal{F}}$, $\widetilde{\mathcal{C}}$, $\widetilde{\mathcal{P}}$, and $\widetilde{\mathcal{O}}$.

Algorithm 3: Construct a feasible solution of the given instance $\mathcal{I}$.

Input: A dual feasible solution related to $\mathcal{I}^{(1)}$ and sets $\widetilde{\mathcal{F}}$, $\widetilde{\mathcal{C}}$, $\widetilde{\mathcal{P}}$, and $\widetilde{\mathcal{O}}$.

Output: A feasible solution of the given RFLEP instance $\mathcal{I}$.

Step 1 Initialization.

Define $\hat{\mathcal{F}} := \emptyset$, $\hat{\mathcal{C}} := \emptyset$, $\hat{\mathcal{P}} := \emptyset$, $\hat{\mathcal{O}} := \emptyset$, and $\sigma(j, t) := \emptyset$ for each client $(j, t) \in \mathcal{D}$.

Step 2 Select the outliers.

If $|\widetilde{\mathcal{O}}| = q$, update $\hat{\mathcal{O}} := \widetilde{\mathcal{O}}$. Else, we have that $|\widetilde{\mathcal{O}}| < q$. Denote by $(i_{\mathrm{la}}, k_{\mathrm{la}}, s_{\mathrm{la}})$ the last temporarily opened facility that is added to $\widetilde{\mathcal{F}}$ in Step 2 of the Algorithm 2. Arbitrarily choose $q - |\widetilde{\mathcal{O}}|$ clients with the largest connection cost in $N^{\mathrm{wit}}_{i_{\mathrm{la}} k_{\mathrm{la}} s_{\mathrm{la}}}$ and denote these $q - |\widetilde{\mathcal{O}}|$ clients as $\mathcal{O}_{\mathrm{ga}}$. Update $\hat{\mathcal{O}} := \widetilde{\mathcal{O}} \cup \mathcal{O}_{\mathrm{ga}}$ and select all the clients in $\hat{\mathcal{O}}$ as outliers.

Step 3 Lease the facilities.

For each facility $(i, k, s) \in \mathcal{F}$, define

$$N^{\mathrm{con}}_{iks} := \{(j, t) \in \mathcal{D} : w_{iks, jt} > 0\}.$$

Define

$$\widetilde{\mathcal{F}}_{\mathrm{new}} := \widetilde{\mathcal{F}} \setminus \{(i_{\max}, k_{\max}, s_{\max}), (i_{\mathrm{la}}, k_{\mathrm{la}}, s_{\mathrm{la}})\}.$$

According to the lease duration of the facilities in $\widetilde{\mathcal{F}}_{\mathrm{new}}$, order the facilities from the one with the largest duration to the one with the smallest. Set $l_{\mathrm{iter}} := 1$ and $\mathcal{F}_{\mathrm{ind}} := \emptyset$.

While $l_{\mathrm{iter}} \leq |\widetilde{\mathcal{F}}_{\mathrm{new}}|$ **do**

For the l_{iter}th facility (i, k, s) in $\widetilde{\mathcal{F}}_{\mathrm{new}}$, if there exists some facility (i', k', s') in $\mathcal{F}_{\mathrm{ind}}$ satisfying

$$N^{\mathrm{con}}_{iks} \cap N^{\mathrm{con}}_{i'k's'} \neq \emptyset,$$

update $l_{\mathrm{iter}} := l_{\mathrm{iter}} + 1$, else, update $\mathcal{F}_{\mathrm{ind}} := \mathcal{F}_{\mathrm{ind}} \cup \{(i, k, s)\}$, and also update $l_{\mathrm{iter}} := l_{\mathrm{iter}} + 1$. For each facility $(i, k, s) \in \mathcal{F}_{\mathrm{ind}}$, define

$$\hat{\mathcal{F}}_{iks} := \{(i, k, s), (i, k, s + l_k), (i, k, \max\{s - l_k, 1\})\}.$$

Update

$$\hat{\mathcal{F}} := \bigcup_{(i,k,s) \in \mathcal{F}_{\mathrm{ind}}} \hat{\mathcal{F}}_{iks} \cup \{(i_{\max}, k_{\max}, s_{\max}), (i_{\mathrm{la}}, k_{\mathrm{la}}, s_{\mathrm{la}})\}$$

and lease all the facilities in $\hat{\mathcal{F}}$.

Step 4 Penalize the clients.

Update $\hat{\mathcal{P}} := \widetilde{\mathcal{P}} \setminus \bigcup_{(i,k,s) \in \mathcal{F}_{\mathrm{ind}}} N^{\mathrm{con}}_{iks}$ and penalize all the clients in $\hat{\mathcal{P}}$.

Step 5 Connect the clients.

Update $\hat{\mathcal{C}} := \mathcal{D} \setminus (\hat{\mathcal{P}} \cup \hat{\mathcal{O}})$. For each client $(j, t) \in \hat{\mathcal{C}}$, find facility

$$(i, k, s) := \arg \min_{(i', k', s') \in \hat{\mathcal{F}} : t \in I^{k'}_{s'}} c_{i'j},$$

and update $\sigma(j, t) := (i, k, s)$, and connect the client (j, t) to the facility (i, k, s).

Output A feasible solution $(\hat{\mathcal{F}}, \hat{\mathcal{C}}, \hat{\mathcal{P}}, \hat{\mathcal{O}}, \sigma)$ of the given RFLEP instance $\mathcal{I}$.

---

and for each client $(j, t) \in \hat{\mathcal{C}}_2$, define

$$v_{jt}^{\mathrm{LC}} := 0 \text{ and } v_{jt}^{\mathrm{CC}} := v_{jt}.$$

It can be seen that for each client $(j, t) \in \hat{\mathcal{C}}$, we have that

$$v_{jt} := v_{jt}^{\mathrm{LC}} + v_{jt}^{\mathrm{CC}}.$$

The following lemmas are essential to analyze the approximation ratio of the Algorithm 3.

Lemma 2. *The total leasing cost of the feasible solution $(\hat{\mathcal{F}}, \hat{\mathcal{C}}, \hat{\mathcal{P}}, \hat{\mathcal{O}}, \sigma)$ obtained from the Algorithm 3 satisfies that*

$$\mathrm{LC} \leq 3 \sum_{(j,t) \in \hat{\mathcal{C}}_1} v_{jt}^{\mathrm{LC}} + 2 f_{i_{\max}}^{k_{\max}}.$$

Lemma 3. *The total connection cost of the feasible solution $(\hat{\mathcal{F}}, \hat{\mathcal{C}}, \hat{\mathcal{P}}, \hat{\mathcal{O}}, \sigma)$ obtained from the Algorithm 3 satisfies that*

$$\mathrm{CC} \leq \sum_{(j,t) \in \hat{\mathcal{C}}_1} v_{jt}^{\mathrm{CC}} + 3 \sum_{(j,t) \in \hat{\mathcal{C}}_2} v_{jt}^{\mathrm{CC}}.$$

Lemma 4. *The total penalty cost of the feasible solution $(\hat{\mathcal{F}}, \hat{\mathcal{C}}, \hat{\mathcal{P}}, \hat{\mathcal{O}}, \sigma)$ obtained from the Algorithm 3 satisfies that*

$$\mathrm{PC} = \sum_{(j,t) \in \hat{\mathcal{P}}} p_j = \sum_{(j,t) \in \hat{\mathcal{P}}} v_{jt}.$$

Combining Lemmas 2–4 yields our main results.

Theorem 1. *There exists a 3-approximation algorithm for the RFLEP.*

Proof. Recall that for each client $(j, t) \in \hat{\mathcal{C}}$, we have that $v_{jt} := v_{jt}^{\mathrm{LC}} + v_{jt}^{\mathrm{CC}}$, and that for each $(j, t) \in \hat{\mathcal{C}}_2$, we have that $v_{jt}^{\mathrm{CC}} := v_{jt}$, and that $\hat{\mathcal{C}} := \hat{\mathcal{C}}_1 \cup \hat{\mathcal{C}}_2$. From

Lemmas 2–4, we obtain that

$$\text{LC} + \text{CC} + \text{PC} \tag{4}$$

$$\leq 3 \sum_{(j,t)\in\hat{\mathcal{C}}_1} v_{jt}^{\text{LC}} + 2f_{i_{\max}}^{k_{\max}} + \sum_{(j,t)\in\hat{\mathcal{C}}_1} v_{jt}^{\text{CC}} + 3 \sum_{(j,t)\in\hat{\mathcal{C}}_2} v_{jt}^{\text{CC}} + \sum_{(j,t)\in\hat{\mathcal{P}}} v_{jt}$$

$$= \left(3 \sum_{(j,t)\in\hat{\mathcal{C}}_1} v_{jt}^{\text{LC}} + \sum_{(j,t)\in\hat{\mathcal{C}}_1} v_{jt}^{\text{CC}} \right) + 3 \sum_{(j,t)\in\hat{\mathcal{C}}_2} v_{jt}^{\text{CC}} + \sum_{(j,t)\in\hat{\mathcal{P}}} v_{jt} + 2f_{i_{\max}}^{k_{\max}}$$

$$\leq 3 \left(\sum_{(j,t)\in\hat{\mathcal{C}}_1} v_{jt}^{\text{LC}} + \sum_{(j,t)\in\hat{\mathcal{C}}_1} v_{jt}^{\text{CC}} \right) + 3 \sum_{(j,t)\in\hat{\mathcal{C}}_2} v_{jt}^{\text{CC}} + \sum_{(j,t)\in\hat{\mathcal{P}}} v_{jt} + 2f_{i_{\max}}^{k_{\max}}$$

$$= 3 \left(\sum_{(j,t)\in\hat{\mathcal{C}}_1} v_{jt} + \sum_{(j,t)\in\hat{\mathcal{C}}_2} v_{jt} \right) + \sum_{(j,t)\in\hat{\mathcal{P}}} v_{jt} + 2f_{i_{\max}}^{k_{\max}}$$

$$= 3 \sum_{(j,t)\in\hat{\mathcal{C}}} v_{jt} + \sum_{(j,t)\in\hat{\mathcal{P}}} v_{jt} + 2f_{i_{\max}}^{k_{\max}}.$$

From Step 2 of the Algorithm 2 and Step 2 of the Algorithm 3, it can be seen that for each client $(j,t) \in \hat{\mathcal{O}}$, we have that $v_{jt} = \theta$, and that $|\hat{\mathcal{O}}| = q$. Since the sets of $\hat{\mathcal{C}}$, $\hat{\mathcal{P}}$, $\hat{\mathcal{O}}$ are a partition of the set $\mathcal{D}$, we obtain that

$$\sum_{(j,t)\in\hat{\mathcal{C}}} v_{jt} + \sum_{(j,t)\in\hat{\mathcal{P}}} v_{jt} \tag{5}$$

$$= \sum_{(j,t)\in\hat{\mathcal{D}}} v_{jt} - \sum_{(j,t)\in\hat{\mathcal{O}}} v_{jt}$$

$$= \sum_{(j,t)\in\hat{\mathcal{D}}} v_{jt} - q\theta$$

$$\leq \text{OPT}^{(1)}.$$

Combining inequalities 4 and 5, and the fact that $f_{i_{\max}}^{k_{\max}} + \mathrm{OPT}^{(1)} \le \mathrm{OPT}$ yields

$$\mathrm{LC} + \mathrm{CC} + \mathrm{PC}$$

$$\le 3 \sum_{(j,t)\in\hat{\mathcal{C}}} v_{jt} + \sum_{(j,t)\in\hat{\mathcal{P}}} v_{jt} + 2f_{i_{\max}}^{k_{\max}}$$

$$\le 3 \left(\sum_{(j,t)\in\hat{\mathcal{C}}} v_{jt} + \sum_{(j,t)\in\hat{\mathcal{P}}} v_{jt} \right) + 2f_{i_{\max}}^{k_{\max}}$$

$$= 3 \left(\sum_{(j,t)\in\hat{\mathcal{D}}} v_{jt} - q\theta \right) + 2f_{i_{\max}}^{k_{\max}}$$

$$\le 3\,\mathrm{OPT}^{(1)} + 2f_{i_{\max}}^{k_{\max}}$$

$$\le 3\,\mathrm{OPT},$$

completing the proof of this theorem. □

5   Discussions

In this paper, we study the RFLEP, and give its first constant-factor approximation algorithm. Note that the penalty costs we considered in the RFLEP are linear, and there are a lot of works concentrated on solving problems with submodular penalties, such as the work of Li et al. [19]. In the future, we are interested in designing approximation algorithms for the RFLEP considering submodular penalties.

References

1. Anthony, B.M., Gupta, A.: Infrastructure leasing problems. In: Proceedings of the International Integer Programming and Combinatorial Optimization Conference, pp. 424–438 (2007)
2. Charikar, M., Guha, S.: Improved combinatorial algorithms for facility location problems. SIAM J. Comput. **34**(4), 803–824 (2005)
3. Charikar, M., Khuller, S., Mount, D.M., Narasimhan, G.: Algorithms for facility location problems with outliers. In: Proceedings the Annual ACM-SIAM Symposium on Discrete Algorithms, pp. 642–651 (2001)
4. Chen, X., Ji, S., Wu, C., Xu, Y., Yang, Y.: An approximation algorithm for diversity-aware fair k-supplier problem. Theor. Comput. Sci. 114305 (2023)
5. Chen, X., Xu, D., Xu, Y., Zhang, Y.: Parameterized approximation algorithms for sum of radii clustering and variants. In: Proceedings of the AAAI Conference on Artificial Intelligence, vol. 38, no. 18, pp. 20666–20673 (2024)
6. Du, D., Lu, R., Xu, D.: A primal-dual approximation algorithm for the facility location problem with submodular penalties. Algorithmica **63**, 191–200 (2012)
7. Guha, S., Khuller, S.: Greedy strikes back: improved facility location algorithms. J. Algorithms **31**(1), 228–248 (1999)

8. Han, L., Xu, D., Xu, Y., Yang, P.: Approximation algorithms for the individually fair k-center with outliers. J. Global Optim. **87**(2), 603–618 (2023)
9. Hochbaum, D.S., Shmoys, D.B.: A unified approach to approximation algorithms for bottleneck problems. J. ACM **33**(3), 533–550 (1986)
10. Jain, K., Mahdian, M., Markakis, E., Saberi, A., Vazirani, V.V.: Greedy facility location algorithms analyzed using dual fitting with factor-revealing LP. J. ACM **50**(6), 795–824 (2003)
11. Jain, K., Vazirani, V.V.: Approximation algorithms for metric facility location and k-median problems using the primal-dual schema and Lagrangian relaxation. J. ACM **48**(2), 274–296 (2001)
12. Kuehn, A.A., Hamburger, M.J.: A heuristic program for locating warehouses. Manag. Sci. **9**(4), 643–666 (1963)
13. Li, S.: A 1.488 approximation algorithm for the uncapacitated facility location problem. Inf. Comput. **222**, 45–58 (2013)
14. Li, S., Svensson, O.: Approximating k-median via pseudo-approximation. In: Proceedings of the Annual ACM Symposium on Theory of Computing, pp. 901–910 (2013)
15. Li, Y., Du, D., Xiu, N., Xu, D.: Improved approximation algorithms for the facility location problems with linear/submodular penalty. In: Proceedings of the International Computing and Combinatorics Conference, pp. 292–303 (2013)
16. Luo, H., Han, L., Shuai, T., Wang, F.: Approximation and heuristic algorithms for the priority facility location problem with outliers. Tsinghua Sci. Technol. **29**(6), 1694–1702 (2024)
17. Nagarajan, C., Williamson, D.P.: Offline and online facility leasing. Discret. Optim. **10**(4), 361–370 (2013)
18. Shmoys, D.B., Tardos, É., Aardal, K.: Approximation algorithms for facility location problems. In: Proceedings of the Annual ACM Symposium on Theory of Computing, pp. 265–274 (1997)
19. Wang, F., Xu, D., Wu, C.: Combinatorial approximation algorithms for the robust facility location problem with penalties. J. Global Optim. **64**, 483–496 (2016)
20. Xu, G., Xu, J.: An LP rounding algorithm for approximating uncapacitated facility location problem with penalties. Inf. Process. Lett. **94**(3), 119–123 (2005)
21. Xu, G., Xu, J.: An improved approximation algorithm for uncapacitated facility location problem with penalties. J. Comb. Optim. **17**(4), 424–436 (2009)
22. Xu, Y., Chau, V., Wu, C., Zhang, Y., Zissimopoulos, V., Zou, Y.: A semi brute-force search approach for (balanced) clustering. Algorithmica **86**(1), 130–146 (2024)

Randomized Strategyproof Mechanisms for Multi-Stage Facility Location Problem with Capacity Constraints

Chi Kit Ken Fong[1]([✉]), Xingchen Sha[2], Hau Chan[3], Vincent Chau[4], and Wai-Lun Lo[5]

[1] School of Data Science, Division of Artificial Intelligence, Lingnan University, Hong Kong, China
kenfong@ln.edu.hk
[2] City University of Hong Kong, Hong Kong, China
xingchsha2-c@my.cityu.edu.hk
[3] University of Nebraska-Lincoln, Lincoln, USA
hchan3@unl.edu
[4] Southeast University, Nanjing, China
vincentchau@seu.edu.cn
[5] Hong Kong Chu Hai College, Hong Kong, China
wllo@chuhai.edu.hk

Abstract. We consider the multi-stage facility location problem with capacity constraints. In the problem, we seek to locate at most one capacity constrained facility in each stage to serve a subset of agents, who arrive over different stages and are located on a line. Our goal is to design randomized strategyproof mechanisms to elicit agents' true information and locate facilities that minimize the social cost and maximum cost, which are defined to be the sum and the maximum of the agents' costs, respectively. Because of the stages, an agent's cost depends on the agent's distance to their assigned facility and the agent's waiting cost. For different facility capacity settings with waiting cost, we provide randomized strategyproof mechanisms for the considered cost objectives. We also establish lower bounds for the approximation ratios given by any randomized strategyproof mechanisms.

Keywords: Facility location · Mechanism design · Algorithmic game theory

1 Introduction

In recent years, facility location problems have received significant attention in mechanism design and social choice communities because of their ability to model real-world preference aggregation settings (e.g., voting [3,4,14] and site locations [6,15]). In the standard mechanism design study of facility location, a social planner aims to locate facilities to serve strategic agents while eliciting agent preferences and optimizing the social or maximum cost [6].

Most existing mechanism design studies on facility location problem have focused on facilities without any capacity constraints or immobile over time. However, in real-world settings, facilities (e.g., mobile health clinics, mobile blood donation centers, mobile outreach programs) are both capacity constrained [1,2,18] and mobile (i.e., need to be relocated over several periods or stages [8,17,19]). In this paper, we consider the multi-stage facility location problem with capacity constraints (MSFLPs-CC) by merging the facility location problem with capacity constraints and the multi-stage facility location problem (with mobile facilities). The considered facility location problem can be used to model a wide range of multistage settings, including locating shuttle stops, providing mobile health services, scheduling education sessions, and other multi-stage extensions of scenarios (see e.g., [1,2,7,9,16,18]).

Notice that both multi-stage models with [8] or without moving cost [17,19] have been studied in the facility location literature. We focus on the latter settings because the (mobile) facility's moving cost are negligible/independent across each period and the social planner focuses on minimizing the agents' costs rather than their own costs.

For instance, consider an organization that is providing shuttle services from a central station/point (e.g., hotels/workplaces) at the beginning of each stage, where each shuttle will travel to its designated location and take their agents (e.g., clients or employees), who have location preferences and arrive over different time intervals, to a relatively distant destination. Because a shuttle often has a limited seating capacity to serve new agents arriving over different time intervals, the company needs to determine the pick-up points of the shuttle for each time period and how to assign agents to the shuttle at different time intervals to better serve the agents, accounting for agent travel distances to the shuttle locations and, possibly, agent waiting times over different time intervals. Moreover, the moving cost of the facilities between stages is negligible compared to the pre-determined round-trip or operation costs (which are independent and incomparable to the agents' costs). As such, moving cost is not considered and incorporated into the planner's objectives.

In addition, the considered multi-stage problems play an important role in assisting the mobile health services [7,16]. At the beginning of each day, the medical facility (e.g., vaccine vehicles, mobile blood donation centers) travels from a central station/point (e.g., medical clinic) to serve residents of several areas. Due to the storage capacity and vaccination time restrictions, the medical facility can only serve a limited number of residents each day. Therefore, the government needs to decide where to locate the facility and which residents to serve over different periods.

1.1 Our Contributions

We study the multi-stage facility location problem with k capacity constrained facilities on the real line $\mathbb{I} = [0, 1]$, where k is the maximum number of available facilities, and the real line is the most widely studied setting [6] that abstractly represents the facility serving range (e.g., the location range of the mobile vaccine vehicle on a street, and the difficulty scale of the standardized materials).

In this paper, we characterize our mechanism design results (see Tables 1) for minimizing the two cost objectives by the number of agents (i.e., n), the number of facilities (i.e., k), facility capacities (i.e., $c_1, \ldots, c_k$), the last stage with arriving agents (i.e., T), and the penalty coefficient (i.e., d).

We consider two different facility capacity settings with waiting cost, i.e., $d > 0$ (see Table 1). For the equal capacity setting (e.g., when the facility capacity cannot change

Table 1. Summary of our results. Notice that it reduces to the classic single facility location problems when $k = 1$.

Objective	Upper Bound	Lower Bound
Equal Capacity Setting:		
Social Cost	$\frac{n}{2d} + 1, T(n - c) + 1$	$\frac{2}{4d+3} + 1$
Max Cost	$\frac{1}{d} + 1, max(T + k - 2, 2)$	$\frac{1}{4d+2} + 1$
Arbitrary Capacity Setting:		
Social Cost	$\frac{n}{2d} + 1, Tn - T + 1$	$\frac{2}{4d+3} + 1$
Max Cost	$\frac{1}{d} + 1, max(T + k - 2, 2)$	$\frac{1}{4d+2} + 1$

over time, such as shuttle buses) where $c_1 = \cdots = c_k = c$ and $k \cdot c = n$, we first present a randomized strategyproof mechanism, which has a good performance when d is relatively large. It achieves approximation ratios of $\frac{n}{2d} + 1$ for social cost and $\frac{1}{d} + 1$ for maximum cost. We also provide another strategyproof mechanism which performs well when d is small, and notice that this mechanism can also work when there is spare capacity, i.e., $kc > n$. It achieves an approximation ratio of $T(n - c) + 1$ for social cost and $max(T + k - 2, 2)$ for maximum cost. We complement the result by giving lower bounds of $\frac{2}{4d+3} + 1$ for social cost and $\frac{1}{4d+2} + 1$ for maximum cost by any strategyproof mechanism.

For the arbitrary capacity setting (e.g., when the facility capacity can change over time such as a classroom in the education setting), we first provide a strategyproof mechanism with dynamic programming to optimize the waiting cost of agents, which performs well when the waiting cost is large. It achieves an approximation ratio of $\frac{n}{2d} + 1$ for social cost and $\frac{1}{d} + 1$ for maximum cost. Then we provide another strategyproof mechanism with dynamic programming to optimize the distance cost of agents, which has a good performance when d is small. It has approximation ratios of $Tn - T + 1$ for social cost and $max\{T + k - 2, 2\}$ for maximum cost.

The remainder of the paper is organized as follows. We first formally define the problem in Sect. 2. Then, we study different capacity settings with waiting cost in Sect. 3. Finally, we conclude our work and discuss the open questions in Sect. 4.

1.2 Related Work

We focus on studies on mechanism design for facility location problem that are most related to ours. We note that there are optimization studies that consider facility location problem where facilities have capacity constraints (see, e.g., [5, 16]) or are mobile (see, e.g., [9, 16]).

In mechanism design for facility location problem, Moulin [14] first characterized strategyproof mechanisms for the classical single stage facility location problem on a line, where agents have single-peaked preferences. The work of Procaccia and Tennenholtz [15] initiated the study of approximate mechanism design without money and used the facility location problem as a case study. They obtained several approximately optimal (deterministic and randomized) strategyproof mechanisms for the single facility location problem under the social and maximum cost objectives. They also considered two homogeneous facilities or multiple locations per agent. Later numerous studies improved the bounds and complemented the results with k facilities [10–13].

Since then, different variants of the classical facility location problem have been proposed. Aziz et al. [2] first introduced the capacity constrained facility location problem from a mechanism design perspective, where the number of agents is larger than the total capacity of facilities. Later, Aziz et al. [1] provided negative results on several classical mechanisms for the uncapacitated settings in terms of the strategyproofness and the approximation ratio. They also provided a deterministic strategyproof mechanism, the INNERPOINT mechanism for two facilities with equal capacity constraint, and a deterministic strategyproof mechanism for two facilities with arbitrary capacity constraints. They also proved that the corresponding optimization problem with arbitrary capacity constraints is NP-hard. Besides the algorithmic and the mechanism perspective, Walsh [18] showed a strong characterization theorem that the INNERPOINT mechanism is the unique strategyproof mechanism that is both anonymous and Pareto optimal for two facilities location problem with equal capacity constraints.

There are also studies on the multi-stage facility location problem without any capacity constraints where agents arrive dynamically over different stages. The work of De Keijzer and Wojtczak [8] investigated the multi-stage facility reallocation problem on the real line, where the goal is to minimize the sum of distance costs between the facility and agents at all stages, plus the facility's moving cost. The work of Wada et al. [17] studied the dynamic facility location problem from the mechanism design perspective, where agents can decide their participations in each stage. Wang et al. [19] studied the multi-stage facility location problem with transient agents who arrive in arbitrary stage and stay for a number of consecutive stages. They analyze the problems from both algorithmic and mechanism design perspectives. To the best of our knowledge, we are the first to simultaneously consider the multi-stage settings, facility capacity constraints, and waiting time of each agent in facility location problem. For more details, please refer to a recent survey on mechanism design for facility location problem [6].

2   Preliminaries

In this section, we formally define the multi-stage facility location problem with capacity constraints (MSFLPs-CC) and the considered mechanism design problems.

Multi-Stage Facility Location Problem with Capacity Constraints. We are given a collection of agents $N := \{1, \ldots, n\}$, where agents arrive in different stages and $T \geq 1$ is the last stage with arriving agents. Each agent $j \in N$ has location $x_j \in \mathbb{I} = [0, 1]$. We denote the location profile of all agents as $X = (x_1, \ldots, x_n)$. We assume agents are ordered such that $x_1 \leq x_2 \leq \cdots \leq x_n$.

In our setting, we have $k \geq 1$ (mobile) facilities to serve these agents. At each stage, at most one facility can be placed due to resource constraints (such a setting with a single facility at each stage is also motivated and considered by [8,9,16,19]).

In order to serve all agents, it is also possible to place the facility at stages beyond T. Therefore, we consider at most $T + k - 1$ stages when locating the facilities. When $k = 1$, the only way to serve all of the agents is to wait until everyone has arrived and serve them all at stage T, which is equivalent to the standard facility location problem with a single facility. Because $k = 1$ has been well-studied (see e.g., [6, 15]) for social and maximum costs, we focus on the cases of $k \geq 2$.

Each facility is indexed by $i \in [k] = \{1, \ldots, k\}$ with a capacity constraint c_i restricting the number of agents it can serve, and its location is denoted as $f_i \in \mathbb{I} = [0, 1]$. The k facilities' capacities are denoted by $C = (c_1, \ldots, c_k)$.

The facility that serves agent $j \in N$ is indicated by $a_j \in [k]$. We denote $N_i := \{j | a_j = i\}$ as the group of agents that facility i serves in the same stage. For simplicity, denote $L_i := \min_{j \in N_i} x_j$ and $R_i := \max_{j \in N_i} x_j$ as the locations of the leftmost agent and the rightmost agent in group i, respectively. The arrival stage of agent j is denoted as $r_j \leq T$. We denote the arrival stage profile of the agents as $R = (r_1, \ldots, r_n)$. The serving stage of facility $i \in [k]$ (i.e., the stage in which the agents in N_i are served) is denoted as $s_i \leq T + k - 1$. Clearly, for N_i to be valid and feasible, $|N_i| \leq c_i$ and $r_j \leq s_i$ for any $j \in N_i$.

Mechanism Design Problems. We are interested in designing randomized strategyproof mechanisms that minimize the social or maximum cost.

A randomized mechanism is a function F that maps the profile (X, R, C) to $Y := (f_{P_1}, \ldots, f_{P_k})$, $O := (N_{P_1}, \ldots, N_{P_k})$, and $S := (s_{P_1}, \ldots, s_{P_k})$, where each f_{P_i} is a set of probability distributions over $\mathbb{I}$, N_{P_i} is a set of probability distributions over N, and S_{P_i} is a set of probability distributions over $\{1, \ldots, T + k - 1\}$ such that for each pair of values (N_i, s_i) from the distribution (N_{P_i}, s_{P_i}), we have $r_j \leq s_i$ for any $j \in N_i$ and $|N_i| \leq c_i$. Given the penalty coefficient d, the cost of an agent is defined to be $cost(F(X, R, C), x_j, r_j) = E_{f_i \sim f_{P_i}, N_i \sim N_{P_i}, s_i \sim s_{P_i}}[|y_{a_j} - x_j| + d \cdot (s_{a_j} - r_j)]$, which is the expected sum of the distance cost and the waiting cost. We focus on the setting where at least one agent has to wait for at least one stage to get served. Otherwise, it can be solved and reduced to the static setting where each facility can serve all the arriving agents in a stage, and thus, agents will only have distance cost.

Denote (X_{-j}, x'_j) as the tuple X with x'_j in place of x_j, and (R_{-j}, r'_j) as the tuple R with r'_j in place of r_j. Below, we provide a formal definition of strategyproofness.

Definition 1. *A mechanism F is strategyproof if for all X, R, $x'_j \in \mathbb{I}$ and $r'_j \geq r_j$, we have*

$$cost(F(X, R, C), x_j, r_j)$$
$$\leq cost(F((X_{-j}, x'_j), (R_{-j}, r'_j), C), x_j, r_j).$$

Given a mechanism F and a profile (X, R, C), the social cost function is defined as $SC[F(X, R, C)] = \sum_{j \in N} cost(F(X, R, C), x_j, r_j)$ and the maximum cost function is defined as $MC[F(X, R, C)] = \max_{j \in N} cost(F(X, R, C), x_j, r_j)$. A mechanism F achieves an approximation ratio of ρ for the social cost (resp. maximum cost), if for any profile (X, R, C), $SC[F(X, R, C)] \leq \rho \cdot SC[OPT(X, R, C)]$ (resp. $MC[F(X, R, C)] \leq \rho \cdot MC[OPT(X, R, C)]$) where $OPT(X, R, C)$ is the optimal solution that minimizes the social or maximum cost.

3 MSFLPs-CC with Waiting Cost

In this section, we study the randomized strategyproof mechanisms for multi-stage facility location problem with capacity constraints.

3.1 Equal Capacity Setting with Waiting Cost

In this subsection, we assume all facilities have equal capacity such that $c_1 = c_2 = \cdots = c_k = c = \frac{n}{k}$. We first design a randomized strategyproof mechanism that has a better performance if the penalty coefficient (i.e., d) is large. Mechanism 1 places all the facilities at the median of all agents. It will gradually allocate facilities from 1 to k as long as at least c agents have arrived but have not been served in each stage. Suppose the allocations of facilities from 1 to $i - 1$ have been determined. Denote the agents that have arrived before or in stage t but have not been served by facilities at $f_1, f_2, \ldots, f_{i-1}$ as $M_{t,i} = \{j | r_j \leq t, j \in N - \sum_{k=1}^{i-1} N_k\}$.

Mechanism 1. *Let $f_1 = \cdots = f_k = x_{\lceil \frac{n}{2} \rceil}$. Starting from stage 1 and facility $i = 1$, we consider one of the following two cases. A function $q()$ is used to help randomly pick c agents with equal probability[1].*
Case 1. *If $|M_{t,i}| \geq c$, let $N_i = q(M_{t,i})$ and $s_i = t$. We then continue with stage $t + 1$ and facility $(i + 1)$.*
Case 2. *Otherwise, no agents are served in stage t. We continue with stage $t + 1$ and facility i.*

Lemma 1. *Mechanism 1 is strategyproof. It has approximation ratios of $\frac{n}{2d} + 1$ for the social cost and $\frac{1}{d} + 1$ for the maximum cost.*

Proof. It is clear that no agent can decrease their distance cost by misreporting their location when all the facilities are placed at the median of all the agents. Besides, arrived agents are all selected with equal probability, and no agent can be served earlier by misreporting their location or a later arrival stage. We use OPT_d and OPT_w to indicate the target distance cost and the waiting cost of the agents in the optimal solution. Since the facility serves agents as soon as at least c agents have arrived in the current stage, the total waiting cost will not exceed the optimal solution's total waiting cost. And the sum of distance cost is at most $\frac{n}{2}$. Therefore, the approximation ratio for social cost is

$$\frac{Mechanism}{OPT} \leq \frac{\frac{n}{2} + OPT_w}{OPT_d + OPT_w} \leq \frac{n}{2d} + 1.$$

The maximum of distance cost is at most 1. Therefore, the approximation ratio for maximum cost is

$$\frac{Mechanism}{OPT} \leq \frac{1 + OPT_w}{OPT_d + OPT_w} \leq \frac{1}{d} + 1.$$

□

---

[1] One implementation of $q()$ is to permute agents by a random order π, and output the agents $\pi(1), \ldots, \pi(c)$.

Algorithm 1. FindMinLen(X,c)

Input: Agent location profile X and the equal capacity c.

Parameter: Define $length(i, m)$ be the minimum length covering the first i agents with m intervals.

Initialize $length(i, 1) =$

$$\begin{cases} x_i - x_1 & \textbf{if } i \leq c, \\ \infty & \textbf{otherwise.} \end{cases}$$

Output:$length(n, k)$.

1: **for** $i \in \{2, \ldots, n\}, j \in \{i - c, \ldots, i - 1\}, m \in \{2, \ldots, k\}$ **do**
2:      $length(i, m) = \min_{j \in \{i-c,\ldots,i-1\}} \{\max\{length(j, m - 1), x_i - x_{j+1}\}\}$
3: **end for**
4: **return** $length(n, k)$

We also provide an example to show that these bounds are tight.

Example 1. Consider there are two facilities with an equal capacity of c, c agents at 0, and c agents at 1. All the agents at 0 arrive at stage 1, one agent at 1 arrives at stage 2, and $c - 1$ agents at 1 arrive at stage 3. Mechanism 1 places two facilities at the median point of all these agents. Therefore, both facilities are placed at 0. Agents at 0 will all be served in stage 1, and agents at 1 will all be served in stage 3. Assuming the waiting cost of each stage is d, the social cost of Mechanism 1 is $\frac{n}{2} + d$ and the maximum cost is $1 + d$. In the optimal solution, one facility is placed at 0, serving all the agents at 0 in stage 1, and the other facility is placed at 1 to serve all the agents at 1 in stage 3, which has a social cost and maximum cost of d. Hence, the approximation ratio is $\frac{n}{2d} + 1$ for social cost and $\frac{1}{d} + 1$ for maximum cost, which shows our analysis is tight.

We also consider the case when the penalty coefficient (i.e., d) is relatively small, and design a randomized strategyproof mechanism where no agents are served until they all arrived. Mechanism 2 is a nontrivial extension of the Equal-Cost randomized mechanism for the classical setting without any capacity constraints for $k \geq 2$ facilities [11]. Mechanism 2 aims to find at most k disjoint intervals such that each interval contains at most c agents (where c is the capacity of the facility), and has a minimum covering length Len. Notice that in order to minimize the maximum length of these intervals, it must be one of the distances between any two agents a and b, i.e., $Len = |x_a - x_b|$. However, different from the Equal-Cost randomized mechanism where each interval has an identical length, some of these lengths might contain more than c agents in which case we can determine and eliminate through the following algorithm $FindMinLen(X, c)$. Notice that Mechanism 2 achieves the same performance and guarantees strategyproofness even if there is spare capacity, i.e., $kc > n$. $FindMinLen(X, c)$ runs in $O(nkc)$ to determine a minimum covering length for each interval and covers at most c agents. Finally, it will wait until all the agents have arrived, and use a random map function $g : \{1, \ldots, k\} \rightarrow \{0, \ldots, k-1\}$ to generate, with equal probability, a permutation order in which facilities serve their agents.

Mechanism 2. *The mechanism performs the following steps.*

1. Use $FindMinLen(X, c)$ to find a profile of allocations $O = \{N_1, \ldots, N_k\}$ such that N_i contains all the agents in the $i-$th interval from left to right and $Len = \max_{i \in [k]}\{R_i - L_i\}$ is minimized where L_i and R_i are the locations of the leftmost agent and the rightmost agent served by facility i given N_i, respectively.

2. If $L_i + Len \leq 1$, f_i is placed at L_i with probability $1/2$ and $L_i + Len$ with probability $1/2$. Otherwise, f_i is placed at R_i with probability $1/2$ and $R_i - Len$ with probability $1/2$.

3. For each facility in the allocations, we pick an arbitrary stage in $\{T, \ldots, T+k-1\}$ for the facility to serve the assigned agents, such that $S = \{T+g(1), \ldots, T+g(k)\}$.

Lemma 2. *Mechanism 2 is strategyproof. It has approximation ratios of $T(n-c)+1$ for social cost and $\max\{2, T+k-2\}$ for maximum cost.*

Proof. We first show that $FindMinLen(X, c)$ returns minimum Len. Suppose there exists a division of intervals with $Len' < Len$ such that all the agents are covered while each interval contains at most c agents. However, Len' can still satisfy the requirements of $FindMinLen(X, c)$. Thus $FindMinLen(X, c)$ will not return Len, which contradicts our assumption.

Because all agents have the expected distance cost of $\frac{Len}{2}$. Any agent may only benefit by reducing the minimum cover length Len. If an agent misreports their location so that the minimum cover length becomes Len', $Len' < L$, the expected distance cost for him will increase to $\frac{Len + (Len - Len')}{2}$. Therefore, agents will not benefit by misreporting. The optimal social distance cost is at least Len since Len is the minimum distance that can cover all the agents in each stage. The total distance cost of Mechanism 2 is at most $\frac{n}{2} \cdot Len$. We use OPT_d and OPT_w to indicate the target distance cost and the waiting cost of the agents in the optimal solution. The worst case in terms of the total waiting cost is when $n-c$ agents arrive in stage 1 and c agents arrive in stage T. The sum of waiting cost of Mechanism 2 is at most $OPT_w + T(n-c)d \leq (T(n-c)+1) \cdot OPT_w$. The approximation ratio for social cost is

$$\frac{Mechanism}{OPT} \leq \frac{(T(n-c)+1) \cdot OPT_w + \frac{n}{2} \cdot OPT_d}{OPT_w + OPT_d}$$
$$\leq T(n-c)+1.$$

The optimal maximum distance cost is at least $\frac{Len}{2}$, and the expected maximum cost of Mechanism 2 is Len. The maximum of waiting cost is at most $(T+k-2)d \leq (T+k-2)\dot{O}PT_w$. Therefore, the approximation ratio for maximum cost is

$$\frac{Mechanism}{OPT} \leq \frac{(T+k-2)\dot{O}PT_w + 2OPT_d}{OPT_d + OPT_w} \leq \max(2, T+k-2).$$

$\square$

Lemma 3. *For $k \geq 2$, any randomized strategyproof mechanism has an approximation ratio of at least $\frac{2}{4d+3} + 1$ for social cost.*

Proof. We construct the first configuration such that $\frac{kc}{2} + 1$ agents locate at 0 and $\frac{kc}{2} - 1$ agents locate at 1, $k \geq 2, c \geq 3$. There are k (k is an even number) facilities with an equal capacity of c. We first focus on the approximation ratio of the distance cost. Now one agent at 0 is moved to $\frac{c-2}{4(c-1)}$. Let $p(x)$ be the probability density function of the probability that the facility serving the agent at $\frac{c-2}{4(c-1)}$ is placed at x in the new configuration. Denote $\bar{x}_1$, $\bar{x}_2$ and $\bar{x}_3$ as the expected facility locations in the intervals $[0, \frac{c-2}{4(c-1)})$, $[\frac{c-2}{4(c-1)}, \frac{c-2}{2(c-1)}]$ and $(\frac{c-2}{2(c-1)}, 1]$ respectively. Let $P_1 = \int_0^{\frac{c-2}{4(c-1)}} p(x)dx$, $P_1 \cdot (\frac{c-2}{4(c-1)} - \bar{x}_1) = \int_0^{\frac{c-2}{4(c-1)}} p(x)x dx$, $P_2 = \int_{\frac{c-2}{4(c-1)}}^{\frac{c-2}{2(c-1)}} p(x)dx$, $P_2 \cdot (\frac{c-2}{4(c-1)} + \bar{x}_2) = \int_{\frac{c-2}{4(c-1)}}^{\frac{c-2}{2(c-1)}} p(x)x dx$, $P_3 = \int_{\frac{c-2}{2(c-1)}}^{1} p(x)dx$, $P_3 \cdot (\frac{c-2}{4(c-1)} + \bar{x}_3) = \int_{\frac{c-2}{2(c-1)}}^{1} p(x)x dx$. Suppose the approximation ratio of a randomized strategyproof mechanism for distance cost is α. In the first configuration, the distance cost of an agent at 0 is at most $(1-P) \cdot \frac{c-2}{4(c-1)} + P \cdot (1 - \frac{c-2}{4(c-1)})$, and the social cost is $P \cdot (\frac{kc}{2} + 1) \le \alpha$, where P is the probability that agents at 0 are served by a facility placed at 1 and the optimal social cost in the first configuration is 1. By strategyproofness, the agent at $\frac{c-2}{4(c-1)}$ in the new configuration cannot benefit by misreporting to 0. Therefore, we have

$$P_1 \cdot \bar{x}_1 + P_2 \cdot \bar{x}_2 + P_3 \cdot \bar{x}_3 \le \frac{\alpha}{\frac{kc}{2}+1} \cdot (1 - \frac{c-2}{4(c-1)}) + (1 - \frac{\alpha}{\frac{kc}{2}+1}) \cdot \frac{c-2}{4(c-1)}$$
$$= \frac{\alpha}{\frac{kc}{2}+1} + (1 - \frac{2\alpha}{\frac{kc}{2}+1}) \cdot \frac{c-2}{4(c-1)}.$$

The optimal social cost of the new configuration is $\frac{3c-2}{4(c-1)}$. If the facility serving the agent at $\frac{c-2}{4(c-1)}$ is placed somewhere in $[0, \frac{c-2}{2(c-1)}]$, the optimal allocation is to serve $c-1$ agents at 0 with the agent at $\frac{c-2}{4(c-1)}$ together. If the facility serving the agent at $\frac{c-2}{4(c-1)}$ is placed in $(\frac{c-2}{2(c-1)}, 1]$, the optimal allocation is to serve $c-1$ agents at 1 and the agent at $\frac{c-2}{4(c-1)}$ together. Therefore, for the social cost of any randomized mechanism, we have

$$\alpha \cdot \frac{3c-2}{4(c-1)} \ge P_1 \cdot ((c-1)(\frac{c-2}{4(c-1)} - \bar{x}_1) + \bar{x}_1 + 1)$$

$$+ P_2 \cdot ((c-1)(\frac{c-2}{4(c-1)} + \bar{x}_2) + \bar{x}_2 + 1)$$

$$+ P_3 \cdot ((c-1)(1 - \frac{c-2}{4(c-1)} - \bar{x}_3) + \bar{x}_3)$$

$$\ge (P_1 + P_2 + P_3)\frac{c+2}{4} + P_3 \cdot \frac{c-2}{2} + P_2 c \bar{x}_2 - (P_1 \bar{x}_1 + P_3 \bar{x}_3)(c-2)$$

$$\ge \frac{c+2}{4} + P_2 c \bar{x}_2 + (c-2)(P_2 \bar{x}_2 - \frac{\alpha}{\frac{kc}{2}+1} - (1 - \frac{2\alpha}{\frac{kc}{2}+1}) \cdot \frac{c-2}{4(c-1)})$$

$$\ge \frac{5c-6}{4(c-1)} - \frac{c(c-2)\alpha}{(kc+2)(c-1)}.$$

Thus, we have $\alpha \ge \frac{5kc^2 - (6k-10)c - 12}{(3k+4)c^2 - (2k+2)c - 4}$. Suppose there exists a randomized strategyproof mechanism with an approximation ratio of β $(1 \le \beta < \frac{5}{3})$ for distance cost. Let

$k = c^2$ and $c > \frac{6+4\beta}{5-3\beta} \geq 5$. We have $\frac{5kc^2-(6k-10)c-12}{(3k+4)c^2-(2k+2)c-4} > \frac{5c^4-6c^3}{3c^4+4c^3} > \beta$, which contradicts the previous statement. Therefore, the sum of distance costs is at least $\frac{5}{3} \cdot OPT_d$. The sum of waiting cost is at least $OPT_w = d$. Let $k = c^2$ and c be as large as possible. According to this instance, the approximation ratio is at least

$$\frac{Mechanism}{OPT} \geq \frac{\frac{5}{3} \cdot OPT_d + OPT_w}{OPT_d + OPT_w} \to \frac{2}{4d+3} + 1. \tag{1}$$

$\square$

Lemma 4. *For $k \geq 2$, any randomized strategyproof mechanism has an approximation ratio of at least $\frac{1}{4d+2} + 1$ for maximum cost.*

Proof. We extend the proof of the lower bound of randomized strategyproof mechanisms in single facility location problem in [15] to this setting. Consider a configuration where we place $2(k-1)$ agents at 0, one agent at $1 - 2\epsilon$, and one agent at $1 - \epsilon$. The capacity of each facility is 2 and all the agents arrive at the same stage. We assume ϵ is small enough such that the agent at $1 - \epsilon$ and the agent at 1 are served by the same facility due to the approximation ratio. It is clear that at least one agent has the expected distance cost of at least $\frac{\epsilon}{2}$. Without loss of generality, we suppose the agent j at $1 - \epsilon$ has the expected distance cost of $E(cost(j)) \geq \frac{\epsilon}{2}$. Now we move the agent at $1 - \epsilon$ to 1, due to the strategyproofness, the expected distance between the facility and point $1 - \epsilon$ in the new configuration should be at least $\frac{\epsilon}{2}$. Therefore, in the new setting, the expected maximum distance cost is at least $\frac{\epsilon}{2} + \epsilon$ while the optimal maximum distance cost is ϵ, which proves a minimum approximation ratio of $\frac{3}{2}$ for distance cost. The sum of waiting cost is at least $OPT_w = d$. Let $k = 2$ and move one agent at $1 - 2\epsilon$ to point 0. According to this instance, the approximation ratio is at least

$$\frac{Mechanism}{OPT} \geq \frac{\frac{3}{2} \cdot OPT_d + OPT_w}{OPT_d + OPT_w} \to \frac{1}{4d+2} + 1.$$

$\square$

3.2 Arbitrary Capacity with Waiting Cost

In this subsection, we consider k facilities with arbitrary capacities such that $n = \sum_{i=1}^{k} c_i$, and provide two randomized strategyproof mechanisms for both social cost and maximum cost. Observe that the equal capacity setting is a special case of this setting. Therefore, we can inherit all the lower bounds in Subsect. 3.1.

Similar to Mechanism 1, serving agents earlier can achieve a better performance when d is large. Thus, we present Mechanism 3 which will serve agents with an optimal allocation that minimizes the waiting cost. It can be proved by a similar method that finding an optimal allocation (for the total or the maximum waiting cost) in the arbitrary capacity setting is also NP-hard [1]. We introduce a dynamic programming algorithm $MinWaiting(R, C)$ to determine the allocation that minimizes the total (resp. maximum) waiting cost. The sub-routine is similar to $FindMinCover(X, C)$ by keeping track of an allocation that minimizes the target cost up to the i-th arriving agents (see the pseudo-code in our supplementary materials).

Algorithm 2. MinWaiting(R,C)

Input: Agent arrival stage profile R in an increasing order and facilities' capacities C

Parameter: Initialize an array Waiting$[n][2^k]$ to $n \cdot T$ and an array of empty tuples Cap$[n][2^k]$, where Waitng$[i][w]$ stores the minimum target waiting cost for agents $\{1, 2, \ldots i\}$ using a set of facilities F, and Cap$[i][w]$ stores the allocation used for agents $\{1, 2, \ldots i\}$ with F, where $w = \sum_{f_j \in F} 2^{j-1}$. Denote $cost(N)$ as the target waiting cost function, i.e., sum or maximum.

Output: An allocation Cap that achieves the minimum target waiting cost.

Mechanism 3   1: **for** $i \in \{1, \ldots, n\}, j \in \{1, \ldots, k\}, w \in \{1, \ldots, 2^k\}$ **do**

2: **if** $i = c_j$ and $w = 2^{j-1}$ **then**

3:      Waiting$[i][w] \leftarrow cost(1, \ldots, i)$

4:      Cap$[i][w] \leftarrow (c_j)$

5: **else if** $c_j \notin$ Cap$[i - c_j][w - 2^{j-1}]$ **then**

6:      Temp$\leftarrow cost($Waiting$[i - c_j][w - 2^{j-1}], i - c_j + 1, \ldots, i)$

7:      **if** Waiting$[i][w]$>Temp **then**

8:          Waiting$[i][w] \leftarrow$ Temp

9:          Cap$[i][w] \leftarrow$ Cap$[i - c_j][w - 2^{j-1}].append(c_j)$

10:     **end if**

11: **end if**

12: **end for**

13: **return** Cap$[n][2^k - 1]$

Let $Cap \leftarrow MinWaiting(R, C)$, and $f_1 = \cdots = f_k = x_{\lceil \frac{n}{2} \rceil}$. Starting from stage 1 and facility $i = 1$, we consider one of the following two cases. A function $q(Cap[i])$ is used to help randomly pick $Cap[i]$ agents with equal probability.

Case 1. If $|M_{t,i}| \geq Cap[i]$, $N_i = q(M_{t,i})$ and $s_i = t$. We then continue with stage $t + 1$ and facility $(i + 1)$.

Case 2. Otherwise, no agents are served in stage t. We continue with stage $t + 1$ and facility i.

Lemma 5. *Mechanism 3 is strategyproof. It has approximation ratios of $\frac{n}{2d} + 1$ for the social cost, and $\frac{1}{d} + 1$ for the maximum cost.*

Proof. Notice that for any feasible solution, the amount of waiting cost decrease due to the misreporting is the same. The total waiting cost of Mechanism 3 will not exceed the optimal solution's total waiting cost. The sum of distance cost is at most $\frac{n}{2}$, and the maximum distance cost is at most 1. By a similar proof of Lemma 1, it is clear that the approximation ratios for the social cost and the maximum cost are $\frac{n}{2d} + 1$ and $\frac{1}{d} + 1$ respectively. □

Notice that finding an (asymptotically) optimal allocation is necessary to achieve bounded approximation ratios when d is small.

Example 2. Consider two facilities with capacities 1 and 2, and three agents at $\{0, x, 1\}$. If we do not always use the optimal one, there are three other possible allocations. The approximation ratio using any one of the three can be arbitrarily large i.e., $x/(1 - x)$.

Thus, we then introduce a dynamic programming algorithm, $FindMinCover$ (X, C), used by Mechanism 4, to determine the allocations and the facilities' locations.

Algorithm 3. FindMinCover(X,C)

Input: Agent location profile X and facilities' capacities C
Parameter: Initialize an array Len[n][2^k] to 1 and an array of empty tuples Cap[n][2^k], where Len[i][w] stores the minimum covering length for agents $\{1, 2, \ldots i\}$ using a set of facilities F, and Cap[i][w] stores the allocation used for agents $\{1, 2, \ldots i\}$ with F, where $w = \sum_{f_j \in F} 2^{j-1}$
.

Output: Minimum length Len covering agents in different groups and its corresponding allocations Cap.

```
1: for i ∈ {1, ..., n}, j ∈ {1, ..., k}, w ∈ {1, ..., 2^k} do
2:    if i = c_j and w = 2^{j-1} then
3:       Len[i][w] ← x_i − x_1
4:       Cap[i][w] ← (c_j)
5:    else if c_j ∉ Cap[i − c_j][w − 2^{j-1}] then
6:       Temp← max{Len[i − c_j][w − 2^{j-1}], x_i − x_{i−c_j+1}}
7:       if Len[i][w]>Temp then
8:          Len[i][w] ← Temp
9:          Cap[i][w] ← Cap[i − c_j][w − 2^{j-1}].append(c_j)
10:      end if
11:   end if
12: end for
13: return Len[n][2^k − 1], Cap[n][2^k − 1]
```

In $FindMinCover(X, C)$, we use an array Len[n][2^k] to store the minimum covering length and an array of tuples Cap[n][2^k] to store the capacities of the allocated facilities based on their locations from left to right. Within the algorithm, we represent each facility i as a binary vector with value 2^i and w is a value (sum of powers of two) to record which facilities have been used and fully occupied so that we will not allocate them again. Mechanism 4, will wait until all the agents have arrived, and use a random map function $g : \{1, \ldots, k\} \rightarrow \{0, \ldots, k-1\}$ to generate, with equal probability, a permutation order in which facilities serve their agents.

Mechanism 4. *Let* $(Len, Cap) \leftarrow FindMinCover(X, C)$. $N_i = \{1 + \sum_{w=1}^{i-1} Cap[w], \ldots, Cap[i] + \sum_{w=1}^{i-1} Cap[w]\}$. *If* $L_i + Len \leq 1$, *the facility* i *is placed at* L_i *with probability* $\frac{1}{2}$ *and* $L_i + Len$ *with probability* $\frac{1}{2}$. *Otherwise,* f_i *is placed at* R_i *with probability* $\frac{1}{2}$ *and* $R_i - Len$ *with probability* $\frac{1}{2}$. *For each facility in the allocations, we pick an arbitrary stage in* $\{T, \ldots, T + k - 1\}$ *for the facility to serve the assigned agents, such that* $S = \{T + g(1), \ldots, T + g(k)\}$.

Lemma 6. *Mechanism 4 is strategyproof, which achieves approximation ratios of* $Tn - T + 1$ *for social cost and* $\max\{2, T + k - 2\}$ *for maximum cost.*

Proof. The sum of distance cost of Mechanism 4 is at most $\frac{n}{2} \cdot OPT_d$. The total waiting cost is at most $OPT_w + T(n-1) \cdot d \leq (Tn - T + 1) \cdot OPT_w$. The approximation ratio for social cost is

$$\frac{Mechanism}{OPT} \leq \frac{(Tn - T + 1) \cdot OPT_w + \frac{n}{2} \cdot OPT_d}{OPT_w + OPT_d}$$
$$\leq Tn - T + 1.$$

The maximum cost of Mechanism 4 is at most $(T + k - 2)d + 2OPT_d$. Therefore, the approximation ratio for maximum cost is

$$\frac{Mechanism}{OPT} \leq \frac{(T + k - 2)\dot{O}PT_w + 2OPT_d}{OPT_d + OPT_w} \leq max(2, T + k - 2).$$

$\square$

4 Conclusion

We initiate the study of the multi-stage facility location problem with capacity constraints from a mechanism design perspective. For settings with various capacity constraint configurations with waiting times, we provide randomized strategyproof mechanisms with approximation guarantees as well as lower bounds to the corresponding settings. There are several directions for future work. It will be intriguing to extend results beyond one dimension to other complex structures like trees and networks. Another approach to extend this work is to consider opening multiple facilities at each stage. Finally, one can also consider a setting where agents have preferences over heterogeneous facilities.

Acknowledgements. We gratefully acknowledge funding from Research Grants Council of the Hong Kong Special Administrative Region, China (Project No. UGC/FDS13/E01/20), and by NSFC No.62202100. Hau Chan is supported by the National Institute of General Medical Sciences of the National Institutes of Health [P20GM130461], the Rural Drug Addiction Research Center at the University of Nebraska-Lincoln, and the National Science Foundation under grant IIS:RI #2302999. The content is solely the responsibility of the authors and does not necessarily represent the official views of the funding agencies.

References

1. Aziz, H., Chan, H., Lee, B., Li, B., Walsh, T.: Facility location problem with capacity constraints: algorithmic and mechanism design perspectives. In: Proceedings of the AAAI Conference on Artificial Intelligence, vol. 34, pp. 1806–1813 (2020)
2. Aziz, H., Chan, H., Lee, B.E., Parkes, D.C.: The capacity constrained facility location problem. Games Econ. Behav. **124**, 478–490 (2020)
3. Black, D.: The decisions of a committee using a special majority. Econometrica: J. Econometric Soc., 245–261 (1948)
4. Blin, J.M., Satterthwaite, M.A.: Strategy-proofness and single-peakedness. In: Public Choice, pp. 51–58 (1976)
5. Brimberg, J., Korach, E., Eben-Chaim, M., Mehrez, A.: The capacitated p-facility location problem on the real line. Int. Trans. Oper. Res. **8**(6), 727–738 (2001)
6. Chan, H., Filos-Ratsikas, A., Li, B., Li, M., Wang, C.: Mechanism design for facility location problems: a survey. In: Zhou, Z.H. (ed.) Proceedings of the Thirtieth International Joint Conference on Artificial Intelligence, IJCAI-21, pp. 4356–4365. International Joint Conferences on Artificial Intelligence Organization (2021). https://doi.org/10.24963/ijcai.2021/596
7. Chen, D.Q., et al.: Efficient and equitable deployment of mobile vaccine distribution centers. In: Proceedings of the Thirty-Second International Joint Conference on Artificial Intelligence, pp. 64–72 (2023)

8. De Keijzer, B., Wojtczak, D.: Facility reallocation on the line. In: Algorithmica, pp. 1–28 (2022)
9. Demaine, E.D., Hajiaghayi, M., Mahini, H., Sayedi-Roshkhar, A.S., Oveisgharan, S., Zadimoghaddam, M.: Minimizing movement. ACM Trans. Algor. (TALG) 5(3), 1–30 (2009)
10. Fotakis, D., Tzamos, C.: Winner-imposing strategyproof mechanisms for multiple facility location games. In: Saberi, A. (ed.) WINE 2010. LNCS, vol. 6484, pp. 234–245. Springer, Heidelberg (2010). https://doi.org/10.1007/978-3-642-17572-5_19
11. Fotakis, D., Tzamos, C.: Strategyproof facility location for concave cost functions. In: Proceedings of the Fourteenth ACM Conference on Electronic Commerce, pp. 435–452 (2013)
12. Fotakis, D., Tzamos, C.: On the power of deterministic mechanisms for facility location games. ACM Trans. Econ. Comput. (TEAC) 2(4), 1–37 (2014)
13. Lu, P., Sun, X., Wang, Y., Zhu, Z.A.: Asymptotically optimal strategy-proof mechanisms for two-facility games. In: Proceedings of the 11th ACM Conference on Electronic Commerce, pp. 315–324 (2010)
14. Moulin, H.: On strategy-proofness and single peakedness. Public Choice 35(4), 437–455 (1980)
15. Procaccia, A.D., Tennenholtz, M.: Approximate mechanism design without money. ACM Trans. Econ. Comput. (TEAC) 1(4), 1–26 (2013)
16. Raghavan, S., Sahin, M., Salman, F.S.: The capacitated mobile facility location problem. Eur. J. Oper. Res. 277(2), 507–520 (2019)
17. Wada, Y., Ono, T., Todo, T., Yokoo, M.: Facility location with variable and dynamic populations. In: AAMAS, pp. 336–344 (2018)
18. Walsh, T.: Strategy proof mechanisms for facility location with capacity limits. In: Raedt, L.D. (ed.) Proceedings of the Thirty-First International Joint Conference on Artificial Intelligence, IJCAI-22, pp. 527–533. International Joint Conferences on Artificial Intelligence Organization (2022). https://doi.org/10.24963/ijcai.2022/75
19. Wang, X., Chau, V., Chan, H., Fong, C.K.K., Li, M.: Multi-stage facility location problems with transient agents. In: Proceedings of the AAAI Conference on Artificial Intelligence (2023)

From Evolutionary Game Dynamics to Non-negative Matrix Factorization: Acceleration with Hessian Geometry

Huili Liang[1], Xiao Wang[1,3($\boxtimes$)], Yechao Wei[1], and Pingfan Wu[2]

[1] Shanghai University of Finance and Economics, Shanghai, China
liang.huili@mail.shufe.edu.cn, wangxiao@sufe.edu.cn
[2] North East Normal University, Changchun, China
wupingfan994@nenu.edu.cn
[3] Key Laboratory of Interdisciplinary Research of Computation and Economics, Shanghai, China

Abstract. Optimization on Riemannian manifolds has attracted the attention of many in the machine learning community due to its potential in solving constrained optimization problems. However, high cost of geodesic-based retraction algorithms often limits the applications of Riemannian optimization algorithm in practical problems that are intensely resource consuming. We focus on the positive orthant constrained optimization problems, a special type of constraint that are present in many application areas from evolutionary game theory to non-negative matrix factorization. We propose an accelerated natural gradient descent algorithm modified from evolutionary game dynamics and show that our algorithm converges to second-order stationary points. Our theoretical analysis and experiments verify that a geodesic free acceleration scheme has the potential to balance the convergence guarantee and computational efficiency in constrained non-convex optimization problems like non-negative matrix factorization.

1 Introduction

Constrained non-convex optimization has been studied extensively due to its applications in Machine Learning and related field, e.g., training neural networks, computing Nash equilibrium in multi-agent games, factorizing a non-negative matrix into the product of several matrices. One major obstacle of non-convex optimization is the existence of saddle points, which can outnumber and be significantly worse than local minima. Thus studying the second-order convergence or saddle avoidance property of algorithms has attracted much attention. Despite the accumulated results in second-order convergence of a variety of first-order methods, the second-order convergence of accelerated algorithm, especially in constrained settings, is less studied. In this paper, we focus on a non-convex optimization problem with a special constraint that is closely related to many areas, say the constraint of positive orthant in Euclidean space. Formally, the optimization problem we study is the following,

$$\text{minimize } f(\boldsymbol{x}) \tag{1}$$

$$\text{subject to } \boldsymbol{x} \in \mathbb{R}_+^d, \tag{2}$$

B. Li et al. (Eds.): IJTCS-FAW 2024, LNCS 14752, pp. 225–248, 2025.
https://doi.org/10.1007/978-981-97-7752-5_18

where $\mathbb{R}_+^d = \{x \in \mathbb{R}^d : x_i > 0, i \in [d]\}$ and $f(x)$ is a non-convex function whose global optima is in the interior of $\mathbb{R}_+^d$.

Table 1. Comparison to related results

	SOSP	ODE solver needed	Non geodesically convex	Types of Manifolds
Momentum method (Alimisis et al., 2021)	No	Yes	Yes	General
Accelerated MD (W.Krichene et al., 2015)	No	No	No	Convex set
RAGD (Zhang and Sra, 2018)	No	Yes	No	General
NMF (Lee and Seung, 2000)	No	No	No	Positive orthant
ANGD and NMF (this work)	Yes	No	Yes	Hessian, positive orthant

Motivation. Studying the non-convex optimization with this type of constraint using a Riemannian geometric approach. is well motivated by its applications in many areas, which are briefly discussed below.

- *Non-negative matrix factorization.* Non-negative matrix factorization (NMF) is a technique widely used in various fields, including unsupervised learning (Lee and Seung, 1996), data mining (Yang and Leskovec, 2013) and bioinformatics.(Aonishi et al., 2022), (Zhao et al., 2022), (Robert et al., 2022), (Sadeghi et al., 2022). Given a matrix V with non-negative entries, we need to find two non-negative matrices W and H to approximate $V \approx WH$. Typical approach is to solve the minimization problem for proper loss functions such as $\|V - WH\|_F^2$ or $D(V\|WH)^1$.
- *Evolutionary game theory.* A differential geometric approach in modeling the evolution of a genetic system has been the object of extensive studies since the 1970s.(Hofbauer and Sigmund, 1998; Mertikopoulos and W.H. Sandholm, 2018; Shahshahani, 1979). The positive orthant, together with a Riemannian metric $\sum_{i=1}^d \frac{|x|}{x_i} dx_i \otimes dx_i$ defined on it, is the geometrized phase space of the evolution dynamical system. Naturally, under this setting, finding the equilibrium of discrete evolution game dynamics is often considered an optimization problem constrain in the $\mathbb{R}_+^d$.
- *(Unconstrained) Geometric and signomial programming.* Geometric programming is an optimization paradigm that generalizes linear programming and has applications in, e.g. aircraft design (Hoburg and Abbeel, 2014) and communication network design (Chiang, 2006; Kandukuri and Boyd, 2002). The standard formulation of geometric programming is to minimize a objective function $f(x)$ subject to $x \in \mathbb{R}_+^d$ and satisfy inequality and equality constraints. Signomial programming is a relaxation of geometric programming so that $f(x)$ contains polynomials with negative coefficients. Due to its computational efficiency and extensibility, signomial programming has proven to be a powerful tool for engineering design optimization. Recent study by (Bürgisser et al., 2020) on unconstrained geometric programming shows its power in solving operator or matrix scaling problems.

---

[1] $D(A\|B)$ is a divergence used in measuring distance between matrices, defined as $\sum_{ij} \left(A_{ij} \log \frac{A_{ij}}{B_{ij}} - A_{ij} + B_{ij} \right)$.

Despite accumulated progress in the aforementioned areas independently, e.g., first-order convergence of (Lee and Seung, 2000), there lacks a unified framework that guarantees second-order convergence, which leave the second-order or global optima convergence of NMF algorithms and important open question. Apart from this issue, current accelerated optimization algorithms on Riemannian manifold heavily relie on computing exponential and logarithmic map (Zhang and Sra, 2018), or parallel transport (Alimisis et al., 2021), which inevitably involves numerical approximation in solving ordinary differential equations of geodesics or parallel transport, and the latter step might consume computational resources intensely. A natural question stimulated by current development of aforementioned areas is the following:

Question: Is there a class of algorithms that scales well with the geometry of constraints, can be implemented without numerical approximation of any Riemannian geometric objects, and guarantees a second-order convergence?

Our Contribution. In this paper, we borrow ideas from information geometry and natural gradient descent, and propose an algorithm that has comparable convergence rate with classic algorithm of (Lee and Seung, 2000) in non-negative matrix factorization. Especially, our algorithm is provably convergent to interior second-order stationary point. Furthermore, as a side result, our experiments comparing (2) and (Lee and Seung, 2000) indicates that the classic algorithm actually converges to second-order stationary points.

Mostly motivated by evolutionary game dynamics and non-negative matrix factorization, we propose a practical geometry based algorithm to solve the problem (1) without numerical approximation of exponential map or logarithmic maps. We combine the idea of natural gradient descent that was initially used in information geometry and neural network approximations, with the acceleration scheme in the sense of Nesterov's accelerated gradient descent. Comparison of our results to most related ones in the literature is provided in Table 1.

Related Works. Convergence of accelerated gradient descent on Riemannian manifolds with geodesic convexity assumption has been studied by (Alimisis et al., 2021; Han et al., 2023; Kim and Yang, 2022; Zhang and Sra, 2018). Non-convex (in the sense of geodesic) accelerated methods on manifold have been studied by (C. Criscitiello and N. Boumal, 2022; Feng et al., 2022) where (Feng et al., 2022) proves an asymptotic saddle avoidance results with locally geodesically convex assumption. In applications of non-negative matrix factorization, (Lee and Seung, 2000) proposes a multiplicative update rule that is easy to implement for this problem, and this method converges to stationary point (Lin, 2007a). (Paatero, 1999) proposes an alternating method for this problem, which decomposes the primal problem into two sub-problems. Although the alternating method can be proved to convergent to a stationary point, it is not simple to implement. (Lin, 2007b) suggests projected gradient descent method to solve the NMF problem with theoretical guarantee of converging to stationary point. Besides, other methods including active set method (Kim and Park, 2008), optimal gradient method (Guan et al., 2012) and block coordinate descent method (Kim et al., 2014), either lack convergence guarantees or only have convergence guarantee for first-order stationary point.

2   Preliminaries

We review some basic notions from Riemannian geometry and dynamical systems that are needed in our analysis. Readers are referred to some standard textbook for a complete treatment, for instance (Petersen, 2006) and (Perko, 2001).

Riemannian Metric and Geodesics. A smooth manifold M is a topological space that is locally equivalent to a Euclidean space, i.e., for each point $x \in M$, there exists a neighborhood that is diffeomorphic to an open subset of a Euclidean space. A Riemannian manifold (M, g) is real, smooth manifold M equipped with a Riemannian metric g. For each $x \in M$, let $T_x M$ denote the tangent space at x. The metric g induces a inner product $\langle \cdot, \cdot \rangle_x : T_x M \times T_x M \rightarrow \mathbb{R}$. We call a curve $\gamma(t) : [0, 1] \rightarrow M$ a geodesic if it satisfies

- The curve $\gamma(t)$ is parametrized with constant speed, i.e., $\left\| \frac{d}{dt} \gamma(t) \right\|_{\gamma(t)}$ is constant for $t \in [0, 1]$.
- The curve is locally length minimized between $\gamma(0)$ and $\gamma(1)$.

Despite the existence and uniqueness of geodesic is guaranteed by the fundamental theorem of ordinary differential equations, we are often involved in numerical approximation of the geodesic equation in the following form

$$\frac{d^2 \gamma^k}{dt^2} + \Gamma_{ij}^k \frac{d\gamma^i}{dt} \frac{\gamma^j}{dt} = 0$$

where $\gamma(t)$ is the curve we try to solve for in local coordinates and Γ_{ij}^k are Christoffel symbols.

Exponential and Logarithmic Maps. The exponential map $\text{Exp}_x(v)$ maps $v \in T_x M$ to $y \in M$ such that there exists a geodesic γ with $\gamma(0) = x$, $\gamma(1) = y$ and $\gamma'(0) = v$. For $x \in M$, let Log_x denote the logarithmic map at x,

$$\text{Log}_x(y) = \underset{u \in T_x M}{\text{argmin}} \text{ subject to } \text{Exp}_x(u) = y,$$

with domain such that this is uniquely defined.

Vector Fields. Vector fields on a Riemannian manifolds are sections of the tangent bundle. Suppose M is a Riemannian manifold, a vector field u in M is a smooth map $u : M \rightarrow TM$, where TM is the tangent bundle, such that $p \circ u$ is the identity map on M. p is the projection map from TM to M.

Riemannian Gradient and Hessian. For differentiable function $f : M \rightarrow \mathbb{R}$, $\text{grad} f(x) \in T_x M$ denotes the Riemannian gradient of f that satisfies $\frac{d}{dt} f(\gamma(t)) = \langle \gamma'(t), \text{grad} f(x) \rangle$ for any differentiable curve $\gamma(t)$ passing through x. The local coordinate expression of gradient is useful in our analysis.

$$\text{grad} f(x) = \left(\sum_j g^{1j}(x) \frac{\partial f}{\partial x_j}, ..., \sum_j g^{dj}(x) \frac{\partial f}{\partial x_j} \right)$$

where $g^{ij}(\boldsymbol{x})$ is the ij-th entry of the inverse of the metric matrix $\{g_{ij}(\boldsymbol{x})\}$ at each point.

The Hessian of f is the covariant derivative of the gradient vector field: $\mathrm{Hess}f(\boldsymbol{x})[\boldsymbol{u}] = \nabla_{\boldsymbol{u}}\mathrm{grad}f(\boldsymbol{x})$ for any vector field $\boldsymbol{u}$ on M. With Riemannian metric $\langle \cdot, \cdot \rangle_g$, the Hessian operator has a matrix form that is defined by $\mathrm{Hess}f(X, Y) = \langle \nabla_X \mathrm{grad}f, Y \rangle_g$, for two vector fields X, Y on M.

Strict Saddle Points. Similar to functions on Euclidean spaces, saddle points of a function $f : M \to \mathbb{R}$ is defined to be the critical points whose Hessian has negative curvature. A strict saddle point $\boldsymbol{x}$ of $f : M \to \mathbb{R}$ satisfies

$$\|\mathrm{grad}f(\boldsymbol{x})\| = 0 \quad \text{and} \quad \lambda_{\min}(\mathrm{Hess}f(\boldsymbol{x})) < 0.$$

Dynamical Systems. Let $\mathbb{T} = \mathbb{R}$ or $\mathbb{N}$. A smooth dynamical system on a manifold M is a continuous differentiable function $\phi : \mathbb{T} \times M \to M$, where $\phi(t, \boldsymbol{x}) = \phi_t(\boldsymbol{x})$ satisfies

- $\phi_0 : M \to M$ is the identity function.
- The composition $\phi_t \circ \phi_s = \phi_{t+s}$ for each $t, s \in \mathbb{T}$.

Limit, Stable and Unstable Set. A point $\boldsymbol{y} \in M$ is an ω-limit point for the solution through $\boldsymbol{x}$ if there is a sequence $t_n \to \infty$ such that $\lim_{n \to \infty} \phi_{t_n}(\boldsymbol{x}) = \boldsymbol{y}$. The set of all ω-limit points of solution through $\boldsymbol{x}$ is the ω-limit set of $\boldsymbol{x}$, denoted by $\omega(\boldsymbol{x})$. The α-limit points and α-limit set $\alpha(\boldsymbol{x})$ are defined by replacing $t_n \to \infty$ with $t_n \to -\infty$ in above definition. By a limit set we mean a set of form $\omega(\boldsymbol{x})$ or $\alpha(\boldsymbol{x})$. We define the stable set of a point $\boldsymbol{x}$ to be

$$W^s(\boldsymbol{x}) = \{\boldsymbol{z} \in M : \lim_{n \to \infty} d(\phi^n(\boldsymbol{z}), \phi^n(\boldsymbol{x})) \to 0\}$$

and unstable set of a point $\boldsymbol{x}$ to be

$$W^u(\boldsymbol{x}) = \{\boldsymbol{z} \in M : \lim_{n \to \infty} d(\phi^{-n}(\boldsymbol{z}), \phi^{-n}(\boldsymbol{x})) \to 0\}$$

3   Accelerated Natural Gradient Descent

In this section, we develop a Riemannian algorithm that is motivated by backtracking line search in Euclidean space and natural gradient descent in information geometry. To avoid complicated computation in most Riemannian geometry based algorithms, such as calculation of inverse of Riemannian metric matrix, parallel transport, exponential and logarithmic maps, we propose a geometry-based algorithm that is not accurately geometric in each step, but still converges to second-order stationary points, provably.

Information Geometry and Natural Gradient. We review the connection between natural gradient descent, information geometry, and Shahshahani geometry, this will make the derivation of accelerated Riemannian gradient descent especially working in positive orthant an intuitive and direct in solving optimization problems constrained on

---

Algorithm 1. Natural Gradient Descent

---

input : $\boldsymbol{x}_0$,
for $t > 0$ **do**
    $\boldsymbol{x}_{t+1} = \boldsymbol{x}_t - \eta \mathrm{grad} f(\boldsymbol{x}_t)$
end for

Algorithm 2. Accelerated Natural Gradient Descent

---

input : $\boldsymbol{x}_0, \boldsymbol{v}_0, \{a_t\}_{t=0}^{\infty}, \{\beta_t\}_{t=0}^{\infty}$ such that inf $a_t > 0$ and inf $\beta_t > 0$.
for $t > 0$ **do**
    $\boldsymbol{y}_t = \boldsymbol{v}_t + \beta_t(\boldsymbol{x}_t - \boldsymbol{v}_t)$
    $\boldsymbol{x}_{t+1} = \boldsymbol{y}_t - \eta \mathrm{grad} f(\boldsymbol{y}_t)$
    $\boldsymbol{v}_{t+1} = \boldsymbol{v}_t - a_{t+1} \mathrm{grad} f(\boldsymbol{y}_t)$
end for

---

positive orthant. The set of categorical distributions on n variables forms a Riemannian manifold via the Fisher information metric

$$g_{ij}(\boldsymbol{x}) = \mathbb{E}\left[\frac{\partial \log p}{\partial x_i} \frac{\partial \log p}{\partial x_j}\right].$$

In information geometry, the replicator equation is known as the *natural gradient*. The gradient flow of the Fisher information metric is the replicator equation (Hofbauer and Sigmund, 1998). The Fisher information metric can be obtained from the Hessian of the KL divergence. The metric is known in evolutionary game theory as the Shahshahani metric and the manifold identified with its embedding into the reals as the $(n-1)$-dimensional simplex. On the other hand, the approach of Shahshahani (Shahshahani, 1979) considers the simplex with such information metric structure as the submanifold of the positive orthant endowed with the metric $\sum_{i=1}^{n} \frac{|\boldsymbol{x}|}{x_i} dx_i \otimes dx_i$. The natural gradient descent (NGD), given by following update rule,

$$\boldsymbol{\theta}_{t+1} = \boldsymbol{\theta}_t - \eta \mathrm{grad} f(x_t),$$

where $\mathrm{grad} f(x_t) = g^{-1}(\boldsymbol{\theta}_t)\nabla_{\theta_t} f(\boldsymbol{\theta}_t)$, is an optimization method initially derives from the perspective of information geometry, which was developed by Amari and his collaborators (Amari and Douglas, 1998). The descent direction of NGD is based on the Riemannian metric (i.e., Fisher Information Matrix,FIM) in a Riemannian manifold, defined as the inverse of FIM times gradient, so the algorithm can make the most of Riemannian structure of parameter space consisting of weights and biases, and works in learning, avoiding plateaus (Amari, 1998),(Amari et al., 2019). As mentioned above, natural gradient descent is motivated from developing a method of implementing steepest descent in the space of realizable distributions (i.e., distributions that correspond to some setting of the model's parameters) instead of the space of parameters (Martens, 2020), where distance in the distribution space is measured by a Riemannian metric which depends on the properties of distributions and not their parameters (Amari and Nagaoka, 2007).

Geometry v.s. Optimization. There are two standard but different schemes of designing accelerated gradient descent on Riemannian manifold, say the logarithmic map based and the parallel transport based. The typical logarithmic acceleration is (Zhang and Sra, 2018) and the parallel transport acceleration is (Alimisis et al., 2021). In principle, the main difference is the following: the logarithmic acceleration algorithm requires calculating logarithmic map in updating the augmented variable $\boldsymbol{v}$, i.e.,

$$\boldsymbol{v}_{t+1} = \mathrm{Exp}_{\boldsymbol{y}_t}\left(\frac{(1-s_t)\gamma_t}{\bar{\gamma}_t}\mathrm{Log}_{\boldsymbol{y}_t}(\boldsymbol{v}_t) - \frac{s_t}{\bar{\gamma}_t}\mathrm{grad}f(\boldsymbol{y}_t)\right)$$

where the parameters are from (Zhang and Sra, 2018). Note that solving logarithmic map is in general a non-trivial minimization problem. In contrast, the parallel transport acceleration requires one to calculate the parallel transport in updating the augmented variable $\boldsymbol{v}$, i.e.,

$$\boldsymbol{v}_{t+1} = \mathrm{Exp}_{\boldsymbol{v}_t}(-a_{t+1}\Gamma_{\boldsymbol{y}_t}^{\boldsymbol{v}_t}\mathrm{grad}f(\boldsymbol{y}_t)),$$

where the main challenge comes from the numerical approximation of the ODE of parallel transport. The advantage of both approaches is the clear geometric meaning in algorithm design and analysis. However, neither computing logarithmic map nor solving parallel transport equation is practically efficient in applications. We would like to find an accelerated Riemannian gradient descent that exhibits good convergence guarantee (e.g. second-order stationary) and easy to implement in applications. In this paper, we extend our scope of natural gradient descent from statistical manifold to general open subset of Euclidean space endowed with a Riemannian metric, for example, positive orthant or Hessian manifold whose metric is defined by barrier functions. Formally, the general benchmark natural gradient has the form given by Algorithm 1, whose second-order stationary convergence is a consequence of using Stable-Manifold Theorem. We will provide a concrete bound for the step-size η for both Algorithm 1 and 2 that suffices to guarantee second-order stationary convergence. Despite sacrificing geometric insight to some extent, our algorithm is proven efficient in experiments.

4 Analysis on Second-Order Stationary Convergence

The Algorithm 2 consists of three steps updating the states $(\boldsymbol{x}, \boldsymbol{v})$ in the augmented state space $M \times M$. The intermediate step updating the state $\boldsymbol{y}$ can be seen as a connecting step in the composition of mappings on $M \times M$. Based on this observation, we denote the mapping induced by Algorithm 2 in the following way.

$$F(t, \boldsymbol{x}, \boldsymbol{v}) = (F_1(t, \boldsymbol{x}, \boldsymbol{v}), F_2(t, \boldsymbol{x}, \boldsymbol{v}))$$

where

$$\begin{aligned}F_1(t, \boldsymbol{x}, \boldsymbol{v}) &= \boldsymbol{y} - \eta\mathrm{grad}f(\boldsymbol{y}) \\ &= \boldsymbol{v} + \beta_t(\boldsymbol{x} - \boldsymbol{v}) - \eta\mathrm{grad}f(\boldsymbol{v} + \beta_t(\boldsymbol{x} - \boldsymbol{v})) \\ &= \beta_t\boldsymbol{x} + (1 - \beta_t)\boldsymbol{v} - \eta\mathrm{grad}f(\beta_t\boldsymbol{x} + (1 - \beta_t)\boldsymbol{v})\end{aligned}$$

and

$$F_2(t, \boldsymbol{x}, \boldsymbol{v}) = \boldsymbol{v} - a_{t+1}\mathrm{grad}f(\boldsymbol{y})$$
$$= \boldsymbol{v} - a_{t+1}\mathrm{grad}f(\beta_t\boldsymbol{x} + (1 - \beta_t)\boldsymbol{v})$$

Remark 1. In the above expression, we ignore the time index t on the state variables $(\boldsymbol{x}_t, \boldsymbol{v}_t)$ in order to emphasize that the update rule can be written as a time-dependent mapping on $M \times M$. We will keep this notation in the rest of the paper if no ambiguity is created.

Assumption 1. *Let ∇ be the Levi-Civita connection. Assume that $\|\nabla\mathrm{grad}f\|_{op} \le L$.*

An immediate consequence of the assumption on Riemannian Hessian is the convergence to second order stationary points for Algorithm 1, which leverages the Center-stable manifold theorem.

Proposition 1. *Let $f : M \to \mathbb{R}$ be a smooth function defined on an open subset of Euclidean space that endowed with a Riemannian metric. The set of initial points that converges to strict saddle points under Algorithm 1 is of measure zero with respect to the volume measure induced by the Riemannian metric.*

The proof (in appendix) is based on a direct computation of $\nabla\mathrm{grad}$ for the Levi-Civita connection ∇ acting on the Riemannian gradient $\mathrm{grad}f$. The step-size η chosen according to the operator norm of $\nabla\mathrm{grad}f$ ensures that the natural gradient descent is a local diffeomorphim, and furthermore, this step-size η will ensure the accelerated natural gradient descent to be local diffeormophism.

As a differentiable mapping on the product manifold $M \times M$, Algorithm 2 induces a differentiable dynamical system whose asymptotic behavior depends on the structure of stationary points of the objective function f defined on M. The local analysis focuses on the spectrum of the differential of the update rule, i.e., the map $F(t, \boldsymbol{x}, \boldsymbol{v})$. The main result of the local characterization of the fixed points and the stationary points of f are presented.

Theorem 1. *If $\boldsymbol{x}^*$ is a second-order stationary point of $f : M \to \mathbb{R}$, it is an attracting fixed point of the dynamical system defined by $F(t, \boldsymbol{x}, \boldsymbol{v})$. If $\boldsymbol{x}^*$ is a strict saddle point of f, it is an unstable fixed point of the dynamical system defined by $F(t, \boldsymbol{x}, \boldsymbol{v})$.*

Proof (sketched). Recall that we have written the Algorithm 2 as the differentiable mapping $F(t, \boldsymbol{x}, \boldsymbol{v})$. Since the mapping itself is defined on an open set of Euclidean space, the differential can be computed in this local coordinate system and then it is a matrix. Concretely, we have

$$DF(t, \boldsymbol{x}, \boldsymbol{v}) = \begin{bmatrix} D_{\boldsymbol{x}}F_1(t, \boldsymbol{x}, \boldsymbol{v}) & D_{\boldsymbol{v}}F_1(t, \boldsymbol{x}, \boldsymbol{v}) \\ D_{\boldsymbol{x}}F_2(t, \boldsymbol{x}, \boldsymbol{v}) & D_{\boldsymbol{v}}F_2(t, \boldsymbol{x}, \boldsymbol{v}) \end{bmatrix}.$$

The detailed analysis of the eigenvalues of $DF(t, \boldsymbol{x}, \boldsymbol{v})$ at fixed points is left in the appendix. The key observation is that as long as $\boldsymbol{x}^*$ is a strict saddle point of f, then the operator $Df(t, x^*, v^*)$ has at least one repelling direction, i.e., the dynamical system is unstable. On the other hand, if x^* is a local minima of f, i.e., the Hessian matrix of f at x^* has all positive eigenvalues, then the corresponding fixed point of the dynamical system induced by the algorithm is stable.

So far we have proved that Algorithm 2 converges to second order stationary points if the initial point lies in a small neighborhood of the stationary point. To see that the initial points that converges to all second order stationary points belongs to a measure zero set, it suffices to prove the following corollary based on the main theorem.

Corollary 1. *Let $f : M \to \mathbb{R}$ be a smooth function and. Then the set of initial points converging to second-order stationary points of f belong to a set of measure zero (measure induced by the Riemannian volume measure).*

5   Experiments

Test Functions. Experiments are conducted to show that our accelerate natural gradient really accelerate as well as to verify the ability to escape saddle points of our algorithms. We choose two special families of functions with multiple saddle points. One of them is the Ackley function, and the other is the multiplied-triangular function (Fig. 1).

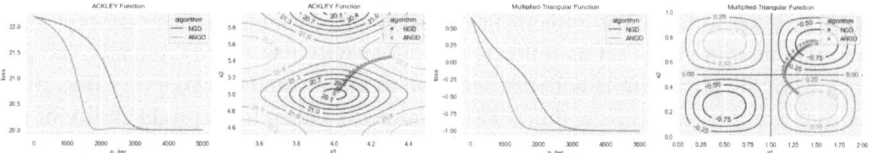

Fig. 1. Loss curves and trajectories for test functions

The Ackley function,

$$f(x) = -a \exp\left(-b\sqrt{\frac{1}{d}\sum_{i=1}^{d} x_i^2}\right) - \exp\left(\frac{1}{d}\sum_{i=1}^{d} \cos cx_i\right) + a + \exp(1)$$

, is widely used for testing optimization algorithms. In its two-dimensional form, it is characterized by a nearly flat outer region, and a large hole at the centre. As recommended commonly, we set the parameters as follows: $d = 2, a = 20, b = 0.2, c = 2\pi$. The initial point $x_0 = (4.3, 5.45)$ is randomly generated uniformly on the support of $[0, 8]$. The stepsize of NGD is 1×10^{-5}, the first stepsize of ANGD is also 1×10^{-5} and the accelerate stepsize is 3×10^{-5}. In the loss curve chart, we can find that the Accelerated Natural Gradient Descent does accelerate compare with the normal Natural Gradient Descent algorithm. In applications, if one attempt to avoid the negative generated by Algorithm 2, it is convenient to update the components $x_i(t)$ by using

$$x_i^{t+1} = y_i^t \exp(-\eta \sum_j g^{ij} \frac{\partial f}{\partial x_j})$$

and

$$v_i^{t+1} = v_i^t \exp(-a_t \sum_j g^{ij} \frac{\partial f}{\partial x_j}).$$

In the contour graph, we can see the optimization trajectories. As shown, there is no tendency of both of the algorithms to get trapped in the saddle points, and verified our theoretical results.

The other text function with multiple saddle points is $f(x) = \cos(\alpha_1(x_1 - \phi_1))$ $\cos(\alpha_2(x_2 - \phi_2))$. We set $\alpha_1 = \pi, \alpha_2 = 2\pi, \phi_1 = \phi_2 = 0.5$. The initial point is generated uniformly randomly on the support $[0, 2] \times [0, 1]$ and in the illustrated case, $x_0 == (1.3, 0.35)$. The stepsize of NGD is 1×10^{-4}, the first stepsize of ANGD is also 1×10^{-4} and the accelerate stepsize is 3×10^{-4}. The loss curve and the optimization trajectories with the contour are consistent with the result for the Ackley function, showing that the accelerate natural gradient descent really converge fast and well.

Applications in NMF. We conduct several experiments to show the effectiveness of our algorithms in practise. We do the non-negative matrix factorization on the CIFAR-10 and the MNIST dataset. The CIFAR-10 dataset is consisted of 60000 colorful pictures with size 32×32. This is considered a high-quality and classical dataset in the machine learning community. The mnist dataset, on the other hand, is a classical picture dataset in the pattern recognition and artificial intelligence. The width of the decomposition matrix, r, is set to be 7. We implement our accelerated natural gradient descent algorithm with learning rates set as follows, without heavily fine-tuned (Actually, the performance of the algorithm is not very sensitive to those parameters), $\eta = 1.0 \times 10^{-5}$, $a_t = 3.0 \times 10^{-5}$. We stops within 5000 iterations. To keep the reproducibility of our results, we just consider the first 64 pictures in the CIFAR-10 datasets firstly. As shown in the Fig. 2 and 3, 4, 5, the reconstructed figures highly agree with the original ones, which reflects the effectiveness of our algorithm.

Fig. 2. Recovered pictures with our algorithm   **Fig. 3.** Origin first 64 pictures in CIFAR-10

Figure 6 is the quantitative performance analysis for the CIFAR-10 dataset. The basic setting is the same as aforementioned. There are 192 matrices because we take 64 pictures and there are 3 matrices for a picture representing the RGB pixels. We use the

Frobenius norm $||V - WH||_F$ as the loss to measure the performance of the recovery for a single matrix and take the average of the 192 matrices to compare performance of algorithms.

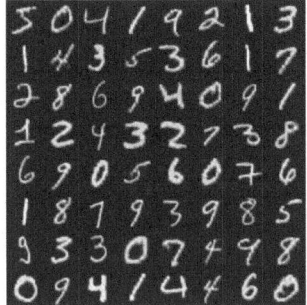

Fig. 4. Recovered pictures with our algorithm

Fig. 5. Origin first 64 pictures in MNIST

About parameters setting, the width of the decompose matrix is, $r = 7$. And in our natural gradient descent algorithm, the stepsize is set to 1.0×10^{-5}. In the accelerate version of the manifold algorithm, the stepsize parameters are, $\eta = 1.0 \times 10^{-5}$, $a_t = 3.0 \times 10^{-5}$.

Although our algorithms do not beat the multiplicative Update algorithm(Lee-Seung) on average, they achieve comparable performance. The key is that our algorithm is quite general, and can be applied in different nonconvex problems besides NMF, while the multiplicative Update algorithm is highly tailed for NMF. Besides, there are some pictures where the ANGD and NGD outperform the Lee-Seung algorithm actually (Fig. 8, 9, 10). And similar results are achieved on the MNIST dataset (Fig. 7).

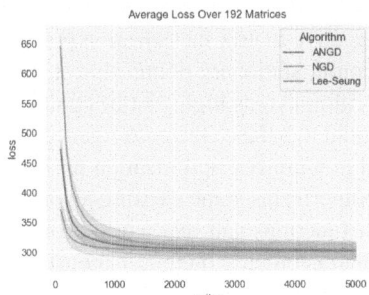

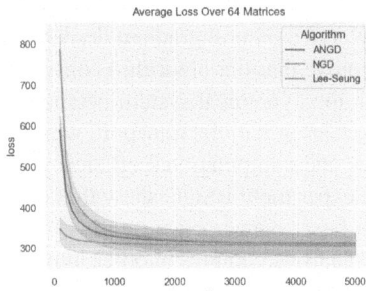

Fig. 6. Quantitative performance analysis for the CIFAR10 dataset.

Fig. 7. Quantitative performance analysis for the MNIST dataset. Similar results are achieved.

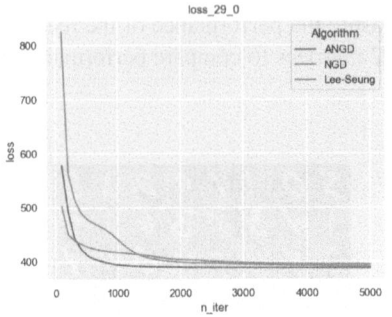

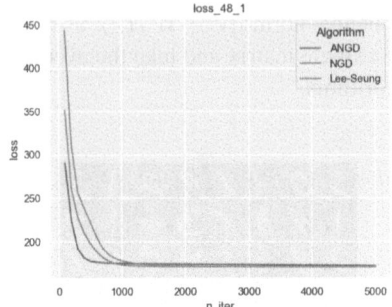

Fig. 8. Case 1 where the NGD, ANGD outperform the Lee-Seung

Fig. 9. Case 2 where the NGD, ANGD outperform the Lee-Seung

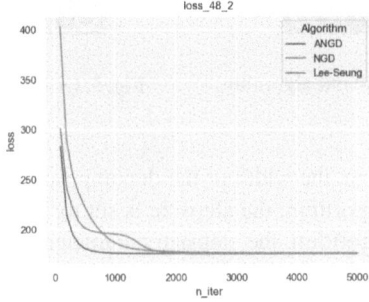

Fig. 10. Case 3 where the NGD, ANGD outperform the Lee-Seung

6   Conclusion

In this paper, We employ the framework of optimization on Riemannian manifolds to leverage geometric information of constrained set and propose an algorithm called accelerated natural gradient descent for the non-negative matrix factorization problem. We prove that our algorithm converges to second order stationary points almost surely from the dynamical system perspective. Our algorithm avoids high cost of computing geodesics or parallel transport, which has the potential to balance the convergence guarantee and computational efficiency in constrained non-convex optimization problems. The experiment results show that our algorithm outperforms some existing algorithms with faster convergence rate in both synthetic test functions and real world problems. Our analysis captures another important Riemannian geometric method in constrained optimization, so called Hessian manifold, which is usually used in convex constraint with same dimension of the ambient space.

A   Positive Orthant as an Hessian Manifold

The problem we focus on in this paper is a special case of the following problem:

$$\text{minimize } f(\boldsymbol{x}) \tag{3}$$

$$\text{subject to } A\boldsymbol{x} > \boldsymbol{b}, \tag{4}$$

In this formulation, $A = (a_{ij}) \in \mathbb{R}^{n \times d}$, $\boldsymbol{x}, \boldsymbol{b} \in \mathbb{R}^d$. The log barrier function in $M = \{\boldsymbol{x} : A\boldsymbol{x} > \boldsymbol{b}\}$ is

$$\phi(\boldsymbol{x}) = -\sum_{p=1}^{n} \log(A_p^T \boldsymbol{x} - b_p),$$

where A_p^T is the p-th row vector of A.

In local coordinate system $x(p) = (x_1, x_2, ..., x_d)$ at point p, given barrier function ϕ, Hessian metric is written as

$$g_{ij} = \frac{\partial^2 \phi}{\partial x_i \partial x_j} = \sum_{p=1}^{n} \frac{a_{pi} a_{pj}}{(A_p^T x - b_p)^2},$$

and the corresponding inverse matrix is (g^{ij}). The $(i, j) - th$ component of Hessian of f can be written as

$$\text{Hess}(f)(i, j) = (\text{Hess}(f))(\frac{\partial}{\partial x_i}, \frac{\partial}{\partial x_j})$$

$$= \frac{\partial^2 f}{\partial x_i \partial x_j} - \sum_{k=1}^{d} \Gamma_{ij}^k \frac{\partial f}{\partial x_k}$$

$$= \frac{\partial^2 f}{\partial x_i \partial x_j} - \sum_{k=1}^{d} \sum_{l=1}^{d} \frac{1}{2} g^{kl} \left(\frac{\partial g_{il}}{\partial x_j} + \frac{\partial g_{lj}}{\partial x_i} - \frac{\partial g_{ij}}{\partial x_l} \right) \frac{\partial f}{\partial x_k} \tag{5}$$

$$= \frac{\partial^2 f}{\partial x_i \partial x_j} - \sum_{k=1}^{d} \sum_{l=1}^{d} \frac{1}{2} g^{kl} \frac{\partial^3 \phi}{\partial x_i \partial x_j \partial x_l} \frac{\partial f}{\partial x_k}$$

$$= \frac{\partial^2 f}{\partial x_i \partial x_j} + \sum_{k=1}^{d} \sum_{l=1}^{d} g^{kl} \sum_{p=1}^{n} \frac{a_{pi} a_{pj} a_{pl}}{(A_p^T x - b_p)^3} \frac{\partial f}{\partial x_k}$$

where Γ_{ij}^k is Christoffel symbol and g^{kl} is the (k, l)-th component of the inverse of matrix (g_{kl}). Also see for details in (Petersen, 2006).

In particular, when M is the positive orthant, we can get the explicit form of Hessian operator from (5) as follows:

$$\text{Hess}(f)(\boldsymbol{x}) = \begin{pmatrix} \frac{\partial^2 f(\boldsymbol{x})}{\partial x_1^2} + \frac{1}{x_1}\frac{\partial f(\boldsymbol{x})}{\partial x_1} & \frac{\partial^2 f(\boldsymbol{x})}{\partial x_1 \partial x_2} & \cdots & \frac{\partial^2 f(\boldsymbol{x})}{\partial x_1 \partial x_d} \\ \frac{\partial^2 f(\boldsymbol{x})}{\partial x_1 \partial x_2} & \frac{\partial^2 f(\boldsymbol{x})}{\partial x_2^2} + \frac{1}{x_2}\frac{\partial f(\boldsymbol{x})}{\partial x_2} & \cdots & \frac{\partial^2 f(\boldsymbol{x})}{\partial x_2 \partial x_d} \\ \vdots & \vdots & & \vdots \\ \frac{\partial^2 f(\boldsymbol{x})}{\partial x_1 \partial x_d} & \frac{\partial^2 f(\boldsymbol{x})}{\partial x_2 \partial x_d} & \cdots & \frac{\partial^2 f(\boldsymbol{x})}{\partial x_d^2} + \frac{1}{x_d}\frac{\partial f(\boldsymbol{x})}{\partial x_d} \end{pmatrix}$$

$$= \begin{pmatrix} \frac{\partial^2 f(\boldsymbol{x})}{\partial x_1^2} & \frac{\partial^2 f(\boldsymbol{x})}{\partial x_1 \partial x_2} & \cdots & \frac{\partial^2 f(\boldsymbol{x})}{\partial x_1 \partial x_d} \\ \frac{\partial^2 f(\boldsymbol{x})}{\partial x_1 \partial x_2} & \frac{\partial^2 f(\boldsymbol{x})}{\partial x_2^2} & \cdots & \frac{\partial^2 f(\boldsymbol{x})}{\partial x_2 \partial x_d} \\ \vdots & \vdots & & \vdots \\ \frac{\partial^2 f(\boldsymbol{x})}{\partial x_1 \partial x_d} & \frac{\partial^2 f(\boldsymbol{x})}{\partial x_2 \partial x_d} & \cdots & \frac{\partial^2 f(\boldsymbol{x})}{\partial x_d^2} \end{pmatrix} + \begin{pmatrix} \frac{1}{x_1}\frac{\partial f(\boldsymbol{x})}{\partial x_1} & & & \\ & \frac{1}{x_2}\frac{\partial f(\boldsymbol{x})}{\partial x_2} & & \\ & & \ddots & \\ & & & \frac{1}{x_d}\frac{\partial f(\boldsymbol{x})}{\partial x_d} \end{pmatrix} \tag{6}$$

According to the property of operator norm, we can obtain:

$$\|\nabla \mathrm{grad} f\|_{op} \geq \lambda_{max}(\text{Hess}(f)(\boldsymbol{x})),$$

where $\lambda_{max}(A)$ is the biggest eigenvalue of matrix A. Then by Assumption 1,

$$\lambda_{max}(\text{Hess}(f)(\boldsymbol{x})) \leq L.$$

We consider the natural manifold gradient descent method with constant step size α:

$$\boldsymbol{x}_{t+1} = \boldsymbol{x}_t - \alpha g^{-1}\nabla f(\boldsymbol{x}_t) \tag{7}$$

where g^{-1} is the inverse matrix of metric g and $\nabla f(\boldsymbol{x})$ is Euclidean gradient.

When the step size $\alpha \leq \frac{1}{L}$, similar to analysis in (Lee et al., 2019), the natural gradient descent algorithm will avoid saddle points almost surely.

Remark 2. In fact, we can assume that

$$\left\|\nabla^2 f(\boldsymbol{x})\right\|_2 + \max_{i=1,\ldots,d}\{\frac{1}{x_i}\frac{\partial f(\boldsymbol{x})}{\partial x_i}\} \leq L.$$

According to (6), we have:

$$\lambda_{max}(\text{Hess}(f)(\boldsymbol{x})) \leq \left\|\nabla^2 f(\boldsymbol{x})\right\|_2 + \max_{i=1,\ldots,d}\{\frac{1}{x_i}\frac{\partial f(\boldsymbol{x})}{\partial x_i}\} \leq L.$$

Thus the natural manifold gradient descent algorithm with step size $\alpha \leq \frac{1}{L}$ can avoid saddle points almost surely. Compared with Assumption 1, this assumption is easier to verify.

B   Variants of Algorithm 2

Despite Algorithm 2 works well in the experiments of NMF, it is possible that the updates fail staying in the positive orthant. To resolve this issue, we can easily add a modification with exponential function.

Algorithm 3. Exp-ANGD

---

input : $\boldsymbol{x}_0, \boldsymbol{v}_0, \{a_t\}_{t=0}^\infty, \{\beta_t\}_{t=0}^\infty$ such that $\inf a_t > 0$ and $\inf \beta_t > 0$.
for $t > 0$ **do**
    $\boldsymbol{y}_t = \boldsymbol{v}_t + \beta_t(\boldsymbol{x}_t - \boldsymbol{v}_t)$
    for $i \in [n]$ **do**
        $x_{t+1}^i = y_t^i \exp\left(-\eta \sum_j g^{ij}(\boldsymbol{y}_t)\frac{\partial f}{\partial x^j}(\boldsymbol{y}_t)\right)$
        $v_{t+1}^i = v_t^i \exp\left(-a_{t+1} \sum_j g^{ij}(\boldsymbol{y}_t)\frac{\partial f}{\partial x^j}(\boldsymbol{y}_t)\right)$
    end for
end for

C   Proof of Proposition 1

Before proving the claim, we present the preliminary theorem that will be used in the proof, say the Center-Stable Manifold Theorem.

Theorem 2. *Let 0 be a fixed point for the C^r local diffeomorphism $\phi : U \to E$, where U is a neighborhood of 0 in the Banach space E. Suppose that $E = E_s \oplus E_u$, where E_u is the span of the eigenvectors corresponding to eigenvalues less than or equal to 1 of $D\phi(0)$, and E_u is the span of eigenvalues greater than 1 of $D\phi(0)$. Then there exist a C^r embedded disk W_{loc}^{cs} that is tangent to E_s at 0 called the local stable center manifold. Moreover, there exist a neighborhood B of 0 such that $\phi(W_{loc}^{cs}) \cap B \subset W_{loc}^{cs}$, and $\cap_{k=0}^\infty \phi^{-k}(B) \subset W_{loc}^{cs}$.*

Now we are ready to prove Proposition 1.

Proof. Recall that an affine connection is a way of differentiate sections in vector bundle. It is defined to be a mapping from sections of a vector bundle $\Gamma(E)$ to a sections in tensor bundle $\Gamma(T^*M \otimes E)$, S.S.Chern and W.H.Chen (1999). A connection on a vector bundle E is a map

$$\nabla : \Gamma(E) \to \Gamma(T^*M \otimes E)$$

which satisfies

1) For any $s_1, s_2 \in \Gamma(E)$,

$$\nabla(s_1 + s_2) = \nabla s_1 + \nabla s_2.$$

2) For $s \in \Gamma(E)$ and any $\alpha \in C^\infty(M)$,

$$\nabla(\alpha s) = d\alpha \otimes s + \alpha \nabla s.$$

Furthermore, if $\{\partial_\alpha\}_{\alpha=1}^n$ is a field of basis of the vector bundle E, the connection can be written by Christoffel symbols:

$$\nabla \partial_\alpha = \sum_{i,\beta} \Gamma_{\alpha i}^\beta du_i \otimes \partial_\beta.$$

Since grad$f$ is a vector field whose local expression is $\sum_{i,j} g^{ij} \frac{\partial f}{\partial x_i} \partial_i$, we can differentiate it to create a section in $\Gamma(T^*M \otimes TM)$. Using Einstein convention for summations, we have

$$
\begin{aligned}
\nabla \mathrm{grad} f &= \nabla \left(g^{ij} \frac{\partial}{\partial x_j} \partial_i \right) \\
&= d \left(g^{ij} \frac{\partial f}{\partial x_j} \right) \otimes \partial_i + g^{ij} \frac{\partial f}{\partial x_j} \nabla \partial_i \\
&= \frac{\partial}{\partial x_k} \left(g^{ij} \frac{\partial f}{\partial x_j} \right) dx_k \otimes \partial_i + g^{ij} \frac{\partial f}{\partial x_j} \Gamma^m_{ik} dx_k \otimes \partial_m \\
&= \left(\frac{\partial g^{ij}}{\partial x_k} \frac{\partial f}{\partial x_j} + g^{ij} \frac{\partial^2 f}{\partial x_k \partial x_j} \right) dx_k \otimes \partial_i + g^{ij} \frac{\partial f}{\partial x_j} \Gamma^m_{ik} dx_k \otimes \partial_m
\end{aligned}
$$

where g^{ij} is the (i,j)-th entry of the inverse of metric matrix g_{ij}. On the other hand, since the natural gradient descent is a mapping $F(x) = x - \eta \mathrm{grad} f(x)$, its differential can be computed as follows

$$
D_x F = I - \eta D_x((g^{ij})\nabla f(x))
$$

where

$$
\begin{aligned}
D_x((g^{ij})\nabla f(x)) &= \begin{bmatrix} \frac{\partial}{\partial x_i}(g^{1j}\frac{\partial}{\partial x_j}) & \cdots & \frac{\partial}{\partial x_d}\left(g^{1j}\frac{\partial f}{\partial x_j}\right) \\ \vdots & & \vdots \\ \frac{\partial}{\partial x_1}\left(g^{dj}\frac{\partial f}{\partial x_j}\right) & \cdots & \frac{\partial}{\partial x_d}\left(g^{dj}\frac{\partial f}{\partial x_j}\right) \end{bmatrix} \\
&= \begin{bmatrix} \frac{\partial g^{1j}}{\partial x_1}\frac{\partial f}{\partial x_j} & \cdots & \frac{\partial g^{1j}}{\partial x_d}\frac{\partial f}{\partial x_j} \\ \vdots & & \vdots \\ \frac{\partial g^{dj}}{\partial x_1}\frac{\partial f}{\partial x_j} & \cdots & \frac{\partial g^{dj}}{\partial x_d}\frac{\partial f}{\partial x_j} \end{bmatrix} + \begin{bmatrix} g^{1j}\frac{\partial^2 f}{\partial x_1 \partial x_2} & \cdots & g^{1j}\frac{\partial^2 f}{\partial x_d \partial x_j} \\ \vdots & & \vdots \\ g^{dj}\frac{\partial^2 f}{\partial x_1 \partial x_j} & \cdots & g^{dj}\frac{\partial^2 f}{\partial x_d \partial x_j} \end{bmatrix}.
\end{aligned}
$$

Since the operator norm of $\nabla \mathrm{grad} f$ is assumed to be bounded by a number L, and $\nabla \mathrm{grad} f$ is nothing but an operator acting on TM point-wisely, which means it is an matrix. Matrix inequality implies that $\|\nabla \mathrm{grad} f\|_F \leq \sqrt{d}\|\nabla \mathrm{grad} f\|_2 \leq \sqrt{d}L$. If we further assume that $g^{ij}\frac{\partial f}{\partial x_j}\Gamma^m_{ik}$ are bounded, we can have that the Frobenius norm of $\left(\frac{\partial g^{ij}}{\partial x_k}\frac{\partial f}{\partial x_j} + g^{ij}\frac{\partial^2 f}{\partial x_k \partial x_j}\right)$ is bounded. Note that this is exactly the matrix $D_x(g^{ij}\nabla f(x))$, thus the operator norm of $D_x(g^{ij}\nabla f(x))$ is also bounded. Therefore, if we tune the step-size η properly, the determinant of $D_x F$ can be positive, which ensures that F is a local diffeomorphism. Further application of Center-stable manifold theorem enables us to conclude the saddle avoidance result of Algorithm 1.

D Proof of Theorem 1

Proof. We start with calculations of the differential operator $DF(t, x, v)$. This is done by computing the differentials $D_x F_1$, $D_v F_1$, $D_x F_2$ and $D_v F_2$ respectively. We have

$$D_x F_1(t, \boldsymbol{x}, \boldsymbol{v}) = D_x \left(\beta_t \boldsymbol{x} + (1 - \beta_t) \boldsymbol{v} - \eta \mathrm{grad} f(\beta_t \boldsymbol{x} + (1 - \beta_t) \boldsymbol{v}) \right) \tag{8}$$
$$= \beta_t I - \eta \mathrm{Dgrad} f(\beta_t \boldsymbol{x} + (1 - \beta_t) \boldsymbol{v}) D_x (\beta_t \boldsymbol{x} + (1 - \beta_t) \boldsymbol{v}) \tag{9}$$
$$= \beta_t I - \eta \beta_t \mathrm{Dgrad} f(\beta_t \boldsymbol{x} + (1 - \beta_t) \boldsymbol{v}) \tag{10}$$

$$D_v F_1(t, \boldsymbol{x}, \boldsymbol{v}) = D_v \left(\beta_t \boldsymbol{x} + (1 - \beta_t) \boldsymbol{v} - \eta \mathrm{grad} f(\beta_t \boldsymbol{x} + (1 - \beta_t) \boldsymbol{v}) \right) \tag{11}$$
$$= (1 - \beta_t) I - \eta (1 - \beta_t) \mathrm{Dgrad} f(\beta_t \boldsymbol{x} + (1 - \beta_t) \boldsymbol{v}) \tag{12}$$

$$D_x F_2(t, \boldsymbol{x}, \boldsymbol{v}) = D_x \left(\boldsymbol{v} - a_{t+1} \mathrm{grad} f(\beta_t \boldsymbol{x} + (1 - \beta_t) \boldsymbol{v}) \right) \tag{13}$$
$$= -a_{t+1} \mathrm{Dgrad} f(\beta_t \boldsymbol{x} + (1 - \beta_t) \boldsymbol{v}) D_x (\beta_t \boldsymbol{x} + (1 - \beta_t) \boldsymbol{v}) \tag{14}$$
$$= -a_{t+1} \beta_t \mathrm{Dgrad} f(\beta_t \boldsymbol{x} + (1 - \beta_t) \boldsymbol{v}) \tag{15}$$

$$D_v F_2(t, \boldsymbol{x}, \boldsymbol{v}) = D_v \left(\boldsymbol{v} - a_{t+1} \mathrm{grad} f(\beta_t \boldsymbol{x} + (1 - \beta_t) \boldsymbol{v}) \right) \tag{16}$$
$$= I - a_{t+1} \mathrm{Dgrad} f(\beta_t \boldsymbol{x} + (1 - \beta_t) \boldsymbol{v}) D_v (\beta_t \boldsymbol{x} + (1 - \beta_t) \boldsymbol{v}) \tag{17}$$
$$= I - a_{t+1} (1 - \beta_t) \mathrm{Dgrad} f(\beta_t \boldsymbol{x} + (1 - \beta_t) \boldsymbol{v}) \tag{18}$$

and then the differential $DF(t, \boldsymbol{x}, \boldsymbol{v})$ is the following matrix,

$$DF(t, \boldsymbol{x}, \boldsymbol{v}) = \begin{bmatrix} D_x F_1 & D_v F_1 \\ D_x F_2 & D_v F_2 \end{bmatrix} \tag{19}$$

$$= \begin{bmatrix} \beta_t I - \eta \beta_t \mathrm{Dgrad} f(\boldsymbol{y}) & (1 - \beta_t) I - \eta (1 - \beta_t) \mathrm{Dgrad} f(\boldsymbol{y}) \\ -a_{t+1} \beta_t \mathrm{Dgrad} f(\boldsymbol{y}) & I - a_{t+1}(1 - \beta_t) \mathrm{Dgrad} f(\boldsymbol{y}) \end{bmatrix} \tag{20}$$

where $\boldsymbol{y} = \beta_t \boldsymbol{x} + (1 - \beta_t) \boldsymbol{v}$. Note that if $\boldsymbol{y}$ is a stationary point, i.e., in Euclidean sense, $\nabla f(\boldsymbol{y}) = 0$, then the differential of the Riemannian gradient is a matrix whose eigenspace of positive and negative eigenvalues is isomorphic to the eigenspace of the Euclidean Hessian $\nabla^2 f(\boldsymbol{y})$. The local dynamical system is determined by the dimension of positive and negative eigenvalues of the Hessian matrix (either in Riemannian or in Euclidean sense), it suffices to focus on the eigenspace decomposition with respect to the positive and negative eigenvalues. Under a change of coordinate system, $\mathrm{Dgrad} f(\boldsymbol{y})$ can be changed to a diagonal matrix whose eigenvalues are the same as the eigenvalues of $\nabla^2 f(\boldsymbol{y})$. We will denote H the diagonal matrix whose diagonal entries are the eigenvalues of $\nabla^2 f(\boldsymbol{y}^*)$ for a fixed point $\boldsymbol{y}^*$. The following matrix provides enough information understanding the local behavior of the dynamical system induced by algorithm NGL.

$$\tilde{D} = \begin{bmatrix} \beta_t I - \eta \beta_t H & (1 - \beta_t) I - \eta (1 - \beta_t) H \\ -a_{t+1} \beta_t H & I - a_{t+1}(1 - \beta_t) H \end{bmatrix} \tag{21}$$

$$= I_{2d} - \begin{bmatrix} (1 - \beta_t) I + \eta \beta_t H & \eta (1 - \beta_t) H - (1 - \beta_t) I \\ a_{t+1} \beta_t H & a_{t+1}(1 - \beta_t) H \end{bmatrix} \tag{22}$$

where I_{2d} is an identity matrix of $2d \times 2d$. Again if there is no ambiguity created we can ignore the dimension indicator $2d$ for convenience. We claim that the main result of this theorem follows from the spectral analysis of the matrix

$$\begin{bmatrix} (1 - \beta_t)I + \eta\beta_t H & \eta(1 - \beta_t)H - (1 - \beta_t)I \\ a_{t+1}\beta_t H & a_{t+1}(1 - \beta_t)H \end{bmatrix}.$$

The determinant of the above matrix can be computed as follows,

$$p(\lambda) = \det \left(\begin{bmatrix} (1 - \beta_t)I + \eta\beta_t H - \lambda I & \eta(1 - \beta_t)H - (1 - \beta_t)I \\ a_{t+1}\beta_t H & a_{t+1}(1 - \beta_t)H - \lambda I \end{bmatrix} \right) \tag{23}$$

$$= \det(((1 - \beta_t)I + \eta\beta_t H - \lambda I)(a_{t+1}(1 - \beta_t)H - \lambda\,I) \tag{24}$$
$$- (\eta(1 - \beta_t)H - (1 - \beta_t)I)a_{t+1}\beta_t H)$$

$$= \prod_{i=1}^{d} (\lambda^2 - (1 - \beta_t + \eta\beta_t\mu_i + a_{t+1}(1 - \beta_t)\mu_i) + a_{t+1}(1 - \beta_t)\mu_i) \tag{25}$$

where μ_i ($i \in [d]$) are eigenvalues of H. The last equality holds due to the fact that the matrix involved are all diagonal. To solve the eigenvalues of $I_{2d} - \tilde{D}$, we let $p(\lambda) = 0$, and then the roots of $p(\lambda)$ corresponds to the roots of quadratic polynomials

$$\lambda^2 - (1 - \beta_t + \eta\beta_t\mu_i + a_{t+1}(1 - \beta_t)\mu_i) + a_{t+1}(1 - \beta_t)\mu_i.$$

Since the quadratic polynomial has roots of the following form,

$$\lambda_1 = \frac{1}{2}(1 - \beta_t + \eta\beta_t\mu_i + a_{t+1}(1 - \beta_t)\mu_i)$$
$$+ \frac{1}{2}\sqrt{(1 - \beta_t + \eta\beta_t\mu_i + a_{t+1}(1 - \beta_t)\mu_i)^2 - 4a_{t+1}(1 - \beta_t)\mu_i}$$

and

$$\lambda_2 = \frac{1}{2}(1 - \beta_t + \eta\beta_t\mu_i + a_{t+1}(1 - \beta_t)\mu_i)$$
$$- \frac{1}{2}\sqrt{(1 - \beta_t + \eta\beta_t\mu_i + a_{t+1}(1 - \beta_t)\mu_i)^2 - 4a_{t+1}(1 - \beta_t)\mu_i}.$$

The local behavior of the dynamical system induced by the algorithm can be highly regular if the parameters are finely tuned. Note that there are three parameters controlling the behavior of the dynamical system, i.e., the step-size η, sequence β_t, and the step-size sequence a_t. In order to make the dynamical system regular, we assume that the sequence a_t is bounded and small enough or even constant in practical uses. The sequence β_t is also assumed to be bounded away from 0 and less that 1. This is a reasonable assumption that captures specific application scenarios such as Nesterov accelerated gradient descent where $\beta_t = \frac{t}{t+1}$. A final assumption is on the step-size η, it is assumed to be as small as we want to make the dynamical system behave regularly. With all these settings, the first term of the eigenvalue λ_i is positive if the parameters are tuned such that the absolute values of $\eta\beta_t\mu_i$ and $a_{t+1}(1 - \beta_t)\mu_i$ are small enough (regardless of the sign of μ_i) and has no impact on the sign when summing with $1 - \beta_t$ which is lower bounded by a positive constant.

Another observation is the following, the discriminant of the quadratic equation of λ is positive if the parameters are tuned properly. To see that, we have

$$\Delta = (1 - \beta_t)^2 + 2(1 - \beta_t)\eta\beta_t\mu_i + \eta^2\beta_t\mu_i^2 \tag{26}$$

$$+ 2a_{t+1}(1 - \beta_t + \eta\beta_t\mu_i)(1 - \beta_t)\mu_i \tag{27}$$

$$+ a_{t+1}^2(1 - \beta_t)^2\mu_i^2 \tag{28}$$

$$- 4a_{t+1}(1 - \beta_t)\mu_i, \tag{29}$$

and since $a_{t+1}^2(1 - \beta_t)^2\mu_i^2$ is higher order of $4a_{t+1}(1 - \beta_t)\mu_i$, the positivity is determined by the difference between the other terms. Let $B = a_{t+1}(1 - \beta_t)\mu_i$, we have

$$\Delta = (1 - \beta_t + \eta\beta_t\mu_i)^2 + 2(1 - \beta_t + \eta\beta_t\mu_i)B + B^2 - 4B \tag{30}$$

$$= (1 - \beta_t + \eta\beta_t\mu_i)^2 + 2(1 - \beta_t + \eta\beta_t\mu_i - 2)B + B^2. \tag{31}$$

Since the non-existence of real eigenvalue is only related to positive μ_i, so we consider the case when $\mu_i > 0$, which implies $B > 0$. For a fixed or bounded a_{t+1}, Δ is a function of η. Therefore, $\Delta_{\eta=0} = (1 - \beta_t)^2 + 2(-1 - \beta_t)B + B^2 > 0$ can be guaranteed by a stronger condition $(1 - \beta_t)^2 - 2(1 + \beta_t)B > 0$. Suppose that

$$(1 - \beta_t)^2 - 2(1 + \beta_t)B = (1 - \beta_t)^2 - 2(1 + \beta_t)a_{t+1}(1 - \beta_t)\mu_i > 0,$$

and this implies a condition that the sequence a_{t+1} is expected to satisfy,

$$a_{t+1} < \frac{1 - \beta_t}{2(1 + \beta_t)\mu_i}.$$

With this condition, we can conclude that there exists some $\eta > 0$ such that $p(\lambda)$ has only real eigenvalues. Therefore, if at certain stationary point $\boldsymbol{y}^*$, H has only positive eigenvalues, it follows that $p(\lambda)$ also has only positive eigenvalues, and this implies that the corresponding stationary point is attracting in all directions, i.e., the second order stationary point. The case when there exists some negative eigenvalue μ_i of H is less transparent, the main idea is to show that the initial points that can be transported to this stationary point has measure zero.

Lemma 1. *Suppose $\boldsymbol{y}^*$ is a stationary point such that $\operatorname{grad} f(\boldsymbol{y}^*) = 0$, and at least one eigenvalue of $\nabla \operatorname{grad} f(\boldsymbol{y}^*)$ is negative. Then in a neighborhood of $\boldsymbol{y}^*$, the initial points that converge to $\boldsymbol{y}^*$ is of measure zero, where the measure is defined by Riemannian volume form.*

Proof. Previous analysis asserts that if the sequence a_{t+1} satisfies $a_{t+1} < \frac{1-\beta_t}{2(1+\beta_t)\mu_{\max}}$ where $\mu_{\max}$ is the maximum of positive eigenvalues, there exists a stepsize $\eta > 0$, such that all eigenvalues of $\tilde{D}$ are real numbers, and this makes $\tilde{D}$ diagonalizable. The analysis of local behavior of NGL boils down to that of a topological conjugated dynamical system whose differential of the mapping is of the form of $I - \mathcal{L}$, where $\mathcal{L}$ is a diagonal matrix whose eigenvalues are time-dependent functions. It is obvious from the expression of λ_i that the eigenvalues satisfy the following property,

- If $\mu_i > 0$, by tuning step-size η and a_{t+1}, the corresponding eigenvalues λ_i are positive and less than 1, i.e., $0 < c_i \leq \lambda_i < 1$.
- If $\mu_i < 0$, the corresponding eigenvalues λ_i satisfies $\lambda_i \leq c_2 < 0$.

The algorithm can be expanded at a neighborhood of fixed point, and in order to proceed the local stability analysis, we will consider a more general dynamical system, based on the above observation of the eigenvalues. The dynamical system we investigate is of the following form,

$$x_{t+1} = (I - \mathcal{L}(t))x_t + R(t, x_t)$$

where $\mathcal{L}(t)$ is diagonal and the diagonal entries satisfies the above two conditions, and $R(t, x)$ is a perturbation satisfying Lipschitz type condition. We will assume that such dynamical system is defined on $\mathbb{R}^n$ and for $1 \leq i \leq s$, $\lambda(t) > 0$) and for $s < i \leq n$, $\lambda(t) < 0$. Thus the matrix $I - \mathcal{L}(t)$ can be decomposed into two blocks,

$$I - \mathcal{L}(t) = \begin{bmatrix} I - L_1(t) & 0 \\ 0 & I - L_2(t) \end{bmatrix}$$

where $L_1(t) = \mathbf{diag}\{\lambda_i(t)\}_{i \in [s]}$ and $L_2(t) = \mathbf{diag}\{\lambda_i(t)\}_{i \in [n-s]}$. Then we have the update rule with initial point x_0) as a closed form given as follows,

$$x_{t+1} = \begin{bmatrix} \prod_{i=0}^{t}(I - L_1(i)) & 0 \\ 0 & \prod_{i=0}^{t}(I - L_2(i)) \end{bmatrix} x_0 \tag{32}$$

$$+ \sum_{i=0}^{t} \begin{bmatrix} \prod_{s=i+1}^{t}(I - L_1(s)) & 0 \\ 0 & \prod_{s=i+1}^{t}(I - L_2(s)) \end{bmatrix} R(i, x_i). \tag{33}$$

The block $\prod_{i=0}^{t}(I - L_1(i))$ and $\prod_{i=0}^{t}(I - L_2(i))$ are called stable and unstable operators respectively. According to this stable-unstable decomposition of operator, which enables use to have a stable-unstable decomposition of the vector space at the fixed point being considered, we can split the dynamical system into stable-unstable components as follows,

$$x_{t+1}^+ = \prod_{i=0}^{t}(I - L_1(i))x_0^+ + \sum_{i=0}^{t} \prod_{s=i+1}^{t}(I - L_1(s))R^+(i, x_i)$$

and

$$x_{t+1}^- = \prod_{i=0}^{t}(I - L_2(i))x_0^+ + \sum_{i=0}^{t} \prod_{s=i+1}^{t}(I - L_2(s))R^-(i, x_i).$$

Since we are interested in the behavior of x_t as $t \to \infty$ expecting x_t converging to an unstable point of the dynamical system, we just let $x_t \to 0$. Note that it is necessarily $x_{t+1}^- \to 0$ despite the linear operator acting on the unstable initial component x_0^- is expanding. So the unstable component x_0^- of the initial point x_0 satisfies

$$x_0^- = -\sum_{i=1}^{\infty} \left(\prod_{s=0}^{i-1}(I - L_2(s))\right)^{-1} R^-(i-1, x_{i-1}).$$

Plugging above expression of x_0^- into the expression of x_{t+1}^- we have that

$$x_{t+1}^- = -\sum_{i=0}^{\infty} \left(\prod_{s=t+1}^{t+1+i} (I - L_2(s)) \right)^{-1} R^-(t+1+i, x_{t+1+i}).$$

Combining with the expression of x_{t+1}^+, we can consider the expression of the transformed $x_{t+1}^+ \oplus x_{t+1}^-$ as a result of an operator T acting on another sequence x_t, i.e.,

$$(Tx)_{t+1} = \left(\prod_{i=0}^{t}(I - L_1(i))x_0^+ + \sum_{i=0}^{t} \prod_{s=i+1}^{t} (I - L_1(s))R^+(i, x_i) \right) \tag{34}$$

$$\oplus \left(-\sum_{i=0}^{\infty} \left(\prod_{s=t+1}^{t+1+i} (I - L_2(s)) \right)^{-1} R^-(t+1+i, x_{t+1+i}) \right). \tag{35}$$

The main idea of showing the local measure zero result is to establish the existence and uniqueness of so called local stable manifold of the dynamical system. Since a manifold is nothing but a graph of certain differentiable mapping defined on Euclidean space, it suffices to show the existence and uniqueness of certain function from stable space to unstable space, where the stable and unstable is from the corresponding decomposition of the diagonal matrix. The existence of such a function follows from the fact that T is a contracting map on sequences converging to 0.

By assumptions on $\lambda_i(t)$, we have the following estimates,

$$\sum_{i=0}^{t} \left\| \prod_{s=i+1}^{t} (I - L_1(s)) \right\| \le \sum_{i=0}^{t} \prod_{s=i+1}^{t} \|I - L_1(s)\| \le \sum_{i=0}^{t} \prod_{s=i+1}^{t} (1 - c_1) = \frac{1}{c_1}. \tag{36}$$

On the other hand, we have

$$\sum_{i=0}^{\infty} \left\| \left(\prod_{s=t+1}^{t+1+i} (I - L_2(s)) \right)^{-1} \right\| \le \sum_{i=0}^{\infty} \prod_{s=t+1}^{t+1+i} \|(I - L_2(s))^{-1}\| \tag{37}$$

$$= \sum_{i=0}^{\infty} \prod_{s=t+1}^{t+1+i} \left\| \mathrm{diag}\{ \frac{1}{1 - \lambda_i(s)} \} \right\| \tag{38}$$

$$\le \sum_{i=0}^{\infty} \prod_{s=t+1}^{t+1+i} \frac{1}{1 - c_2} \tag{39}$$

$$= \sum_{t=0}^{\infty} \left(\frac{1}{1 - c_2} \right)^{i+1} \tag{40}$$

$$= \lim_{N \to \infty} \sum_{i=0}^{N} \frac{1}{(1 - c_2)^{i+1}} \tag{41}$$

$$= \frac{1}{-c_2}. \tag{42}$$

Moreover,

$$\left\|\prod_{s=0}^{t}(I - L_1(s))\right\| \leq (1 - c_1)^t < 1 - c_1 < 1,$$

so as long as $\epsilon > 0$ so that

$$\left\|\prod_{s=0}^{t}(I - L_1(s))\right\| + \epsilon\frac{1}{c_1} + \epsilon\frac{1}{-c_2} \leq 1$$

for example, such ϵ can be chosen as

$$\epsilon = \frac{c_1^2 c_2}{2(c_2 - c_1)}$$

the operator T will be a contracting map on the space of sequences converging to 0. Therefore, we can conclude that if the initial point x_0 converges to 0, which is a fixed point with non-trivial unstable space with respect to the differential of the dynamical system, then the stable-unstable components of x), say (x_0^+, x_0^-) satisfies that there exist a unique function ϕ from the stable space to unstable space such that $x_0^- = \phi(x_0^+)$. An important implication of this result is what we are proving for this lemma, i.e., the initial points that converges to an unstable fixed point is of measure zero. Since it is assumed in the lemma that y^* is a saddle point, which implies, based on the spectral analysis of the differential of the algorithm, the corresponding fixed point (x^*, v^*) of the algorithm is unstable, thus the initial points converging to (x^*, v^*) is of measure zero.

E   Proof of Corollary 1

Define $\tilde{F}(m, n, x, v) = F(m, ..., F(n + 1, F(n, x, v))...)$ for $m > n$. For each $(x^*, v^*)\mathcal{A}^* \times \mathcal{A}^*$, the set of second order stationary points, there is an open neighborhood $U_{(x^*, v^*)}$ depending on the dynamical system we consider. $\cup_{x^* \in \mathcal{A}^*} U_{x^*}$ forms an open cover, and since manifold is second-countable, we can find a countable subcover, so that

$$\cup_{(x^*, v^*)\in\mathcal{A}^* \times \mathcal{A}^*} U_{(x^*, v^*)} = \cup_{i=1}^{\infty} U_{(x_i^*, v_i^*)}.$$

Since $\mathcal{A}^* \times \mathcal{A}^*$ is a global stable set of the algorithm on $M \times M$, we have

$$W^s(\mathcal{A}^* \times \mathcal{A}^*) = \{(x_0, v_0) : \lim_{t\to\infty} \tilde{F}(t, 0, x_0, v_0) \in \mathcal{A}^* \times \mathcal{A}^*\}.$$

Fix a point $(x_0, v_0) \in W^s(\mathcal{A}^* \times \mathcal{A}^*)$, since $F(t, 0, x_0, v_0) \to (x^*, v^*) \in \mathcal{A}^* \times \mathcal{A}^*$, there exists some non-negative integer T and all $t \geq T$, such that

$$\tilde{F}(t, 0, x_0, v_0) \in \bigcup_{(x^*, v^*)\in\mathcal{A}^* \times \mathcal{A}^*} U_{(x^*, v^*)} = \bigcup_{i=1}^{\infty} U_{(x_i^*, v_i^*)\in\mathcal{A}^* \times \mathcal{A}^*}.$$

So $\tilde{F}(t, 0, \boldsymbol{x}_0, \boldsymbol{v}_0) \in U_{(\boldsymbol{x}_i^*, \boldsymbol{v}_i^*)}$ for some $(\boldsymbol{x}_i^*, \boldsymbol{v}_i^*) \in \mathcal{A}^* \times \mathcal{A}^*$ and all $t \geq T$. This is equivalent to $\tilde{F}(T + t, T, T, 0, \tilde{\boldsymbol{x}}_0, \boldsymbol{v}_0) \in U_{(\boldsymbol{x}_i^*, \boldsymbol{v}_i^*)}$ for all $k \geq 0$, and this implies that

$$\tilde{F}(T, 0, \boldsymbol{x}_0, \boldsymbol{v}_0) \in \tilde{F}^{-1}(T + t, T, U_{(\boldsymbol{x}_i^*, \boldsymbol{v}_i^*)})$$

for all $t \geq 0$. And then we have

$$\tilde{F}(T, 0, \boldsymbol{x}_0, \boldsymbol{v}_0) \in \bigcap_{t=0}^{\infty} \tilde{F}^{-1}(T + t, T, U_{(\boldsymbol{x}_i^*, \boldsymbol{v}_i^*)}).$$

Denote $S_{i,T} = \bigcap_{t=0}^{\infty} \tilde{F}^{-1}(T + t, T, U_{(\boldsymbol{x}_i^*, \boldsymbol{v}_i^*)})$ and the above relation is equivalent to $(\boldsymbol{x}_0, \boldsymbol{v}_0) \in \tilde{F}^{-1}(T, 0, S_{i,T})$. Take the union for all non-negative integers T, we have $(\boldsymbol{x}_0, \boldsymbol{v}_0) \in \bigcup_{T=0}^{\infty} \tilde{F}^{-1}(T, 0, S_{i,T})$. Taking union of all i we can conclude

$$(\boldsymbol{x}_0, \boldsymbol{v}_0) \in \bigcup_{i=1}^{\infty} \bigcup_{T=0}^{\infty} \tilde{F}^{-1}(T, 0, S_{i,T})$$

which implies $W^s(\mathcal{A}^* \times \mathcal{A}^*) \subset \bigcup_{i=1}^{\infty} \bigcup_{T=0}^{\infty} \tilde{F}^{-1}(T, 0, S_{i,T})$. Since $S_{i,T}$ is a subset of measure zero so it has measure zero as well, with respect to the volume measure of the Riemannian metric. On the other hand, the image of measure zero set under diffeomorphism is of measure zero, and countable union of measure zero set is still of measure zero, we can conclude that $W^s(\mathcal{A}^* \times \mathcal{A}^*)$ is of measure zero.

References

Alimisis, F., Orvieto, A., Bécigneul, G., Lucchi, A.: Momentum improves optimization on Riemannian manifolds. In: AISTATS (2021)

Amari, S.I.: Natural gradient works efficiently in learning. Neural Comput. **10**(2), 251–276 (1998)

Amari, S.I., Douglas, S.C.: Why natural gradient? In: Proceedings of ICASSP 1998, pp. 1213–1216 (1998)

Amari, S.I., Karakida, R., Oizumi, M.: Fisher information and natural gradient learning of random deep networks. In: Proceedings of AISTATS, vol. 89, pp. 694–702 (2019)

Amari, S.I., Nagaoka, H.: Methods of Information Geometry, vol. 191. American Mathematical Society (2007)

Aonishi, T., Maruyama, R., Ito, T., Miyakawa, H., Murayama, M., Ota, K.: Imaging data analysis using non-negative matrix factorization. Neurosci. Res. **179**, 51–56 (2022)

Bürgisser, P., Li, Y., Nieuwboer, H., Walter, M.: Interior-piont methods for unconstrained geometric programming and scaling problems (2020). https://arxiv.org/abs/2008.12110

Criscitiello, C., Boumal, N.: An accelerated first-order method for non-convex optimization on manifolds. Foundations of Computational Mathematics (2022)

Chiang, M.: Geometric programming for communication systems. Foundations and Trends in Communications and Information Theory (2006)

Feng, Y., Panageas, I., Wang, X.: Accelerated multiplicative weights update avoids saddle points almost always. In: IJCAI (2022)

Guan, N., Tao, D., Luo, Z., Yuan, B.: Nenmf: an optimal gradient method for nonnegative matrix factorization. IEEE Trans. Signal Process. **60**(6), 2882–2898 (2012)

Han, A., Mishra, B., Jawanpuria, P., Gao, J.: Riemannian accelerated gradient methods via extrapolation. In: AISTATS (2023)

Hoburg, W., Abbeel, P.: Geometric programming for aircraft design optimization. AIAA J. (2014)

Hofbauer, J., Sigmund, K.: Evolutionary Games and Population Dynamics. Cambridge University Press (1998)

Kandukuri, S., Boyd, S.: Optimal power control in interference-limited fading wireless channels with outage-probability specifications. IEEE Trans. Wirel. Commun. 1 (2002)

Kim, H., Park, H.: Nonnegative matrix factorization based on alternating nonnegativity constrained least squares and active set method. SIAM J. Matrix Anal. Appl. **30**(2), 713–730 (2008)

Kim, J., He, Y., Park, H.: Algorithms for nonnegative matrix and tensor factorizations: a unified view based on block coordinate descent framework. J. Global Optim. **58**, 285–319 (2014)

Kim, J., Yang, I.: Accelerated gradient methods for geodesically convex optimization: tractable algorithms and convergence analysis. In: ICML (2022)

Lee, D., Seung, H.S.: Unsupervised learning by convex and conic coding. In: Advances in Neural Information Processing Systems, vol. 9 (1996)

Lee, D., Seung, H.S.: Algorithms for non-negative matrix factorization. In: Advances in Neural Information Processing Systems, vol. 13 (2000)

Lee, J.D., Panageas, I., Piliouras, G., Simchowitz, M., Jordan, M.I., Recht, B.: First-order methods almost always avoid strict saddle points. Math. Program. **176**, 311–337 (2019)

Lin, C.J.: On the convergence of multiplicative update algorithms for nonnegative matrix factorization. IEEE Trans. Neural Netw. **18**(6), 1589–1596 (2007)

Lin, C.J.: Projected gradient methods for nonnegative matrix factorization. Neural Comput. **19**(10), 2756–2779 (2007)

Martens, J.: New insights and perspectives on the natural gradient method. J. Mach. Learn. Res. **21**(1), 5776–5851 (2020)

Mertikopoulos, P., Sandholm, W.H.: Riemannian game dynamics. J. Econ. Theory **177** (2018)

Paatero, P.: The multilinear engine—a table-driven, least squares program for solving multilinear problems, including the n-way parallel factor analysis model. J. Comput. Graph. Stat. **8**(4), 854–888 (1999)

Perko, L.: Differential Equations and Dynamical Systems. Springer, Cham (2001)

Petersen, P.: Riemannian Geometry, vol. 171. Springer, Cham (2006)

Robert, C., Patel, R., Blostein, N., Steele, C.J., Chakravarty, M.M.: Analyses of microstructural variation in the human striatum using non-negative matrix factorization. Neuroimage **246**, 118744 (2022)

Sadeghi, S., Lu, J., Ngom, A.: A network-based drug repurposing method via non-negative matrix factorization. Bioinformatics **38**(5), 1369–1377 (2022)

Shahshahani, S.: A New Mathematical Framework for the Study of Lineage and Selection, vol. 17. American Mathematical Society (1979)

Chern, S.S., Chen, W.H.: Lectures on Differential Geometry. World Scientific (1999)

Krichene, W., Bayen, A., Bartlett, P.: Accelerated mirror descent in continuous and discrete time. In: NIPS (2015)

Yang, J., Leskovec, J.: Overlapping community detection at scale: a nonnegative matrix factorization approach. In: Proceedings of the Sixth ACM International Conference on Web Search and Data Mining, pp. 587–596 (2013)

Zhang, H., Sra, S.: Towards Riemannian accelerated gradient methods. In: COLT (2018)

Zhao, Y., Deng, F., Pei, J., Yang, X.: Progressive deep non-negative matrix factorization architecture with graph convolution-based basis image reorganization. Pattern Recogn. **132**, 108984 (2022)

A Case for Copeland: from Theory to Practice

Michelle Le[1] (ID), Chloe Nguyen[1], Leo Claney[1], Krishh Tipnis[1], Brian MacSweeney[3],
Eric Huber[2], and Christine Chung[1(✉)] (ID)

[1] Department of Computer Science, Connecticut College, New London, CT 06320, USA
{nle6,cnguyen2,lclaney,ktipnis,cchung}@conncoll.edu
[2] Raytheon Corp, Woburn, MA 01801, USA
[3] Carnegie Mellon University, Pittsburgh, PA 15213, USA
bmacswee@andrew.cmu.edu

Abstract. We consider the Copeland voting rule, a classical and simple voting rule that takes a set of voters' rankings over a set of candidates and outputs the candidate that wins the most pair-wise match-ups against other candidates. We examine Copeland in the metric distortion model of voting, with a focus on low-dimensional spaces. We show that Copeland voting on the line metric has a metric distortion of 3, which is a lower metric distortion than its ratio of 5 for general metrics, and also better than the distortion on the line of other common voting rules like Plurality and Borda. It is also lower than the best-known bound for STV, known to the general public as "Ranked Choice Voting" (RCV), which is increasingly being adopted by state and local governments in the United States. We then run simulations in which randomly generated voters and candidates are placed in a Euclidean space. For each random trial, we compare the Copeland rule to Plurality, Borda, Single Transferable Vote (STV/RCV), and Plurality Veto, the novel rule recently proposed that achieves the best possible metric distortion of 3 for general metrics. We show in our empirical study that the Copeland rule outperforms all others, including Plurality Veto and STV/RCV, with respect to distortion as well as the rate at which the optimal candidate is elected. We also test the same voting rules for the satisfaction rate of desirable criteria, such as the Independence of Irrelevant Alternatives, and in these respects, Copeland still outperforms the other voting rules.

1 Introduction

In recent years, as computing has increasingly permeated decision-making and operations in every industry, there has been a rising demand for transparent and explainable computing models and algorithms. In social choice settings, transparency has always been important, and it is especially called for in political applications of voting rules. Since various governing bodies have slowly come to grips with the numerous and flagrant flaws of the de facto Plurality voting rule, a movement in the United States has succeeded in shifting an increasing number of state and local governments to "Ranked

The authors would like to thank Barbara Anthony of Southwestern University for her input on early drafts of portions of this work. We are also grateful to the anonymous reviewers for their valuable feedback.

Choice Voting" (RCV) [34], which is a social choice method known in the voting theory literature as Single Transferable Vote (STV). Yet STV/RCV and other voting rules are somewhat slow to be accepted, partly due to their perceived opacity by the mainstream public [22,24,33].[1] Meanwhile, recent exciting progress has been made in the literature on the metric distortion of voting rules [4], with elegant and novel procedures, like the Plurality Veto rule, achieving optimal distortion ratios [20,25]. While these new voting methods are appealing to researchers and theorists, the procedures can still be difficult to explain concisely and clearly to a lay audience.

In this work, we investigate the Copeland voting rule [36], a straightforward, classical, yet effective voting method that determines a winner by comparing candidates in a pairwise fashion. According to standard assumptions for social choice functions, each voter first submits a ballot ranking all the candidates. Copeland then pits each candidate head-to-head against every other candidate, and the candidate with the most wins across all pairwise comparisons is the winner. Forms of the Copeland rule are used in popular round-robin tournament settings like sports (e.g., soccer) or other competitions (e.g. chess) [37]. This makes it approachable, familiar, and simple enough to be easily understood by the general public, making it an attractive option for practical use.

Metric distortion [2,3] has recently become an influential measure that captures how many times worse the candidate chosen by a voting rule is than the optimal or "correct" candidate. It was already shown by [3] that Copeland has a metric distortion of 5, in contrast with Plurality, Borda, STV, and others, which all have super-constant distortion. In this work, we show that assuming a one-dimensional metric, the Copeland rule has a worst-case distortion of 3, the lowest possible for any deterministic voting rule. In contrast, the metric distortion of STV/RCV on the line is only known thus far to have an upper bound of 15 [1], while Borda and Plurality are both known to have a metric distortion of $\Omega(m)$, where m is the number of candidates, even on the line [3].

Recent works [1,13,38] have suggested that low-dimensional metrics are important to consider, and we believe even modeling voters and candidates along a one-dimensional metric space has practical relevance, as it captures the everyday reference of a basic ideological axis, ranging from "liberal" to "conservative." It has also been long suggested that "single-issue" voters are quite common [11,23,27], and a one-dimensional metric neatly models the spectrum of positions a voter may have on a single issue.[2]

We also empirically study the Copeland rule with simulations of candidates and voters in low-dimensional metric spaces, and find that Copeland outperforms other commonly studied voting rules, including STV/RCV and Plurality Veto. Our experiments evaluate the voting rules both in terms of distortion and accuracy (i.e. how often the optimal candidate is chosen). We further find in our experiments that, compared to other voting rules, Copeland excels in adhering to a certain commonly studied social choice criteria, the Condorcet criterion and Independence of Irrelevant Alternative criterion.

---

[1] This is despite the fact that STV is already the rule used for the national elections of Australia, Ireland, and India [1].

[2] A recent national poll indicated that only 21 percent of voters did not have a single issue that would determine their support for or against a candidate [6].

1.1   Metric Distortion and Related Work

Voting rules, or social choice functions, aggregate the preferences of voters over a set of candidates and output a candidate as the "winner," who ideally represents the collective preference of the voters. Traditionally, voting rules have been functions of the voters' ordinal rankings, wherein decisions are made based solely on each voter's preference ordering of the candidates. An ordinal voting rule may however fall short of optimizing social utilities, resembling an approximation algorithm whose worst-case performance is known as its *distortion*.

The notion of *distortion* in voting mechanisms was introduced by Procaccia and Rosenschein [35], who defined it to be the worst-case ratio between the social welfare achieved by a specific ordinal voting rule and the optimal social welfare. Later, Anshelevich et al. [3] introduced the metric distortion framework, which became another influential step in quantifying the quality of a voting rule. Within this framework, a set of n voters and m candidates is mapped into a shared metric space. The distances between a voter and all candidates are consistent with the voter's preference over the candidates. This approach assumes that each candidate embodies a specific standpoint on crucial issues, akin to points in an ideological space. Similarly, each voter holds a standpoint on these issues, presumably reflected in their ballot. The distance between a voter and a candidate in this metric space serves as the voter's (cardinal) metric cost. It is reasonable to assume that voters would prefer candidates with standpoints closer to their own, leading to a lower metric cost. The *social cost* of a candidate is the sum of the distances of all n voters to the candidate.

Hence, *metric distortion* of a voting rule is defined as the worst-case ratio of the social cost of the candidate selected by the voting rule to the social cost of the optimal candidate, over all possible metric spaces and all induced ordinal preference profiles. The initial study by Anshelevich et al. [2,3] established a lower bound of 3 for the distortion of any deterministic voting rule. They further demonstrated that choosing any candidate from the *uncovered set*—which the Copeland rule does - results in a constant distortion of 5. Subsequent research aimed to narrow this gap. Munagala and Wang [31] made the first improvement by introducing a weighted variant of the uncovered set, achieving a distortion of $2 + \sqrt{5} \approx 4.236$. They also proved that selecting from a novel *matching uncovered set* ensures a distortion upper bound of 3. Gkatzelis et al. [20] then established a distortion of 3 for their proposed Plurality Matching rule, conclusively closing the gap for deterministic social choice functions. Building on this achievement, Kizilkaya and Kempe [25] more recently proposed a refined and more practical voting rule known as Plurality Veto, which also achieves the optimal distortion of 3, providing an elegant alternative to Plurality Matching.

Despite this recent progress and activity in the academic literature on the metric distortion of voting rules, in real-world elections, progress has gone in a different direction. STV/RCV is a voting rule that eliminates candidates with the fewest votes one round at a time, transferring the votes of those whose favorite candidate was eliminated to the remaining candidates in each round. In contrast with mechanisms like Copeland and Plurality Veto that have constant distortion bounds, STV/RCV has a metric distortion of $\Omega(\sqrt{\ln m})$ [38]. Notably, while STV/RCV and Plurality Veto are simple enough for a technical audience to find appealing, neither have proven to be easy for a mainstream audience to digest.

The Copeland rule has also been found in past work to have various desirable properties. Faliszewski et al. [14] has shown that the Copeland election system exhibits notable resistance to various forms of bribery and procedural control, surpassing what was known for any natural election system where winners can be computed in polynomial time. Copeland voting also satisfies the desirable Condorcet criterion, where a candidate who defeats all other candidates in head-to-head match-ups is always chosen as the winner, while many other rules that satisfy the Condorcet criterion cannot be computed in polynomial time [15].

It is already known that candidates and voters in a line metric always admit a Condorcet winner [9], and that Condorcet winners have distortion at most 3 on a line [2]. However, the Condorcet-consistency of Copeland only applies to *strong* Condorcet winners (candidates who are preferred by a strict majority of the voters when compared head-to-head against any other candidate). Copeland does not always elect a *weak* Condorcet winner (even assuming one exists) [16,17]. Our result in Sect. 4.2 effectively shows that when restricted to the line metric, Copeland not only satisfies the Condorcet criterion for strong Condorcet winners, but for weak ones as well.

Later, Goel [21] studied the Copeland rule in the context of quantifying the "fairness" of social choice rules, and proposed the "fairness ratio" as a generalization on the standard definition of metric distortion. Their work showed that Copeland achieves a fairness ratio of 5, amounting to a constant-factor approximation for a large class of convex objectives, and matching its previously established metric distortion bound for the standard sum-of-distances objective [2].

Regarding prior empirical studies related to ours, the work of [13] evaluates various multi-winner voting rules using elections randomly generated in a two-dimensional Euclidean space. Their work used statistical properties of the elected winners to determine the most suitable voting rules for three applications of multi-winner voting. To our knowledge, little other prior empirical work has been done with respect to simulating a spatial model of voting, especially for sake of comparing the cost of the candidate elected against the cost of an optimal candidate.

1.2 Our Contributions

In this work, we show that while rules like Plurality and Borda have distortion $2m - 1$, even on a line [3], Copeland achieves a worst-case distortion of 3 on the line, the best possible for any deterministic voting rule. Furthermore, we consider simulated settings in which the positions of voters and candidates in a low-dimensional metric space are drawn from standard probability distributions. We test commonly studied voting rules for average distortion, max distortion, and accuracy (the rate at which the optimal candidate is elected). In these simulations, Copeland outperforms STV/RCV, as well as many other well-known voting mechanisms, old and new (e.g. Borda and Plurality Veto). We then build upon the work of previous empirical studies in social choice theory [8,12,19], which evaluate voting rules by considering how often certain desirable criteria are satisfied. Our experiments test six voting rules against two criteria: the Condorcet criterion and the Independence of Irrelevant Alternatives. We find that the Copeland rule is the most likely to satisfy these criteria when compared against Plurality, Borda, STV/RCV, and Plurality Veto.

2   Preliminaries

An *election* is a tuple $\mathcal{E} = (\mathcal{V}, \mathcal{C}, \sigma)$ consisting of a set of n *voters*, $\mathcal{V}$, a set of m *candidates*, $\mathcal{C}$, and a *preference profile* $\sigma = (\sigma_1, \sigma_2, \ldots, \sigma_n)$, where σ_i is the *preference ranking* of voter i represented as a strict total/linear order over $\mathcal{C}$. We will use $X \succ_i Y$ to denote candidate X being preferred over candidate Y by voter i, according to the ordering of candidates in σ_i. We let $XY = \{i \in \mathcal{V} : X \succ_i Y\}$, i.e., the set of voters that prefer X over Y. We say X *defeats* Y when $|XY| > |YX|$ and X *weakly defeats* Y when $|XY| \geq |YX|$.

A *social choice function* or *voting rule* f is a procedure that, given a preference profile σ, returns a candidate $f(\sigma) \in \mathcal{C}$. We refer to $f(\sigma)$ as a *winner* of the election under voting rule f. Given an election $\mathcal{E} = (\mathcal{V}, \mathcal{C}, \sigma)$, we say that a metric d over $\mathcal{V} \cup \mathcal{C}$ is *consistent* with the ranking σ_i of voter i if $d(i, X) < d(i, Y)$ for all $X, Y \in \mathcal{C}$ such that $X \succ_i Y$.[3] We say that d is consistent with σ if it is consistent with the ranking σ_i for all voters $i \in \mathcal{V}$. We use $\mathcal{D}(\sigma)$ to denote the set of metrics consistent with σ.

The *social cost* of a candidate X with respect to a metric d is defined as the candidate's sum of distances to all voters: $SC(X, d) = \sum_{i \in \mathcal{V}} d(i, X)$. A candidate X is *optimal* with respect to the metric d if $X \in argmin_{X \in \mathcal{V}} SC(X, d)$. The *distortion* of a voting rule f, given preference profile σ, denoted by $dist(f, \sigma)$, is the worst-case ratio between the social cost of $f(\sigma)$ and that of an optimal candidate OPT over all metrics in $\mathcal{D}(\sigma)$. That is, $dist(f, \sigma) = \sup_{d \in \mathcal{D}(\sigma)} SC(f(\sigma), d)/SC(OPT, d)$. We overload notation and say that the distortion of a voting rule f is the maximum value of $dist(f, \sigma)$, taken over all possible preference profiles σ. That is,

$$dist(f) = \max_{\sigma} \sup_{d \in \mathcal{D}(\sigma)} \frac{SC(f(\sigma), d)}{SC(OPT, d)}.$$

The following useful Lemma is due to [2], and restated here for later reference.

Lemma 1 (from [2]). *For every preference profile σ, every pseudo-metric $d \in \mathcal{D}(\sigma)$, and every pair of candidates W, X we have $\frac{SC(W,d)}{SC(X,d)} \leq \frac{2n}{|WX|} - 1$.*

2.1   Voting Rules

We consider the following well-known voting rules. In all cases, a set of candidates may emerge as tied for winning the election (and also, in the case of each round of STV, tied for elimination). We assume ties are broken arbitrarily; for our worst-case analysis, any candidate in the set of possible winners may be chosen as the winner.

---

[3] Since agents are allowed to be at distance 0 from each other, d is technically a pseudo-metric; in this paper we use the terms metric and pseudo-metric interchangeably.

Copeland. Candidates are compared head-to-head with every other candidate. The winner of each pairwise match receives one point and the loser receives zero points; in case of a tie both candidates receive α points ($0 \leq \alpha \leq 1$), and a candidate with the most points is the winner. The original variant as proposed by [36] used $\alpha = 1/2$. [4]

Plurality. The score of each candidate X is equal to the number of voters who rank X first. A candidate with the highest score wins the election.

Borda. A candidate X receives m points for each voter that ranks X first, $m - 1$ points for each voter that ranks X second, and so on. The candidate with the most points wins.

Single Transferable Vote (STV). In each round, if a candidate has a majority of votes (more than 50% of the voters rank them first), they are declared to be the winner. If no majority winner exists, an alternative with the lowest Plurality score in that round (one that is ranked first by the fewest voters) is eliminated from the set of candidates; the Plurality rule is then applied again to the remaining candidates, and the process repeats until one of the remaining candidates is ranked first by more than half the voters. [5]

Plurality Veto. Each candidate X starts with a score equal to the number of voters that rank X first. Any candidates with a score of 0 are eliminated. For each voter $i \in \mathcal{V}$, the score of their lowest-ranked candidate among the remaining candidates is decremented, and during this process, whenever the score of any candidate Y reaches 0, Y is eliminated. The last candidate remaining wins.

3   Distortion of Copeland on the Line

In this section, we show that the distortion of the Copeland rule on a line is 3, which is as low as the distortion of any deterministic voting rule, and much lower than that of rules like Plurality and Borda. Note that the following theorem holds for Copeland with any α value, $0 \leq \alpha \leq 1$.

Theorem 1. *The distortion of the Copeland rule on a line metric is at most 3.*

Proof. Consider a profile σ over a set of candidates $\mathcal{C}$ and a metric $d \in \mathcal{D}(\sigma)$. Let W be a Copeland winner for σ and let X be an optimal candidate for σ with respect to d. If W weakly defeats X, then $|WX| \geq n/2$, and from Lemma 1 we immediately have $SC_{\Sigma}(W, d) \leq 3 \cdot SC_{\Sigma}(X, d)$. Hence, for the rest of the proof we may assume X defeats W, i.e.,

$$|XW| > n/2. \tag{1}$$

---

[4] Our theorem in Sect. 3 holds for any $0 \leq \alpha \leq 1$. We found that our experimental results, reported in Sect. 4, were similar for different values of α between 0 and 1, so in Sect. 4 we focus on $\alpha = 1/2$.

[5] We follow the lead of [10, 38] in assuming a winner from any valid sequence of tie-breaks in elimination rounds is possible.

It has been previously established by [3,30] that the Copeland rule always outputs a subset of the *uncovered set*, which is a set comprised of candidates from $\mathcal{C}$ who either weakly defeat every other candidate $Z \in \mathcal{C}$, or weakly defeats another candidate Y who in turn weakly defeats all $Z \in \mathcal{C}$. Since W is defeated by X, then there must be a candidate Y such that W weakly defeats Y, who in turn weakly defeats X. Note that W, Y, and X must all be in distinct locations on the line from one another, since if any pair is co-located, either (1) or another above assumption is immediately violated. We consider three cases based on the possible relative positions of W, Y, and X and show that in all cases a contradiction is reached, so, in fact, the condition (1) cannot occur.

Case 1: Y is between W and X. In this case, for every voter $i \in WY$, it is the case that $d(i, W) < d(i, X)$, so $i \in WX$ and thus $|WX| \geq |WY|$. Because W weakly defeats Y, $|WY| \geq \frac{n}{2}$, so $|WX| \geq \frac{n}{2}$, a contradiction with (1).

Case 2: W is between Y and X. Without loss of generality, let us assume the position of Y is to the "left" of W and X is to the "right" of W. Consider any voter $i \in YX$. If i is located to the left of W, then $d(i, X) = d(i, W) + d(W, X) > d(i, W)$. If i is to the right of W, or co-located with W, then $d(i, X) > d(i, Y) = d(i, W) + d(W, Y) > d(i, W)$. In both cases, $d(i, W) < d(i, X)$, so $i \in WX$, and hence $|WX| \geq |YX| \geq \frac{n}{2}$, again a contradiction with (1).

Case 3: X is between W and Y. Without loss of generality, let us assume W is to the left of X and Y is to the right of X. For every $i \in WY$, it must be the case that $d(i, X) < d(i, Y)$ (due to analogous reasoning as in the previous case). So $i \in XY$ and hence $|XY| \geq |WY| \geq n/2$. Since Y also weakly defeats X, then X and Y must be tied, i.e. $|XY| = |YX|$. Note further that W and Y must also be tied, since if not, W strictly defeats Y, and due to (1), this means X defeats Y, which contradicts the fact that X and Y are tied. So we know $|WY| = |YW|$.

We will proceed to show that in this case W cannot be the Copeland winner since the Copeland score of X must be strictly greater than the Copeland score of W. It suffices to show that for every candidate Z that W defeats (respectively, weakly defeats), Z is also defeated (resp., weakly defeated) by X. Then since (by 1) X also defeats W, X would certainly have a higher Copeland score than W. Let Z be a candidate that W defeats. Note that Z cannot be in the same location as X since X defeats W. We now consider three subcases.

Subcase 3a: Z is to the left of W. For every $i \in XW$, $d(i, X) < d(i, W) < d(i, Z)$, so $i \in XZ$. Using (1), we have $|XZ| \geq |XW| > \frac{n}{2}$. Hence in this case Z is defeated by X. **Subcase 3b: Z lies between W and X.** For each $i \in XW$ in this case, it can be verified that regardless of the voter i is to the left or right of candidate Z, we will have $d(i, Z) \leq d(i, W)$, so $|ZW| \geq |XW| > n/2$, again due to (1). This contradicts the fact that Z was defeated by W, so this case cannot happen. **Subcase 3c: Z lies to the right of X.** It can be verified in this case that for each voter $i \in WZ$, regardless of whether i is left or right of X, $d(i, X) < d(i, Z)$, so $|XZ| \geq |WZ| > n/2$. Hence Z is defeated by X.

To see that every candidate weakly defeated by W is also weakly defeated by X, assume Z is a candidate that is weakly defeated by W and follows the same arguments as in the three subcases above. They all apply as written for weak defeat, with the exception of Subcase 3c, where the final inequality must change slightly to read $|WZ| \geq n/2$

for the case of weak defeat. Further, in the case of weak defeat, Z and W may be in the same location, but due to (1), this case still implies X defeats Z. Since X defeats every candidate that W defeats, X weakly defeats every candidate that W weakly defeats, and X also defeats W, the Copeland score of X is at least 1 greater than the Copeland score of W, contradicting the definition of W.   □

We note that it was previously established that elections on a line always have a Condorcet winner [9], Condorcet winners have distortion at most 3 on a line [2], and that Copeland always elects the Condorcet winner when one exists (i.e., that it is *Condorcet-consistent*). However, this only applies to strong Condorcet winners. A *strong Condorcet winner* (SCW) is a candidate who wins a strict majority of the voters against any other candidate (i.e. more than $n/2$ voters prefer the SCW when the SCW is pitted head-to-head against any other candidate). A *weak Condorcet winner* (WCW) is a candidate that wins or ties against any other candidate head-to-head. Copeland does not always elect a weak Condorcet winner, even when one exists [16, 17]. Our result above effectively shows that when restricted to the line metric, Copeland is not only SCW-consistent, but it is also WCW-consistent.

4   Empirical Results

We generate election instances on one-, two-, and three-dimensional Euclidean spaces, consisting of random points for both candidates and voters drawn from four distributions. The uniform distribution is drawn as floating point values in the range [0,100], the normal distribution is generated with a mean of 50 and standard deviation of 18, the Poisson distribution is generated using $\lambda = 30$, and the bimodal distribution is generated using two normal distributions with means of 30 and 70 and standard deviations of 10. The normal distribution models a population with one dominant idea (e.g., where being moderate is the most popular position, or where a single dominant party exists). The bimodal distribution models a population with polarized or dual cultures. The uniform distribution models voters and candidates that are spread evenly across the ideological spectrum, while the Poisson distribution models situations where the population of voters are more skewed to one side.

For each voting rule[6], and for each number of voters taking on values between 5 and 200, and for each number of candidates taking on values between 5 and 100, and for each number of dimensions in $\{1, 2, 3\}$, and for each probability distribution mentioned above, results from 1000 randomly generated elections are aggregated. Voters rank the candidates by preferring those that are closer in Euclidean distance to them over those that are further away. In the event of two candidates who are equidistant to a voter, one candidate is arbitrarily chosen over another on the voter's ballot to preserve a complete ranking over all candidates. For every voting rule, there is also the possibility of tied outcomes, i.e., more than one possible winner. We break such ties uniformly at random, as is often done in real-life applications [32].

The performance metrics we evaluate in our simulations are: (1) *Distortion*—the ratio between the social cost of the winner W selected by the voting rule and the social

---

[6] Note that the STV implementation used is from an open source GitHub repository [7].

cost of the optimal candidate OPT. Note we used this term in Sect. 2 to refer to worst-case distortion over the space of all possible inputs, but to simplify exposition, we use "distortion" in this section to refer to $SC(W)/SC(OPT)$ on a per-instance basis. (2) *Accuracy*—the rate at which a voting rule correctly selects the optimal candidate OPT as the winner. And (3) *Condorcet/IIA satisfaction*—the rates at which a voting rule satisfies the Condorcet or IIA criteria (these are explained below in Sect. 4.3).

Our experiments show that as the number of voters increases, distortion and accuracy rates both tend to improve across all voting rules. However, beyond 100 voters, the observed changes in distortion and accuracy rates become minimal. For example, see Fig. 5 of Appendix A.1. Thus, to conserve space, we focus on results for 200 voters.

4.1 Accuracy

We use the term *accuracy* to refer to how often a voting rule successfully elects the optimal candidate as the winner in our simulated trials. The optimal candidate, denoted in our figures as OPT, is as defined in Sect. 2 above: the candidate whose total distance to all voters (or social cost) is minimized.

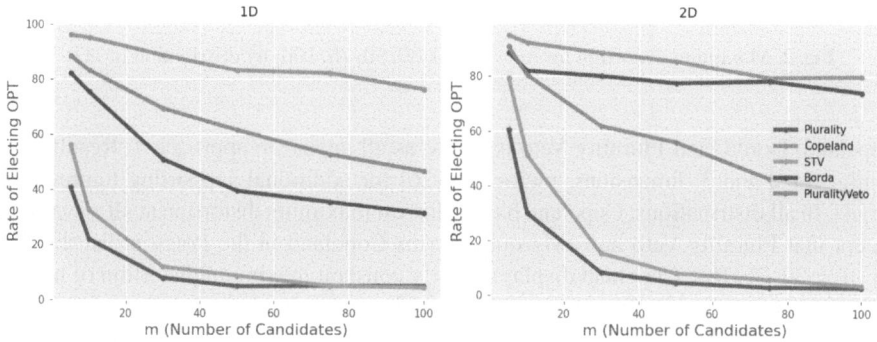

Fig. 1. Accuracy of each voting rule on trials of random instances generated using the normal distribution for $m = [5, 10, 30, 50, 75, 100]$.

Copeland has the highest accuracy across all values of m, as seen in Fig. 1 for the normal distribution. The higher accuracy of Copeland is also seen in other distributions (see Appendix A.2). Copeland has a range of 92–96.4 percent accuracy at $m = 5$ across all distributions. Plurality Veto does reach an accuracy of 89.2 percent at $m = 100$ in the Poisson distribution, but this is still not comparable to Copeland's 99.8 percent accuracy under the same conditions. Experiments in 2D and 3D metric spaces reveal similar results (please see the extended version of our work [26] for more detail).

4.2 Distortion

Figure 2 shows that Copeland has a lower maximum distortion than other voting rules across all distributions. As m increases, the difference in maximum distortion between

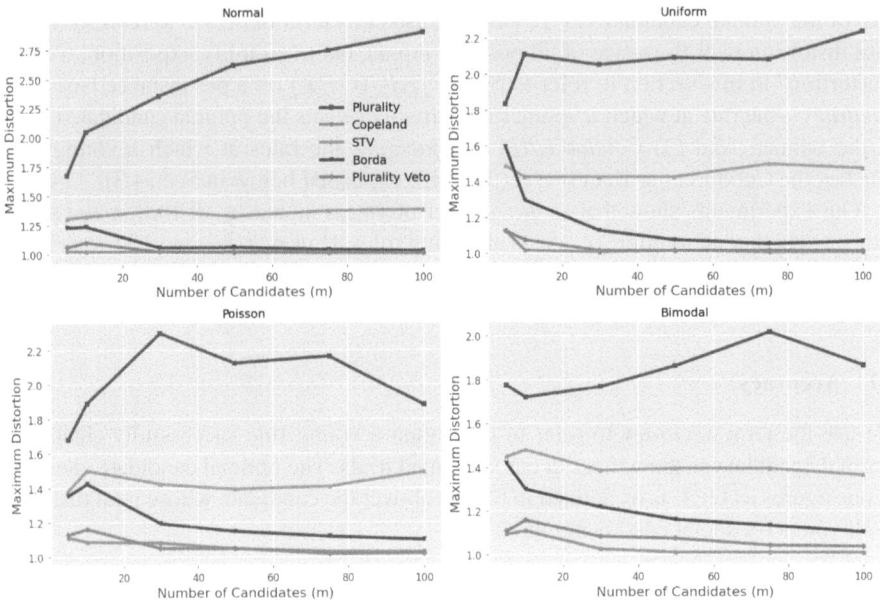

Fig. 2. Maximum distortion for $m = [5, 10, 30, 50, 75, 100]$ by distribution in 1D.

Copeland, Borda, and Plurality Veto narrows, as all appear to approach 1. Results are similar for 2 and 3 dimensions; please see [26] for additional supporting figures and details. In all distributions, Copeland has the lowest maximum distortion at all m values, except that Plurality Veto narrowly outperforms Copeland in the Poisson distribution for $m = 30, 75, 100$. Copeland displays a fairly constant maximum distortion of near 1 with respect to m, while Plurality and STV tend to grow in maximum distortion as m increases, and Borda and Plurality Veto slightly decrease in distortion as m increases. Similar behavior is seen when evaluating average distortion.

As Fig. 7 in Appendix A.1 shows, Copeland also has the lowest average distortion across all m values and distributions in 1D. Results are similar for 2 and 3 dimensions; please see [26] for additional details. Copeland achieves its worst average distortion of 1.0027 at $m = 5$ in the Poisson distribution, while Plurality-Veto achieves a similar 1.0034 average distortion under the same conditions. At $m \geq 50$ in all distributions, Copeland never exceeds an average distortion of 1.0002. Please see Tables provided in an extended version of our work [26] for all referenced distortion values. Copeland and Plurality Veto together widely outperform the other voting rules, and while the difference between Copeland and Plurality Veto's average distortion is near negligible, Copeland still outperforms Plurality Veto in every environment tested.

4.3 Additional Criteria

An alternative avenue for evaluating a voting method is to measure its adherence to certain desirable criteria. Arrow's Impossibility Theorem shows that in the presence of three or more distinct candidates, no preference aggregation rule can satisfy all of the following important and sensible criteria: unrestricted domain, non-dictatorship, Pareto efficiency, and Independence of Irrelevant Alternatives (IIA) [5]. Among the criteria in Arrow's theorem, we are most interested in IIA. The voting rules we consider here always satisfy unrestricted domain and non-dictatorship by definition. Copeland and Borda also always satisfy Pareto efficiency. IIA is the only criterion that all rules violate. High IIA satisfaction is correlated to better resistance to vote-splitting (or the spoiler effect) [28].

Since Arrow's theorem tells us finding an ideal voting system is not possible, and all voting rules do not violate these criteria at the same frequency, we empirically investigate how often each voting rule violates IIA. The IIA criterion effectively states that the outcome of an election should remain unchanged if a losing candidate is removed. While the IIA criterion presents a challenge for most preferred voting rules, its concept remains compelling because it rules out the vulnerability of vote splitting. Vote splitting occurs when candidate X would defeat Y in a head-to-head comparison but loses to Y when Z also runs (as Z diverts some of the votes that would otherwise go to X).

Our evaluation of the IIA criterion begins by running each voting rule on the generated dataset of voters and candidates to find the winner W. Then, successive elections are conducted by removing each of the non-winning candidates in each iteration, repeating this process $m - 1$ times, with m being the number of candidates. The outcome of each iteration is compared with the original winner W. If the winner remains unchanged throughout, the trial is counted as a success for satisfying IIA.

Prior work [39] has suggested that to satisfy IIA for as many voting profiles as possible, a Condorcet method is best, and Condorcet methods are more robust to strategic voting than other methods. A *Condorcet method* is one that always chooses a Condorcet winner if one exists. A *Condorcet winner* is a candidate that pairwise defeats all other candidates (i.e., in head-to-head match-ups). Hence we also test our voting rules in each randomized trial for the Condorcet criterion. Studying the probability that a voting rule violates various criteria has been done using computer simulations by [18] and [12]. The work of [18] finds the Condorcet efficiency of different voting rules under various assumptions, while [12] explores the likelihood of violating the criteria of Arrow's theorem, especially IIA. However, they do not use a spatial-model as in our work. The first simulations for Condorcet efficiency under spatial-model assumptions were performed by Chamberlin and Cohen [8], who discuss in depth the motivation for such simulations. However, their results are limited to a four-dimensional model with four candidates and they do not consider Copeland, STV, and Plurality Veto. To our knowledge, there is little prior work on the probability of satisfying or violating IIA under a spatial model.

Condorcet Criterion Results. For the range of cases we have studied, the five social choice rules can be ordered as follows in their overall ability to select Condorcet winners: (1) Copeland, (2) Plurality Veto, (3) Borda, (4) STV, and (5) Plurality. (Note, however, that in two- and three-dimensional space, Borda satisfies Condorcet criterion more frequently than Plurality Veto.) Since Copeland's rule considers pairwise comparisons and selects the candidate who wins the most of them, it inherently satisfies the

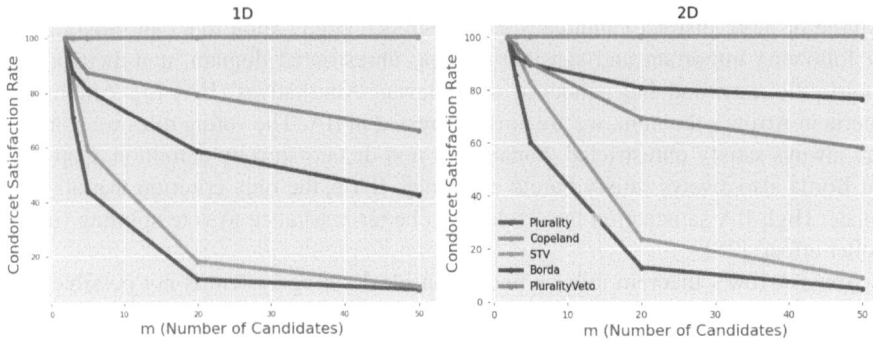

Fig. 3. Condorcet consistency rate with $n = 200$ and $m = [2, 3, 5, 20, 50]$, normal distribution.

Condorcet criterion, achieving the expected 100% satisfaction under all scenarios considered. For all other rules, the satisfaction rate decreases as the number of candidates increases (see Fig. 3).

In a symmetrical distribution, the Condorcet winner typically emerges as the candidate positioned closest to the distribution's center [29]. This candidate usually receives fewer last-place votes due to the relationship between voter preferences and candidate proximity. Consequently, they are more likely to persist through the elimination process under the Plurality Veto method. However, despite their advantageous central position, such a centrist candidate (the Condorcet winner) may receive fewer first-place votes and face elimination in Plurality and run-off systems (like STV) due to being squeezed by surrounding opponents. This expectation is supported by the results of our simulation studies. STV and Plurality are significantly worse at selecting the Condorcet winner, with both satisfaction rates being less than 10% when $m = 50$. Merrill [29] additionally proposes that the Condorcet efficiency can increase with the expansion of the generalized dimension due to the "squeeze effect." Augmenting the spatial dimensions seems to offer greater space for voters and candidates, thereby reducing the likelihood of the Condorcet candidate (typically situated centrally) being squeezed by their opponents. This concept is notably evident for Borda and Plurality Veto, as their Condorcet satisfaction rates increase when the number of spatial dimensions increases to two and three dimensions (see Fig. 3 and Appendix A.3 Fig. 9). It is also noted that as the number of voters increases in Fig. 10 of Appendix A.3, the change in Condorcet satisfaction appears to diminish, which helps to justify our focus here on $n = 200$.

Independence of Irrelevant Alternatives (IIA) Results. None of the voting rules in our study always satisfy the IIA criterion. In trials with 5 candidates, Copeland is the only voting rule that has an IIA satisfaction rate of 100%. It is interesting to note that the probability that Copeland satisfies IIA does not decrease monotonically as the number of voters increases - the satisfaction rate is greater for odd n than even n (please refer to the extended version of our work [26] for supporting figures). This is because ties are possible for an even, but not odd, number of voters. A social ranking created in an even-sized population need not be a strict one, thus increasing the opportunity for an

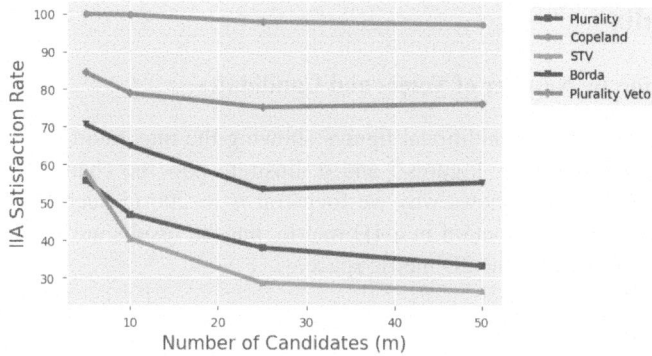

Fig. 4. IIA satisfaction rate, $n = 100$ and $m = [5, 10, 25, 50]$, uniform distribution, 1D metric.

IIA violation. This parity-based observation is consistent with prior findings by [12] on the frequency of IIA violation. In contrast, the IIA satisfactions for Plurality, Borda, and Plurality Veto appear to decrease, then remain relatively unchanged as the number of voters increases.

The difference in IIA performance among voting rules can be quite large as the number of candidates increases. See Fig. 4. For $n = 100$ and $m \geq 25$, with a uniform distribution of points in a one-dimensional metric, Copeland has the highest IIA satisfaction rate among the voting rules at roughly 97%. This result is similar in other distributions, ranging around 95–97%. Plurality Veto and Borda satisfy IIA less frequently, under 75% and 60% of the time, respectively, followed by Plurality with an IIA satisfaction rate of under 40%. STV is the worst performer, violating IIA more than 70% of the time. Results in other distributions follow a similar trend (see, e.g., Fig. 11 of Appendix A.4).

5 Conclusion

When restricted to one dimension, we show that Copeland voting has a worst-case distortion of 3, matching the best possible distortion guarantee of any voting rule on the line. While Copeland's distortion in general metrics is 5, which is higher than the better distortion guarantee of 3 for Plurality Veto, our empirical study shows that interestingly, Copeland has lower distortion and higher accuracy than Plurality Veto in simulated trials. In addition, Copeland has a higher rate of success in satisfying the Condorcet and Independence of Irrelevant Alternatives (IIA) criterion. Despite its widespread use, our experiments confirm traditional Plurality is one of the least effective rules by all measures studied. The recent increasing adoption of the Single Transferable Vote (STV/RCV) method in political elections is an improvement over Plurality, but STV/RCV performs worse than Copeland in all measures tested, and usually worse than other voting rules like Plurality Veto and Borda. Since Copeland is also a simple voting rule that is easily understood by the lay population, we are compelled to conclude from our results that Copeland should be considered above both STV/RCV and Plurality Veto for practical use in real-world social choice settings.

A   Appendix

A.1   Distortion by Number of Voters and Candidates

In this section, we present additional figures showing the maximum and average distortion for each voting rule. Figures 5 and 6 illustrate how the changes in distortion diminish as n grows, and shows why we focus on $n = 200$ in Sect. 4. These results were for the uniform distribution in a 1D metric, but the results are similar for other distributions as well as for the 2D metric space.

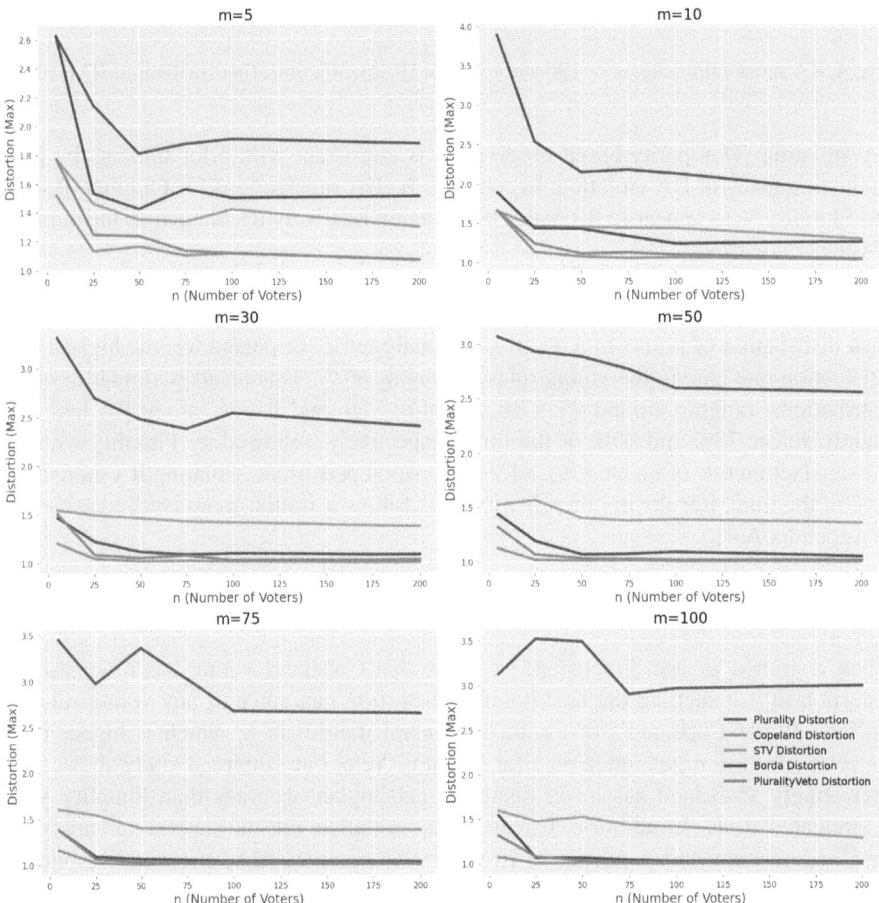

Fig. 5. Uniform distribution, 1D metric.

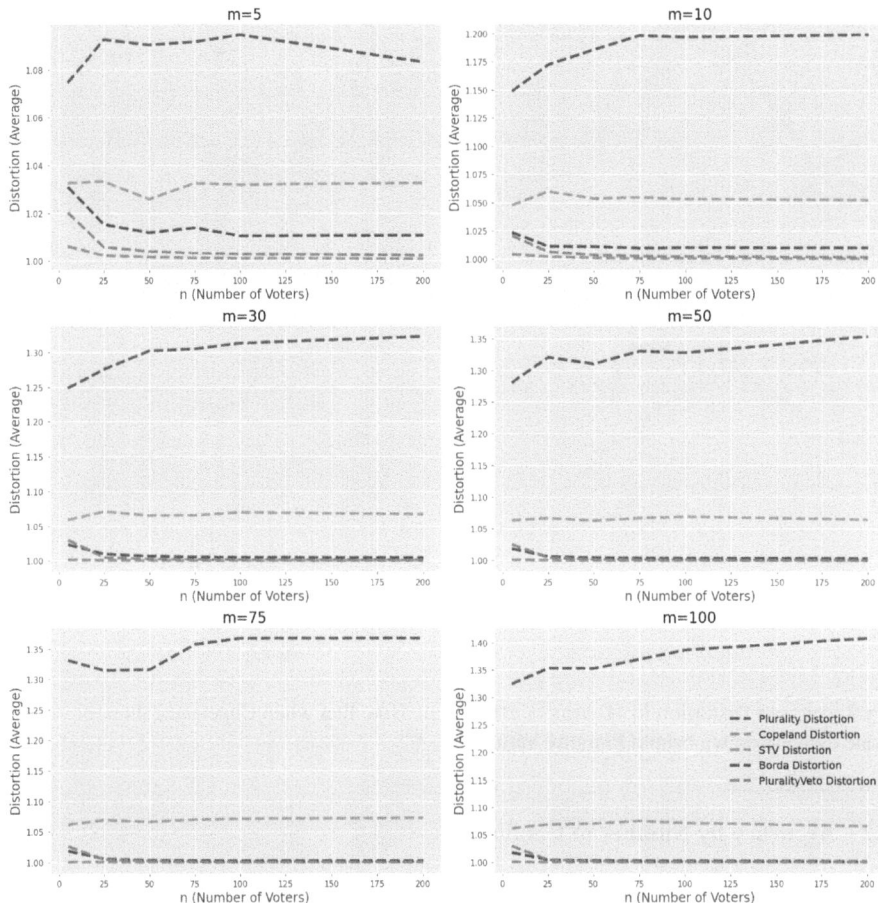

Fig. 6. Average Distortion, Uniform Distribution, 1D metric.

Average Distortion (1D). In this section we provide Fig. 7, referenced by statements in Sect. 4.2, showing the average distortion for each voting rule, grouped by the number of candidates ($n = 200$ voters).

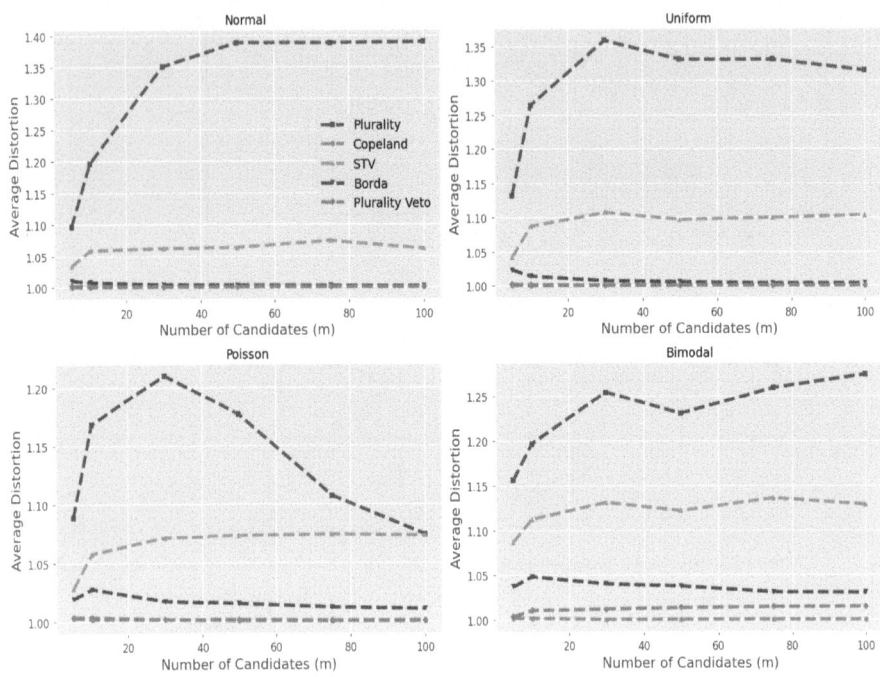

Fig. 7. Average Distortion in 1D at $n = 200$ voters. Note that when Copeland cannot be seen it is due to being drawn behind Plurality Veto.

A.2 Accuracy by Number of Candidates

To support Sect. 4.1, we present a series of plots in Fig. 8 showing the accuracy for each voting rule at $n = 200$ voters, as the number of candidates grows, separated by distribution.

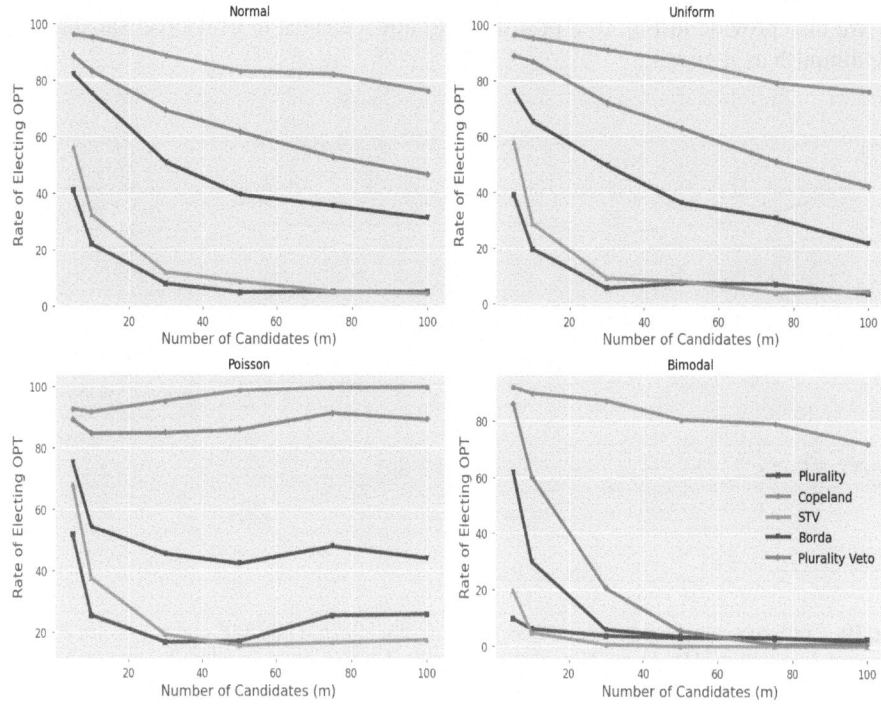

Fig. 8. Accuracy at $n = 200$ in 1D.

A.3   Condorcet Satisfaction Rate

To support results from Sect. 4.3, we first present, in Fig. 9, the Condorcet satisfaction rate for each voting rule as the number of candidates changes using normal distribution in 3-dimensional space.

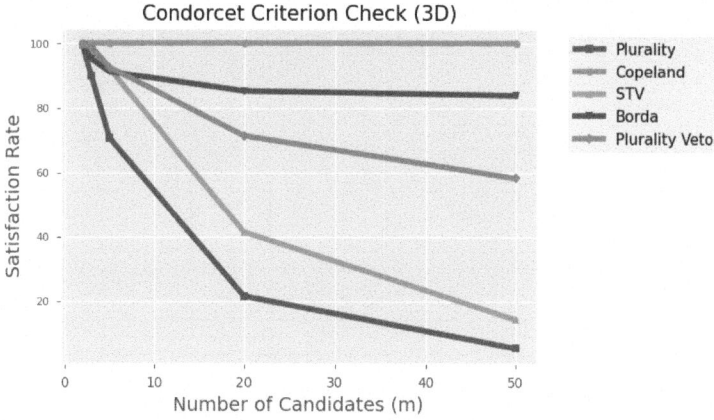

Fig. 9. Normal distribution, 3D metric, n=200

We also provide in Fig. 10 a plot showing how changes in Condorcet satisfaction rate diminish as n grows.

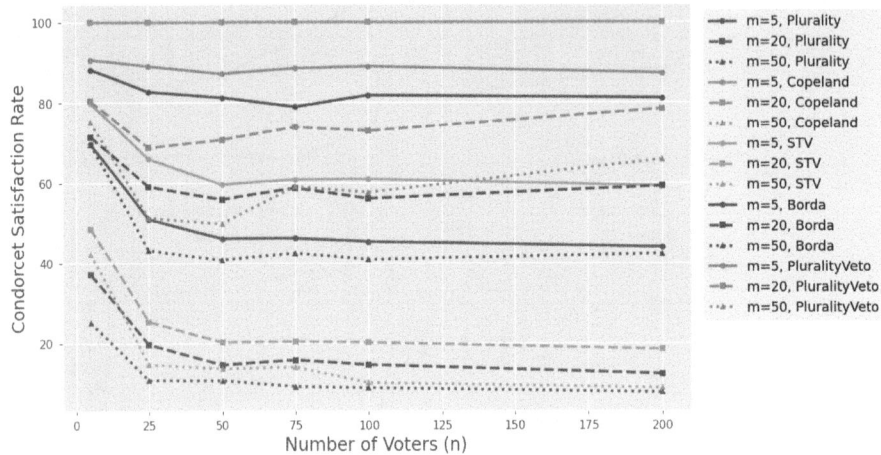

Fig. 10. Cordorcet satisfaction rate across $n = [5, 25, 50, 75, 100, 200]$ and $m = [5, 20, 50]$ (Normal distribution, 1D space). Note that all Copeland curves are coinciding for all values of n and m at 100%, as expected.

A.4   IIA Satisfaction Rate

In this section, we present Fig. 11, referenced from the latter part of Sect. 4.3, showing the IIA criterion satisfaction rate. For additional IIA results, please see [26].

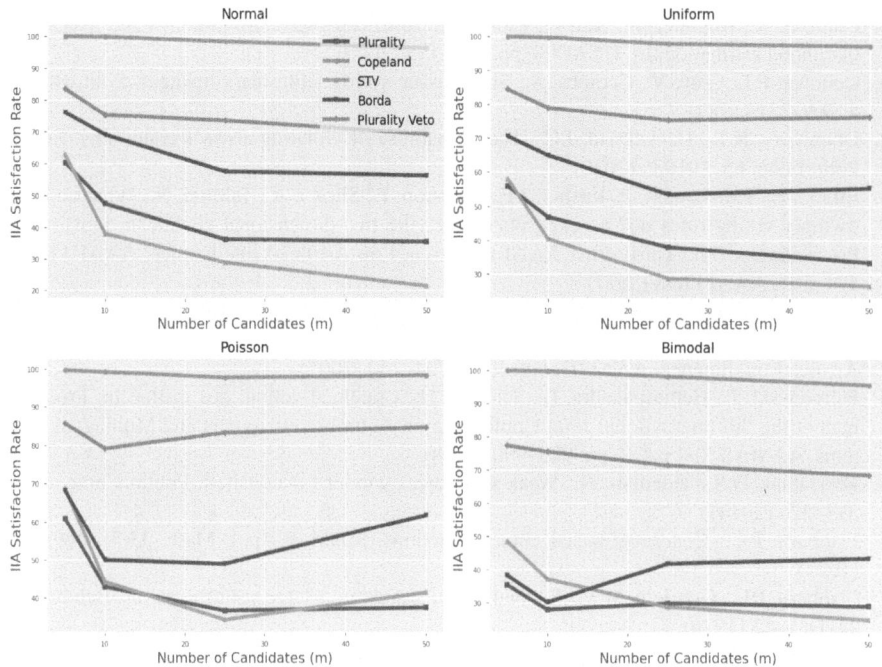

Fig. 11. IIA criterion satisfaction rate when $n = 100$ and $m = [5, 10, 25, 50]$, 1D metric

References

1. Anagnostides, I., Fotakis, D., Patsilinakos, P.: Dimensionality and coordination in voting: the distortion of stv. In: Proceedings of the AAAI Conference on Artificial Intelligence, vol. 36, pp. 4776–4784 (2022). https://doi.org/10.1609/aaai.v36i5.20404
2. Anshelevich, E., Bhardwaj, O., Elkind, E., Postl, J., Skowron, P.: Approximating optimal social choice under metric preferences. Artif. Intell. **264**, 27–51 (2018)
3. Anshelevich, E., Bhardwaj, O., Postl, J.: Approximating optimal social choice under metric preferences. In: Proceedings of the Twenty-Ninth AAAI Conference on Artificial Intelligence, AAAI 2015, pp. 777–783. AAAI Press (2015)
4. Anshelevich, E., Filos-Ratsikas, A., Shah, N., Voudouris, A.A.: Distortion in social choice problems: the first 15 years and beyond. In: 30th International Joint Conference on Artificial Intelligence, pp. 4294–4301 (2021)
5. Kenneth Joseph Arrow: Social Choice and Individual Values. Wiley, New York (1951)
6. Bowman, B.: Poll: Democracy, abortion are top priorities for single-issue voters (2023). https://www.nbcnews.com/meet-the-press/meetthepressblog/
7. Cannon, W.: ranked-choice-voting (2013). https://github.com/wcannon/ranked-choice-voting
8. Chamberlin, J.R., Cohen, M.D.: Toward applicable social choice theory: a comparison of social choice functions under spatial model assumptions. Am. Polit. Sci. Rev. **72**(4), 1341–1356 (1978)
9. Conitzer, V.: Eliciting single-peaked preferences using comparison queries. In: Proceedings of the 6th International Joint Conference on Autonomous Agents and Multiagent Systems, pp. 1–8 (2007)

10. Conitzer, V., Rognlie, M., Xia, L.: Preference functions that score rankings and maximum likelihood estimation. In: IJCAI'09, pp. 109–115 (2009)
11. Conover, P.J., Gray, V., Coombs, S.: Single-issue voting: elite-mass linkages. Polit. Behav. **4**, 309–331 (1982)
12. Dougherty, K.L., Heckelman, J.C.: The probability of violating arrow's conditions. Eur. J. Polit. Econ. **65**, 101936 (2020)
13. Elkind, E., Faliszewski, P., Laslier, J.F., Skowron, P., Slinko, A., Talmon, N.: What do multiwinner voting rules do? an experiment over the two-dimensional euclidean domain. In: Proceedings of the Thirty-First AAAI Conference on Artificial Intelligence, AAAI'17, pp. 494–501. AAAI Press (2017)
14. Faliszewski, P., Hemaspaandra, E., Hemaspaandra, L.A., Rothe, J.: Llull and copeland voting broadly resist bribery and control. In: Proceedings of the 22nd National Conference on Artificial Intelligence, AAAI'07, vol. 1, pp. 724–730 (2007)
15. Faliszewski, P., Hemaspaandra, E., Schnoor, H.: Copeland voting: ties matter. In: Proceedings of the 7th International Joint Conference on Autonomous Agents and Multiagent Systems, AAMAS '08, vol. 2, pp. 983–990 (2008)
16. Felsenthal, D.S., Tideman, N.: Weak Condorcet winner (s) revisited. Public Choice **160**, 313–326 (2014)
17. Fishburn, P.C.: Condorcet social choice functions. SIAM J. Appl. Math. **33**(3), 469–489 (1977)
18. Fishburn, P.C., Gehrlein, W.V.: An analysis of simple two-stage voting systems. Behav. Sci. **21**(1), 1–12 (1976)
19. Gehrlein, W.V.: Condorcet's paradox and the Condorcet efficienty of voting rules. Mathematica Japonica **45**, 173–199 (1997)
20. Gkatzelis, V., Halpern, D., Shah, N.: Resolving the optimal metric distortion conjecture. In: 2020 IEEE 61st Annual Symposium on Foundations of Computer Science (FOCS), pp. 1427–1438 (2020)
21. Goel, A., Krishnaswamy, A., Munagala, K.: Metric distortion of social choice rules: Lower bounds and fairness properties. In: Proceedings of the 2017 ACM Conference on Economics and Computation (2016)
22. Graves, L.: Ranked-choice voting coming to more statewide ballots in 2024 (2024). https://www.pbs.org/wnet/preserving-democracy/2023/12/18
23. Hutter, S.: Single-issue voters emerge ahead of the 2024 election (2023). https://www.laloyolan.com/e2024/single-issue-voters-emerge-ahead-of-the-2024-election/article_637d04d8-3595-11ee-9194-3b421dce4bc4.html
24. Kambhampaty, A.P.: New york city voters just adopted ranked-choice voting in elections. Here's how it works (2019). https://time.com/5718941/ranked-choice-voting/
25. Kizilkaya, F.E., Kempe, D.: Plurality veto: a simple voting rule achieving optimal metric distortion. In: Raedt, L.C. (ed.) Proceedings of the Thirty-First International Joint Conference on Artificial Intelligence, IJCAI 2022, pp. 349–355 (2022)
26. Le, M., Nguyen, C., Claney, L., Tipnis, K., Chung, C.: A case for copeland: from theory to practice (extended version) (2024). http://cs.conncoll.edu/cchung/research/EC24.pdf
27. Longley, R.: What are single issue voters? (2022). https://www.thoughtco.com/single-issue-voters-5214543
28. Maskin, E.: A modified version of arrow's IIA condition. Social Choice Welfare (2020)
29. Merrill, S.: A statistical model for condorcet efficiency based on simulation under spatial model assumptions. Public Choice **47**(2), 389–403 (1984)
30. Moulin, H.: Choosing from a tournament. Social Choice Welfare **3**(4), 271–291 (1986). http://www.jstor.org/stable/41105842
31. Munagala, K., Wang, K.: Improved metric distortion for deterministic social choice rules. In: Proceedings of the 2019 ACM Conference on Economics and Computation (2019)

32. National Conference of State Legislatures. Resolving tied elections for legislative offices (2023). https://www.ncsl.org/elections-and-campaigns/
33. Olsen. T.: Duluth voters reject ranked-choice voting (2018). https://www.duluthnewstribune.com/news/
34. Parks, M.: Ranked choice is "the hot reform" in democracy. Here's what you should know about it (2023). https://www.npr.org/2023/12/13/1214199019
35. Procaccia, A.D., Rosenschein, J.S.: The distortion of cardinal preferences in voting. In: Proceedings of the 10th International Conference on Cooperative Information Agents, CIA'06, pp. 317–331 (2006)
36. Saari, D.G., Merlin, V.R.: The copeland method: I.: Relationships and the dictionary. Econ. Theory **8**(1), 51–76 (1996)
37. Scimia, E.: Scoring systems in chess tournaments (2019). https://www.thesprucecrafts.com/tournament-scoring-systems-611116
38. Skowron, P., Elkind, E.: Social choice under metric preferences: scoring rules and STV. In: Proceedings of the Thirty-First AAAI Conference on Artificial Intelligence, AAAI 2017, pp. 706–712. AAAI Press (2017)
39. Wang, T., Sturm, J., Cuff, P., Kulkarni, S.: Condorcet voting methods avoid the paradoxes of voting theory. In: 2012 50th Annual Allerton Conference on Communication, Control, and Computing, Allerton 2012, pp. 201–203 (2012)

Deterministic and Universal Truthful Mechanism for Fair Matching

Hao Xu[1], Jinshan Zhang[2(✉)], and Feng Wang[1(✉)]

[1] School of Computer Science, Wuhan University, Wuhan, China
{xuhao2002,fengwang}@whu.edu.cn
[2] School of Software Technology, Zhejiang University, Ningbo, China
zhangjinshan@zju.edu.cn

Abstract. This study addresses the bipartite graph weight maximum matching problem with a focus on achieving fairness. We introduce a deterministic truthful mechanism and a universal truthful mechanism tailored for fair matching, striving for an asymptotically lexicographically optimal solution. To tackle this, we employ a transformation function that reduces the original problem to a weight maximum matching problem, alongside an iterative algorithm based on TMHA and RTMHA (TMHA means Truthful Matroid House Allocation and RTMHA means Random Truthful Mechanism for Matroid House Allocation [18], they are both extensions of serial dictatorship and random serial dictatorship to cases where agents have ties.). Our approach yields an approximation ratio of at most 2 for deterministic truthful mechanisms and at most 1.58 for universally truthful mechanisms, while also requiring polynomial time complexity. By offering a balanced solution, our method contributes to equitable outcomes in matching scenarios.

Keywords: Bipartite matching · Mechanism design · Fair matching · Truthful

1 Introduction

The bipartite graph matching problem finds wide application in scenarios such as dormitory assignment, marriage matching, and various other contexts. The matching problem has two aspects: maximum weight matching and maximum cardinality matching [10]. For example, for maximum weight matching, consider a series of dormitories and a group of students, each student has a ranking for dormitories, and it is important for the manager to strive to maximize the rating value of the dormitories allocated to students. Another example of maximum cardinality matching is that each student does not have a ranking for dormitories. Thus, the school manager should try to match students as much as possible.

For the maximum weight matching problem, algorithms such as [7,9,15,23] have demonstrated effectiveness in solving this problem within acceptable time complexities. Regarding the maximum cardinality matching problem, the Hungarian method [19] can solve it in polynomial time. Additionally, algorithms like [5,16,21] exist to address the problem across various scenarios.

In a maximum weight matching problem, an agent may have a preference order for the matched items. In a preference order based maximum matching problem, there are three different conditions: rank maximal matching, maximum cardinality rank-maximal and fair matching [11]. In the rank maximal matching problem, we aim that matching should maximize the number of individuals matched to better outcomes, Irving et al. first propose it and solve it in $O(\min(n + r, r\sqrt{n})m)$ time where n means the number of agents and m means the number of items and r means the maximal rank used in an optimal solution [13], Kavitha et al. solve it in $O(\min(mn, rm\sqrt{n}))$ time [14]. In the fair matching problem, we aim that matching should minimize the number of individuals matched to worse outcomes as much as possible, i.e., let the matching be fairer and more equitable. Mehlhorn et al. were the first to propose and study the fair matching problem [20], and presented an $O(\gamma\sqrt{n}m \log n)$ time algorithm to solve the fair matching problem, where γ is the largest size of all indifference classes. Abraham et al. extended the concept of fair matching to rental markets through value functions and provided several constant-factor approximation algorithms for the assignment problem [2]. Huang et al. solve the problem through an iterative method in $O(rm\sqrt{n} \log n)$ time where r means the maximum rank number [12]. In our work, we aim to achieve fair matching conditions and propose a fair matching algorithm under the constraint of truthfulness.

A truthful mechanism mandates that an agent cannot achieve a superior outcome by misrepresenting its preference order, a detailed concept we will delve into later. In [8], truthful mechanisms are explored across various problem domains, with proofs demonstrating that no straightforward truthful mechanism for normal weight maximum matching can be devised under the no-money condition. In the preference order based maximum matching problem, Chakrabarty et al. introduce a range of randomized truthful definitions such as universal truthful and lexicographical truthful [6,17]. For rank maximal matching, Zhang et al. [24] put forward a truthful matching mechanism under the lexicographical truthful constraint. In our study, we endeavor to propose a truthful mechanism under fair matching conditions.

Fair matching represents a subset of profile-based matching problems, as elucidated in [12], wherein the goal is to ensure that no agent receives a worse outcome. However, research on this problem remains relatively scarce, particularly regarding truthful mechanisms for fair matching. Recent studies have primarily focused on truthful mechanisms for rank maximal matching, predominantly constrained to randomized settings. To our knowledge, there has been minimal exploration of truthful mechanisms for fair matching. Hence, we aim for our work to bridge this gap and contribute to the advancement of truthful mechanisms in fair matching scenarios.

Our Result. We have successfully constructed deterministic truthful and universal truthful algorithms for the fair matching problem and proved that it can solve the problem with polynomial complexity of $O(rn^3\gamma)$, where r means the maximum rank number and n means the number of agents and γ means the largest size of all indifference classes. Additionally, we demonstrate that the approximation ratio of our deterministic algorithm is not better than 2 and the approximation ratio of our universal truthful algorithm is not better than $\frac{e}{e-1}$, which aligns with TMHA and RTMHA.

Technical Overview. Firstly, we will devise a transformation function to reduce the fair matching problem to a weight maximum matching problem. Subsequently, we will demonstrate that the solution to the reduced weight maximum matching problem is equivalent to the solution to the preference order based maximum matching problem (Proposition 1). Next, we will develop an algorithm (Algorithm 1 and Algorithm 2) based on TMHA and RTMHA to solve the preference order-based maximum matching problem with truthful constraint. To enhance the approximation ratio, we will employ an iterative method to approach the asymptotically optimal solution. Finally, we will furnish the upper bound for the algorithm approximation ratio.

2   Preliminaries

2.1   Basic Model

We have a set of agents $N = \{1, 2, \cdots, n\}$ and a set of items $A = \{a_1, a_2, \cdots, a_m\}$. Each agent $i \in N$ possesses a complete preference order P_i over items. Let G be a bipartite graph $G = (N \cup A, E)$ where $E = \{(i, a) \mid i \in N \wedge a \in P_i\}$. We define the matching μ as a subset of E where each pair of vertices of every edge in μ has only one connecting edge.

For each agent's preference order, an indifference class exists wherein an agent may prefer some items equally within the same rank. We denote C_j^i as agent i's jth indifference class, $rank(i, e)$ denote the rank of e to i i.e. $rank(i, e) = j$ means $e \in C_j^i$, and r as the maximum rank among all agents' preference orders and γ be the size of the largest indifference class. Note that the preference weights of items within the same indifference class are equal. We define $a \succ_i b$ to indicate that agent i prefers item a to b, $a \succeq_i b$ to indicate that agent i has the same preference for a and b or i prefers a to b, and $C_j^i \succ_i C_k^i$ to signify that agent i prefers items in C_j^i more than items in C_k^i. We define $a \simeq_i b$ to signify that both a and b are in the same indifference class. So that

$$P_i = \{C_1^i, C_2^i, \cdots, C_r^i\}$$
$$\forall j \in [r], C_j^i \subseteq A$$
$$\forall j, k \in [r] \wedge j < k, C_j^i \prec_i C_k^i$$
$$\forall a, b \in C_j^i, a \simeq_i b$$

Note that indifference class can be $\emptyset$ and $\forall j, k \in [r]$, $C_j^i \cap C_k^i = \emptyset$. We define E_j and G_j so that the weight of each edge in E_j and G_j will not exceed j i.e.

$E_j = \{(i, a) \mid rank(i, a) \leq j\}$, $G_j = (N \cup A, E_j)$ and denote μ_j as the matching on G_j. Let $P = \{P_1, P_2, \cdots, P_n\}$ as the total preference order, P_{-i} as the total preference order except P_i. We denote $I = (N, A, P)$ as the instance for our problem. The preference order based maximum matching problem is that each agent tries to obtain a more favorite item based on the agent's preference order, and the Pareto optimal is that no agent can obtain a better result without decreasing others' profit. Formally, there does not exist a matching μ' that satisfies

$$\forall i \in N, \ \mu'(i) \succeq_i \mu(i)$$
$$\exists j \in N, \ \mu'(j) \succ_j \mu(j)$$

2.2 Fair Matching

A fair matching [12] is that matching should maximize the number of edges; subject to this, matching should minimize the number of matched items ranked r, and subject to that, matching should minimize the number of matched items ranked $r - 1$, and $r - 2$, and so forth.

Let's consider a scenario where only agents can rank items. Here, similar to the definition in [12], we define an edge e with a series of weights from w_1 to w_r. For this case, we define $w_j(e)$ where e connects with i as follows:

$$w_j(e) = \begin{cases} 1, & \text{if } rank(i, e) \leq r - j + 1 \\ 0, & \text{if } rank(i, e) > r - j + 1 \end{cases} \tag{1}$$

The signature of a matching is defined as

$$signature(\mu) = \{w_1(\mu), w_2(\mu), ..., w_r(\mu)\}$$

where $w_j(\mu) = \sum_{e \in \mu} w_j(e)$. We define $signature_j(\mu)$ as the j-th element in vector $signature(\mu)$. The comparative relationship of $signature(\mu)$ is defined by lexicographic order, such that we define $\succ_{sig}$ if $signature(\mu_1) \succ_{sig} signature(\mu_2)$, then there exists $0 \leq k \leq r$ such that when $j < k$, $signature_j(\mu_1) = signature_j(\mu_2)$ and $signature_k(\mu_1) > signature_k(\mu_2)$.

Let OPT be the lexicographically maximal matching for G based on the signature. If a matching μ satisfies that for $i \leq j$, always has $w_i(\mu) = w_i(OPT)$, we call matching μ a j-optimal matching. We define a j-fair matching as a matching that maximizes cardinality while minimizing the number of edges ranked r, then $r - 1$, down to $r - j + 1$.

Theorem 1. *A j-optimal matching is a j-fair matching.*

From Theorem 1 we know that OPT is a r-optimal matching and r-fair matching. So, we denote OPT as lexicographically optimal matching for fair matching.

2.3 Truthful

A *deterministic* truthful matching is that any agent cannot obtain a better item through lying about its preference order. Let P_i be the preference order when i lies and $P' = (P'_i, P_{-i})$ be the lied total preference order, $\mu(P)_i$ as the item obtained by i under matching μ and total preference order P. So a deterministic truthful matching must have

$$\mu(P)_i \succ \mu(P')_i$$

A *universal* truthful matching is a distribution over deterministic truthful mechanism. We define $\phi_{ij}(P)$ as the probability that agent i receives an item in indifference class C^i_j under preference order P. So a universal truthful matching must have

$$\mathbb{E}_j[\phi_{ij}(P)] \succ \mathbb{E}_j[\phi_{ij}(P')]$$

In our work, we aim to achieve a deterministic truthful and a universal truthful fair matching that serves as an asymptotically lexicographically optimal solution.

3 Truthful Mechanism for Fair Matching

3.1 Reduce to Maximum Weight Matching

To transform the fair matching problem to a weight maximum matching problem, we now design a transformation weight function $f : \mathbb{N}^+ \to \mathbb{N}^+$ and $v : E \to \mathbb{N}^+$. The weight function f should map the jth rank to a constant larger than the sum of weights for all edges with weights greater than j. This ensures that function v aggregates all the weights, transforming the original problem into a straightforward weight maximum matching problem

To achieve lexicographical optimal solutions, the function f needs to satisfy the following inequalities:

$$
\begin{aligned}
f(j) &\geq nf(j+1) + nf(j+2) + \cdots + nf(r) \\
&\geq n^2 f(j+1)
\end{aligned}
\tag{2}
$$

It's easy to notice that $f(j) = n^{2(r-j)}$ can satisfy the above inequalities. Therefore, we obtain a new edge weight function:

$$
\begin{aligned}
v(e) &= \sum_{j=1}^{r} w_j(e) f(j) \\
&= \sum_{j=1}^{r} w_j(e) n^{2(r-j)}
\end{aligned}
\tag{3}
$$

In this case, consider the assignment conditions of fair matching, for each indifference class C^i_j, there is a weight

$$v(e) = n^{2(r-1)} + n^{2(r-2)} + \cdots + n^{2(j-1)} \quad \forall e \in C^i_j \tag{4}$$

In this case, we define graph G' that each edge e in G' has a weight $v(e)$ for $\forall e \in G$.

Proposition 1. *The following propositions hold:*

(1) A Pareto optimal solution for preference order based maximum matching in G is an asymptotically weight maximum matching in G'

(2) A weight maximum matching on G' is a lexicographical optimal solution for fair matching on G.

Consider there are significant limitations in constructing a truthful mechanism for weight maximum matching without money [8]. Therefore, from Proposition 1, although limitations exist, we still consider that a truthful preference order based maximum matching algorithm can solve the problem.

3.2 Deterministic Truthful Algorithm for Fair Matching

We will introduce mechanism TMHA (Truthful Matroid House Allocation) [18], which is a truthful mechanism for (Matroid) house allocation for uniform matroid with one single element. We now present it in Algorithm 3 in the appendix.

To improve the lexicographical approximation ratio, we employ an iterative method (see Algorithm 1) based on TMHA to identify the maximum matching.

Algorithm 1: Deterministic Truthful Algorithm for Fair Matching

---

Input: Instance (N, A, P)
Output: A truthful matching μ
Initialize $G_1 \leftarrow \emptyset$, $\mu_0 \leftarrow \emptyset$;
Select an order π;
for $j = 1$ *to* r **do**
> Add all agents' C_j to G_{j-1} obtain G_j with weight 0;
> Add $f(j)$ to all edges in G_j;
> Run TMHA on G_j based on order π since agents in μ_{j-1} are matched;
> Augmenting μ_j from μ_{j-1} with the result of TMHA;

end
Return the result μ_r as μ;

Theorem 2. *Algorithm 1 can obtain an asymptotically lexicographically optimal, truthful matching in polynomial time.*

3.3 Universal Truthful Algorithm for Fair Matching

Based on TMHA, we introduce RTMHA proposed in [18], which constructs a universal truthful mechanism for preference order based weight maximum matching. We present in an Algorithm 4 in the appendix. Note that $g(y) = e^{y-1}$.

Next, we use a method similar to constructing the deterministic truthful mechanism based on an iterative method to approach it. We only need to replace TMHA in Algorithm 1 to RTMHA and obtain Algorithm 2.

Algorithm 2: Universal Truthful Algorithm for Fair Matching

Input: Instance (N, A, P)
Output: A truthful matching μ
Initialize $G_1 \leftarrow \emptyset$, $\mu_0 \leftarrow \emptyset$;
for *each agent i in N* **do**
| Pick $Y_i \in [0, 1]$ uniformly at random;
end
Sort agents in decreasing order of $1 - g(Y_i)$ and obtain order π;
for $j = 1$ *to r* **do**
| Add all agents' C_j to G_{j-1} obtain G_j with weight 0;
| Add $f(j)$ to all edges in G_j;
| Run TMHA on G_j based on order π since agents in μ_{j-1} are matched;
| Augmenting μ_j from μ_{j-1} with the result of TMHA;
end
Return the result μ_r as μ;

Theorem 3. *Algorithm 2 can obtain a universal asymptotically lexicographically optimal truthful matching for fair matching problems in polynomial time.*

4 Approximation Ratio of Fair Matching

For lexicographical approximation, the same as [24], we define the approximation ratio for fair matching as below. Note that we define μ^* as the lexicographical optimal solution for fair matching:

$$\alpha_{lex} = \max_{s \leq r} \frac{\sum_{j=1}^{s} signature_j(\mu^*)}{\sum_{j=1}^{s} signature_j(\mu)} \tag{5}$$

Proposition 2. *The lexicographical optimal solution for fair matching problem is a Pareto optimal solution for preference order based maximum matching problem.*

From Proposition 2, we can use a preference order based maximum matching solution to calculate the approximation ratio so that the lexicographical optimal solution must also satisfy the approximation ratio.

Theorem 4. *Algorithm 1 can provide an approximate ratio not better than 2 for an asymptotically lexicographically optimal solution with deterministic truthful constraint.*

Theorem 5. *Algorithm 2 can provide an approximate ratio not better than 1.58 for an asymptotically lexicographically optimal solution with universal truthful constraint.*

5 Conclusion

In this paper, we construct a TMHA based and RTMHA based algorithms and propose a deterministic truthful mechanism and a universal truthful mechanism for fair matching problem. We prove that our mechanisms are asymptotically lexicographically optimal and polynomially solvable. Also, it is noteworthy that our mechanisms achieve an approximation ratio of 2 for deterministic truthful mechanisms and $\frac{e}{e-1} \approx 1.58$ for universal truthful mechanisms.

In the future, it would be intriguing to explore the construction of a deterministic truthful mechanism for rank maximal matching and maximum cardinality rank-maximal matching problems, as discussed in [11]. Investigating whether our transformation method remains applicable in these scenarios would be particularly insightful. Moreover, further research is warranted to devise strategies for achieving a better approximation ratio in such contexts. These endeavors hold promise for advancing the field of matching theory and algorithm design.

Acknowledgements. This work is supported by the National Natural Science Foundation of China (Grant Nos. 62173258, 61773296), and the National Key Research and Development Program of China (2022YFF0902005). Jinshan Zhang was supported by the Key Research and Development Jianbing Program of Zhejiang Province (2023C01002), Hangzhou Major Project and Development Program (2022AIZD0140) and Yongjiang Talent Introduction Programm (2022A-236-G). We are also grateful for the valuable comments from the anonymous reviewers.

A Algorithms of TMHA and RTMHA

Algorithm 3: Truthful Matroid House Allocation

Input: An instance $I = (N, A, P)$, An order π
Output: A truthful matching μ
Let $G = (N \cup A, E)$, $E \leftarrow \emptyset$, $\mu \leftarrow \emptyset$;
for *each agent i in order π* **do**
    Let $l \leftarrow 1$;
    while $C_l^i \neq \emptyset$ **do**
        $E \leftarrow E \cup \{(i, a) : a \in C_l^i\}$;
        Run any algorithm on G and obtain a maximum cardinality matching μ';
        if $|\mu'| = |\mu| + 1$ **then**
            modify μ to μ';
            Break;
        end
        else
            $E \leftarrow E \setminus \{(i, a) : a \in C_l^i\}$;
            $l \leftarrow l + 1$;
        end
    end
end
Return μ;

Algorithm 4: Random Truthful Mechanism for MHA (RTMHA)

Input: Instance (N, A, P)
Output: A truthful matching μ
for *each agent i in N* **do**
  | Pick $Y_i \in [0, 1]$ uniformly at random;
end
Sort agents in decreasing order of $1 - g(Y_i)$;
Run TMHA according to the above order;
Return the matching μ;

B   Further Related Works

Dughmi et al. [8] and Krysta et al. [18] demonstrate that no deterministic truthful algorithm for weight maximum matching can obtain an approximation ratio better than 2, and a truthful mechanism for weight maximum matching without payment has significant limitations so that researchers can only consider truthful mechanism with a limitation or turn to consider randomized truthful mechanism. Therefore, for truthful weight maximum matching, Reiffenhauser et al. introduce a truthful online weighted bipartite matching problem, leveraging the VCG mechanism, and offer a deterministic truthful algorithm for online matching [22]. In a similar vein, Amanatidis et al. present a deterministic truthful mechanism for offline matching conditions, albeit considering only two agents [4]. Additionally, for offline conditions, Abebe et al. propose a truthful cardinal matching mechanism aimed at optimizing Nash Bargaining [1], which serves as a new benchmark proposed by Abebe et al. [1] based on Nash Social Welfare. For the fair matching problem, Huang et al. [12] propose an algorithm based on iterative methods by combining it with its dual problem but with a trivial preference order situation.

Under the preference order based situation, Krysta et al. propose a truthful matching mechanism called TMHA and RTMHA under matroid constraint, in which agent's preference orders include indifference classes, within $O(n^3\gamma)$ complexity and 2 approximation ratio for deterministic truthful mechanism and $\frac{e}{e-1}$ for randomized truthful mechanism [18]. Zhang et al. employ TMHA to explore the lexicographical truthful mechanism for the rank maximal matching problem with an iterative method and obtain a mechanism called MRAND, ultimately obtaining a Pareto optimal solution with an approximation ratio of $\frac{2\sqrt{e}-1}{2\sqrt{e}-2}$ [24]. Adamcyzk et al. also consider the truthful matching problem in a more complex and diverse preference order structure in increasing the social welfare [3].

C   An Example

Next, we will give an example for Algorithm 1: Let $N = \{a, b, c, d\}$ and $A = \{i_1, i_2, i_3\}$. The preference orders for agents in A are below:

$$P_{i_1} = a \succ_{i_1} b \succ_{i_1} c \succ_{i_1} d$$

$$P_{i_2} = a \succ_{i_2} b \simeq_{i_2} c \succ_{i_2} d$$
$$P_{i_3} = a \succ_{i_3} b \simeq_{i_3} c \simeq_{i_3} d$$

We assume $\pi = \{i_1, i_2, i_3\}$.

In the first iteration, G_1 includes edges $(i_1, a), (i_2, a), (i_3, a)$. After running Algorithm 1, we obtain $\mu_1 = \{(i_1, a)\}$ since i_1 has the highest rank in π. And $(i_2, a), (i_3, a)$ is deleted since they cannot be matched without influencing others.

In the second iteration, G_2 includes edges $(i_1, a), (i_2, b), (i_2, c), (i_3, b)$. After augmenting μ_1, we obtain $\mu_2 = \{(i_1, a), (i_2, b)\}$ and edge (i_3, b) is deleted.

In the third iteration, G_3 includes edges $(i_1, a), (i_2, b), (i_3, b), (i_3, c), (i_3, d)$. Then $\mu_3 = \{(i_1, a), (i_2, b), (i_3, c)\}$.

Then all agents are matched, so the result of Algorithm 1 is

$$\mu = \{(i_1, a), (i_2, b), (i_3, c)\}$$

.

D   Omitted Proof

D.1   Proof of Theorem 1

Proof: Since the edges with a rank less than $r - j + 1$ weight 1 for w_j and the others weight 0 for w_j, then a maximum weight matching on graph G_j matches the maximum number of edges with a rank of less than j. Simultaneously, this implies that the aforementioned matching can also match the minimum number of edges with a rank greater than j.

If $w_j(\mu) = w_j(OPT)$, it indicates that matching μ matches the maximum number of edges with ranks less than $r - j + 1$, which is equivalent to minimizing the number of edges with ranks more than $r - j + 1$. If the matching μ is j-optimal, it implies that the number of edges with rank r is minimized, and this criterion extends to ranks $r - 1$ until $r - j + 1$ (since OPT is the lexicographically maximal matching). Therefore, matching μ qualifies as a j-fair matching.

D.2   Proof of Proposition 1

Proof: For (1), it is easy to notice that the weight function with respect to the rank of indifference class is *monotonic*, that means if $C_j^i \prec C_k^i$, then for $\forall a \in C_j^i$ and $\forall b \in C_k^i$, we must have $v((i, a)) < v((i, b))$. So a Pareto optimal preference based maximum matching in G is an asymptotically weight maximum matching in G'.

For (2), from inequalities (2), we know that $f(j)$ is larger than $\sum_{k=j+1}^{r} nf(k)$, which means even matching all edges with the rank of w greater than j, the weight v obtained will not be greater than the weight of an edge with $w_j = 1$. So a maximum matching based on the preference order is a lexicographical optimal result for fair matching.

D.3   Proof of Theorem 2

Proof: We know that TMHA can obtain an asymptotically lexicographically optimal solution for preference order based maximum matching, from Proposition 1 and Theorem 1, we know that process TMHA can construct an asymptotically fair matching for graph G.

Next, the proof of the truthful condition is similar to the approach in [24]. Assume $\mu(P')_i = b \succ_i \mu(P)_i = a$ and let $b \in C_k^i$. From Algorithm 1, we know that b is allocated to i in the k-th iteration (run TMHA on G_k). Thus, TMHA can allocate b to i since i reports P_i'. This contradicts the truthfulness of TMHA.

Remark that γ is the size of the largest indifference and n is the size of N, the complexity of TMHA is $O(n^3\gamma)$, so we can easily notice that the complexity of Algorithm 1 is $O(rn^3\gamma)$, which satisfies the polynomial requirement.

D.4   Proof of Theorem 3

Proof: Similar to the proof for Theorem 2, according to the truthfulness of RTMHA, we know that it can obtain a universal truthful matching for preference order based maximum matching problem. Similar to Theorem 2, leveraging Proposition 1 and Theorem 1, we can deduce that Algorithm 2 is capable of achieving an asymptotically lexicographically optimal truthful matching for the fair matching problem under a truthful constraint. Let γ be the size of the largest indifference class, the complexity of Algorithm 2 is $O(rn^3\gamma)$, which satisfies the polynomial condition.

D.5   Proof of Proposition 2

Proof: We assume that the lexicographical optimal solution is not a Pareto optimal solution for preference order based maximum matching problem, then there exists a solution that an agent a can obtain a better result without influencing others. From Proposition 1, we know that the weight function with respect to the rank of the indifference class is monotonic, so a can obtain a larger weight without influencing others. That means the solution is not a lexicographical optimal solution, a contradiction to the assumption. That proves the proposition.

D.6   Proof of Theorem 4

Proof: In each iteration, the graph G_j contains indifference classes C_1 to C_j, according to formula 1, thus the $signature_j(\mu)$ is

$$
\begin{aligned}
& signature_j(\mu) \\
&= |\mu_{r-j+1}| \\
&\leq \frac{1}{\alpha_{TMHA}} |\mu_{r-j+1}^*| \; \forall j \in [1, r]
\end{aligned}
$$

For deterministic truthful mechanism, note that the approximate ratio for TMHA is $\alpha_{TMHA} \geq 2$ [18], so in each iteration, we have

$$\frac{\sum_{j=1}^{s} signature_j(\mu^*)}{\sum_{j=1}^{s} signature_j(\mu)}$$

$$= \frac{\sum_{j=1}^{s} |\mu^*_{r-j+1}|}{\sum_{j=1}^{s} |\mu_{r-j+1}|}$$

$$\geq \alpha_{TMHA} \frac{\sum_{j=1}^{s} |\mu^*_{r-j+1}|}{\sum_{j=1}^{s} |\mu^*_{r-j+1}|}$$

$$\geq 2$$

Therefore, we can conclude that

$$\alpha_{lex}$$

$$= \max_{s \leq r} \frac{\sum_{j=1}^{s} signature_j(\mu^*)}{\sum_{j=1}^{s} signature_j(\mu)}$$

$$= \max_{s \leq r} \frac{\sum_{j=1}^{s} |\mu^*_{r-j+1}|}{\sum_{j=1}^{s} |\mu_{r-j+1}|}$$

$$\geq \alpha_{TMHA} \max_{s \leq r} \frac{\sum_{j=1}^{s} |\mu^*_{r-j+1}|}{\sum_{j=1}^{s} |\mu^*_{r-j+1}|}$$

$$\geq \alpha_{TMHA}$$

$$\geq 2$$

D.7   Proof of Theorem 5

Proof: For universal truthful mechanism, the approximation ratio for RTMHA is $\alpha_{RTMHA} \geq \frac{e}{e-1}$ [18]. The same as the deterministic truthful mechanism, we have

$$\alpha_{lex}$$

$$= \max_{s \leq r} \frac{\sum_{j=1}^{s} signature_j(\mu^*)}{\sum_{j=1}^{s} signature_j(\mu)}$$

$$= \max_{s \leq r} \frac{\sum_{j=1}^{s} |\mu^*_{r-j+1}|}{\sum_{j=1}^{s} |\mu_{r-j+1}|}$$

$$\geq \alpha_{RTMHA} \max_{s \leq r} \frac{\sum_{j=1}^{s} |\mu^*_{r-j+1}|}{\sum_{j=1}^{s} |\mu^*_{r-j+1}|}$$

$$\geq \alpha_{RTMHA}$$

$$\geq \frac{e}{e-1}$$

$$\approx 1.58$$

References

1. Abebe, R., Cole, R., Gkatzelis, V., Hartline, J.D.: A truthful cardinal mechanism for one-sided matching. In: Proceedings of the Fourteenth Annual ACM-SIAM Symposium on Discrete Algorithms, pp. 2096–2113. SIAM (2020)
2. Abraham, D., Chen, N., Kumar, V., Mirrokni, V.S.: Assignment problems in rental markets. In: Internet and Network Economics: Second International Workshop, WINE 2006, Patras, Greece, 15–17 December 2006. Proceedings 2, pp. 198–213. Springer (2006)
3. Adamczyk, M., Sankowski, P., Zhang, Q.: Efficiency of truthful and symmetric mechanisms in one-sided matching. In: International Symposium on Algorithmic Game Theory, pp. 13–24. Springer (2014)
4. Amanatidis, G., Birmpas, G., Christodoulou, G., Markakis, E.: Truthful allocation mechanisms without payments: Characterization and implications on fairness. In: Proceedings of the 2017 ACM Conference on Economics and Computation, pp. 545–562 (2017)
5. Azad, A., Buluç, A.: Distributed-memory algorithms for maximum cardinality matching in bipartite graphs. In: 2016 IEEE International Parallel and Distributed Processing Symposium (IPDPS), pp. 32–42. IEEE (2016)
6. Chakrabarty, D., Swamy, C.: Welfare maximization and truthfulness in mechanism design with ordinal preferences. In: Proceedings of the 5th conference on Innovations in Theoretical Computer Science, pp. 105–120 (2014)
7. Duan, R., Su, H.H.: A scaling algorithm for maximum weight matching in bipartite graphs. In: Proceedings of the Twenty-Third Annual ACM-SIAM Symposium on Discrete Algorithms, pp. 1413–1424. SIAM (2012)
8. Dughmi, S., Ghosh, A.: Truthful assignment without money. In: Proceedings of the 11th ACM Conference on Electronic Commerce, pp. 325–334 (2010)
9. Dulmage, A.L., Mendelsohn, N.S.: Coverings of bipartite graphs. Can. J. Math. **10**, 517–534 (1958)
10. Gerards, A.: Matching. Handb. Oper. Res. Manag. Sci. **7**, 135–224 (1995)
11. Huang, C.C., Kavitha, T.: Weight-maximal matchings. Proc. MATCH-UP **12**, 87–98 (2012)
12. Huang, C.C., Kavitha, T., Mehlhorn, K., Michail, D.: Fair matchings and related problems. Algorithmica **74**(3), 1184–1203 (2016)
13. Irving, R.W., Kavitha, T., Mehlhorn, K., Michail, D., Paluch, K.E.: Rank-maximal matchings. ACM Trans. Algorithms (TALG) **2**(4), 602–610 (2006)
14. Kavitha, T., Shah, C.D.: Efficient algorithms for weighted rank-maximal matchings and related problems. In: International Symposium on Algorithms and Computation, pp. 153–162. Springer (2006)
15. Kesselheim, T., Radke, K., Tönnis, A., Vöcking, B.: An optimal online algorithm for weighted bipartite matching and extensions to combinatorial auctions. In: European Symposium on Algorithms, pp. 589–600. Springer (2013)
16. Korula, N., Mirrokni, V.S., Zadimoghaddam, M.: Bicriteria online matching: maximizing weight and cardinality. In: Web and Internet Economics: 9th International Conference, WINE 2013, Cambridge, MA, USA, 11–14 December 2013, Proceedings 9, pp. 305–318. Springer (2013)
17. Krysta, P., Manlove, D., Rastegari, B., Zhang, J.: Size versus truthfulness in the house allocation problem. In: Proceedings of the Fifteenth ACM Conference on Economics and Computation, pp. 453–470 (2014)

18. Krysta, P., Zhang, J.: House markets with matroid and knapsack constraints. In: 43rd International Colloquium on Automata, Languages, and Programming (ICALP 2016). Schloss Dagstuhl-Leibniz-Zentrum fuer Informatik (2016)
19. Kuhn, H.W.: The Hungarian method for the assignment problem. Nav. Res. Logistics Q. **2**(1–2), 83–97 (1955)
20. Mehlhorn, K., Michail, D.: Network problems with non-polynomial weights and applications (2005)
21. Punnen, A.P., Nair, K.: Improved complexity bound for the maximum cardinality bottleneck bipartite matching problem. Discret. Appl. Math. **55**(1), 91–93 (1994)
22. Reiffenhauser, R.: An optimal truthful mechanism for the online weighted bipartite matching problem. In: Proceedings of the Thirtieth Annual ACM-SIAM Symposium on Discrete Algorithms, pp. 1982–1993. SIAM (2019)
23. Schwartz, J., Steger, A., Weißl, A.: Fast algorithms for weighted bipartite matching. In: International Workshop on Experimental and Efficient Algorithms, pp. 476–487. Springer (2005)
24. Zhang, J., Liu, Z., Deng, X., Yin, J.: Improved truthful rank approximation for rank-maximal matchings. In: International Conference on Web and Internet Economics, pp. 637–653. Springer (2023)

Equilibrium Strategies of Carbon Emission Reduction in Agricultural Product Supply Chain Under Carbon Sink Trading

Tingting Meng[1], Yukun Cheng[1(✉)], Xujin Pu[1], and Rui Li[2]

[1] Jiangnan University, Wuxi, China
ykcheng@amss.ac.cn

[2] Suzhou University of Science and Technology, Suzhou, China

Abstract. As global climate change and environmental issues escalate, carbon reduction has emerged as a paramount global concern. Agriculture accounts for approximately 30% of global greenhouse gas emissions, making carbon reduction in this sector crucial for attaining global emission targets. Carbon sink trading serves as a supplementary mechanism to achieve carbon peaking and neutrality, helping to lower the rate of carbon emissions. However, practical projects and research in the field of carbon sink trading are not enough currently. This work aims to thoroughly explore the cooperative models between farmers and retailers within the context of agricultural carbon sink trading, as well as the optimal decisions on the efforts to reduce carbon emission for both parties under different cooperative models. To this end, we delve into three distinct cooperative frameworks: the decentralized, the Stackelberg, and the centralized models, each accompanied by a corresponding differential game model. The Hamilton-Jacobi-Bellman equation is utilized to investigate the equilibrium strategies of each participant under these three cooperative models, respectively. Furthermore, we conducted numerical simulations to analyze the carbon emission reduction efforts of farmers and retailers, the carbon emission reduction level of the agricultural supply chain, and the overall profits of the supply chain. We also compare scenarios with and without carbon sink trading to provide a comprehensive assessment. The numerical results indicate that the centralized model excels in all aspects, followed by the Stackelberg model, with the decentralized model showing the weakest performance. Additionally, carbon sink trading can significantly increase the profits of the participants under each cooperative model.

Keywords: carbon emission reduction · agricultural product supply chain · carbon sink trading · differential game

This work is supported by NSFC (No. 72271109).

1   Introduction

As the global economy continues to grow, the consumption of energy and resources, as well as emissions of greenhouse gases, exert substantial pressure on the ecological environment, leading to intensified climate change. The agricultural system, being one of the primary sources of greenhouse gas emissions, contributes more than 30% to global emissions[1]. Consequently, reducing carbon emissions in agriculture is crucial for achieving global emission reduction goals. In this context, many countries have set targets for carbon emission reduction.

Currently, countries worldwide primarily adopt carbon trading as a means to address carbon emissions issues, which typically occur in the midstream and downstream stages of the supply chain, such as processing and transportation. However, besides carbon trading, carbon sink trading is also important for the carbon market, serving as an effective measure for reducing agricultural carbon emissions with significant advantages. Agricultural carbon sink refers to the process of reducing greenhouse gas concentrations by absorbing carbon dioxide from the atmosphere through agricultural practices, such as crop planting and vegetation restoration. Enhanced carbon sequestration can slow the continued rise in atmospheric carbon dioxide concentrations and can aid in achieving carbon neutrality goals [1]. In carbon sink trading, farmers or agricultural enterprises can generate carbon credits or sinks through sustainable and low-carbon agricultural practices, which they can then sell to individuals or companies needing carbon sinks in the market. Carbon sink trading can effectively motivate farmers to reduce carbon emissions, thereby increasing carbon sequestration and reducing greenhouse gas emissions. Countries, such as the United States, Australia, and the European Union, have initiated agricultural carbon sink trading and achieved notable success. In China, the nationwide carbon market is progressively being established. On January 22, 2024, the ceremony to launch the National Greenhouse Gas Voluntary Emission Reduction Trading Market was held in Beijing, marking the official restart of the china certified emission reduction (CCER) market since its suspension in 2017, presenting a key opportunity for the development of agricultural carbon sink projects in the country[2].

Simultaneously, increasing consumer awareness of low-carbon products has fueled their growing popularity in the market. Nowadays, consumers tend to prefer products with low-carbon properties when making purchasing decisions. The dual impetus of low-carbon policies and consumer demand preferences has incentivized carbon emission reduction in the agricultural product supply chain. On the sales end, many large retailers have begun to set emission reduction targets and adopt promotional strategies, such as Walmart's plan to reduce emissions by one billion tons, actively promoting a low-carbon supermarket image[3]. On the production side, farmers can adopt emission reduction measures, such as reducing fertilizer usage and implementing conservation tillage, to increase soil car-

---

[1] https://www.chinanews.com.cn/gj/2021/03-11/9430282.shtml.

[2] http://paper.people.com.cn/zgnyb/html/2024-01/29/content_26041024.htm.

[3] https://finance.sina.com.cn/roll/2019-04-29/doc-ihvhiqax5710370.shtml.

bon sequestration. They participate in carbon sink trading and earn additional income. The first carbon sink trading platform in Xiamen has been successfully established, facilitating the trade of 3,357 tons of carbon sink from agricultural tea gardens in two villages[4].

However, the main obstacle to carbon reduction within the agricultural product supply chain is the imbalance in the distribution of revenues and costs related to carbon reduction among the participants. Participants including farmers and retailers, encounter different economic and technological constraints, as well as varying levels of awareness and willingness to invest in carbon reduction. So the obstacle lead to the challenge of promoting carbon reduction within the agricultural product supply chain. Specifically, the carbon emissions of agricultural products mainly occur in the upstream production phase. Farmers need to bear greater responsibility for carbon emission reduction and make high-cost investments. Still, these investments often cannot be converted into direct economic benefits in the short term, which undoubtedly increases the burden on farmers and reduces their enthusiasm for participating in carbon emission reduction. Retailers, as downstream participants in the supply chain, can directly benefit from consumer purchasing behavior and market demand, but they rely on upstream farmers to produce low-carbon agricultural products to meet market needs.

To address this issue and assist participants in making informed carbon reduction decisions within the agricultural product supply chain, this paper examines the impact of carbon sink trading and carbon emission reduction levels, constructing three differential game models: decentralized, Stackelberg, and centralized, for the supply chain system that includes the farmer and the retailer. Employing the Hamilton-Jacobi-Bellman equation, the optimal carbon reduction efforts and associated profits for each participant within these three modes of cooperation are explored. Our findings indicate that the centralized model yields the best long-term benefits, with higher levels of carbon reduction and efforts compared to the decentralized and the Stackelberg models. In the Stackelberg game model, the retailer achieves higher overall profits by sharing costs with the farmer compared to the decentralized model. In addition, we found that carbon sink trading can increase the profit of the farmer and the retailer, incentivizing them to engage in carbon emission reduction.

1.1   Related Work

Carbon trading is an effective means to promote carbon reduction. Some researches mainly focus on carbon quota allocation, carbon quota subsidies, and carbon emission reduction to explore the optimal carbon quota allocation mechanism and decision-making of supply chain members under carbon trading policies. Xu et al., [2,3] investigated the production and emission reduction decision-making problem of Order-based supply chain, consisting of manufacturers and retailers under carbon cap and trade rules. Zhang et al., [4] con-

---

[4] http://www.fujian.gov.cn/xwdt/fjyw/202205/t20220506_5903853.htm.

sidered two scenarios, static carbon trading prices and dynamic carbon trading prices. They developed an evolutionary game model involving the government and manufacturers to investigate how government policies influence manufacturer decisions and the carbon trading market. Cai et al., [5] studied a supply chain model consisting of a supplier and a manufacturer, and discussed the optimal pricing and carbon reduction decisions of the supply chain members. In the context of carbon sink trading, Wang et al., [6] considered carbon sinks can fully utilize the characteristics of natural ecosystems and will be an important means in the process of carbon emission reduction. Ke et al., [7] conducted research on forest carbon sink trading. The study assessed the potential of forest carbon sink trading in China. Wang et al., [8] addressed the gap in carbon sink trading in China and, based on a duopoly model, examined the impact of carbon sink mechanisms on the profits of emission trading participants and industry output. The results indicate that when companies exceed their carbon emission limits, they can purchase carbon sinks to compensate for the excess emissions.

In terms of supply chain coordination, many researches have started applying supply chain contracts to agricultural product supply chains. Yu et al., [9] proposed a Stacklberg game model to coordinate pricing and service levels in a three-stage Agricultural Product Supply Chain that includes third-party logistics service providers. The model takes into account both supplier-led and logistics service provider-led scenarios. Furthermore, Yan et al., [10] examined the optimal decision-making challenges in an Internet of Things (IoT) fresh agricultural product supply chain, which comprises manufacturers, distributors, and retailers. They proposed improved revenue-sharing contracts to achieve supply chain coordination. Additionally, Song et al., [11] investigated the decision-making problems in a three-stage supply chain, including fresh e-commerce companies, third-party logistics service providers, and community convenience stores, under both centralized and decentralized model. They introduced cost-sharing contracts and profit-sharing contracts to achieve coordination and maximize the profits of all parties in the supply chain. Ma et al., [12] studied the coordination issues in a three-stage supply chain system consisting of a supplier, a third-party logistics service provider, and a retailer supplying seasonal fresh agricultural products to customers.

In practice, carbon emissions reduction in the agricultural product supply chain is a long-term process [13], where the outcomes of previous stage decisions can impact the subsequent carbon reduction decisions, and the carbon emissions of the final product are influenced by the carbon reduction decisions made by different participants in the supply chain. Therefore, it is necessary to analyze it from a dynamic perspective. Differential game theory is the useful tool to study the dynamic decision problems involving continuous time and continuous actions, in which participants aim to achieve optimal decisions through dynamic optimization. Some researches have employed the approach of differential game theory to investigate carbon emissions reduction in supply chains. Wang et al., [14] developed three differential game models for a supply chain system consisting of two suppliers and a single manufacturer, examining the impact of different carbon emission allocation rules on carbon reduction in the supply chain.

Some studies have explored supply chain decision-making under carbon trading [2–5]. Others have examined the significance of carbon sink trading, primarily focusing on carbon sink assessment and the factors influencing carbon sink trading [7,8]. Notably, several of these studies have been applied in the forestry sector. However, research on agricultural carbon sink trading, particularly in terms of coordinated decision-making within the supply chain, is not too enough. Also, previous research on low-carbon supply chains has primarily focused on carbon emissions during processing and transportation. To enrich the research of carbon sink trading, we mainly investigate carbon emissions during the production of agricultural products and innovatively incorporate carbon sink trading into the agricultural supply chain. We utilize the differential game approach to examine the carbon emission reduction decisions of farmers and retailers within the supply chain.

1.2   Organization

The remaining part of this paper is organized as follows: Sect. 2 provides a detailed description of the model framework along with the necessary assumptions. In Sect. 3, we introduce and explain three different cooperation models, namely decentralized decision model, Stackelberg game model, and centralized decision model, and analyze their respective equilibrium solutions. Numerical analysis and sensitivity analysis are conducted in Sect. 4. Finally, Sect. 5 presents the conclusion, discussion, and outlines potential directions for future research.

2   Problem Description and and Assumptions

2.1   Description of the Problem

Due to the widespread distribution of farmers, it is difficult to account for the carbon sink produced by an individual farmer. Therefore, in practice, it is common to aggregate the overall carbon sinks produced by agricultural activities at the village or township level for the purposes of unified accounting and trading. To simplify the expression and calculation, we treat the farmers of the entire village as a single collective entity, denoted by f. This work focuses on a low-carbon agricultural product supply chain, consisting of one farmer f and one retailer r. The primary responsibility of the farmer is to undertake efforts such as implementing advanced production techniques and enhancing soil quality to achieve the carbon emission reduction targets. In detail, they adopt production technologies such as nitrogen oxide emission reduction in farmland to enhance production efficiency and reduce the emission intensity per unit of product. Additionally, they improve soil quality to enhance carbon sequestration capacity in farmland, including implementing conservation tillage, promoting straw returning to the field. On the other hand, the retailer enhance the environmental image of agricultural product brands through active low-carbon promotion and advertising activities, thereby attracting more consumers. These combined efforts aim to enhance the brand's environmental image, collectively referred to as low-carbon promotional initiatives. Figure 1 demonstrates the process of carbon emission reduction in the agricultural product supply chain.

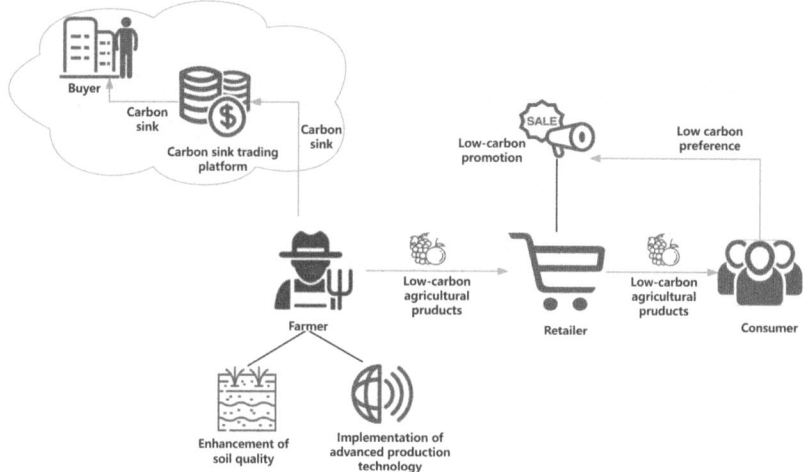

Fig. 1. The process of carbon emission reduction in the agricultural product supply chain

2.2 Problem Assumptions

The following assumptions are necessary for our discussion.

Assumption 1. Generally, products that consume less energy and emit lower amounts of carbon dioxide and other greenhouse gases during their production and consumption processes are referred to as low-carbon products. Given consumers' preference for low-carbon products, the level of carbon emissions reduction H in the agricultural supply chain has become a critical factor that influences the marketing of agricultural products and consumer purchasing decisions. This level H is influenced by two factors: the emission reduction and carbon sequestration efforts $E_f(t)$ of the farmer and low-carbon promotion efforts $E_r(t)$ of the retailer. Similar to [14], we can describe the dynamic process of $H(t)$ in agricultural products using the following differential equation:

$$\dot{H}(t) = \mu_f E_f(t) + \mu_r E_r(t) - \delta H(t), \tag{1}$$

where μ_f and μ_r represent the impact coefficient of the farmer's and the retailer's efforts to reduce carbon emission, on the level of carbon emission reduction, and $\delta > 0$ represents the decay coefficient of the carbon emission reduction level.

Assumption 2. Faced with environmental deterioration, people's awareness of low-carbon living has risen, leading more consumers to pay more for low-carbon products [19]. Additionally, as the carbon emission reduction level of the supply chain improves, the low-carbon attributes of products become increasingly significant, leading to a larger market demand. Also, different cooperation models may lead to information asymmetry between farmers and retailers. This asymmetry may cause the supply of agricultural products to not match market demand. As a result, retailers could face stock shortages. Alternatively, they might experience issues with inventory overstock. Therefore, we assume that the supply $Q(t)$ and demand $D(t)$ of agricultural products are influenced by the level of carbon emission reduction $H(t)$, the price per unit of agricultural product p, and consumers' preferences for low-carbon products θ. Following [5], the supply and demand function of agricultural products are

$$Q(t) = (Q_0 + ap)\theta H(t), \tag{2}$$
$$D(t) = (D_0 - bp)\theta H(t), \tag{3}$$

where $Q_0 \geq 0$ and $D_0 \geq 0$ represent the initial supply and initial demand of agricultural products, respectively; $a \geq 0$ and $b \geq 0$ represent the influence of agricultural product prices on product supply and demand, respectively.

Assumption 3. The carbon emission reduction during the agricultural production process is crucial for the carbon emission reduction of the agricultural product supply chain. By implementing advanced technologies and enhancing soil quality, farmers can lower carbon emissions, and thus increase carbon sink F, a distinctive feature of the agricultural industry. Assuming that the carbon emission reduction level of agricultural products is correlated with the farmer's efforts $E_f(t)$ in emission reduction and carbon sequestration, represented by the coefficient ω [14]. Similar to [15,16], we assume that the farmer's initial carbon sink per unit product is 1. Thus, the total carbon sink $F(t)$ at time t can be expressed as:

$$F(t) = (1 + \omega E_f(t))Q(t), \tag{4}$$

where $Q(t)$ is the supply of the farmer at time t.

Assumption 4. The costs of emission reduction efforts by the farmer and the retailer have a quadratic relationship with $E_f(t)$ and $E_r(t)$, respectively. The form of quadratic function, which is widely adopted to describe the cost pattern in the literature [17,18]. To simplify the problem and highlight the research focus, we assume that the production cost of agricultural products is zero [14]. Therefore, the costs of carbon emission reduction efforts by the farmer and the retailer are

$$C_s(t) = \frac{1}{2}\lambda_f E_f^2(t), \quad C_m(t) = \frac{1}{2}\lambda_r E_r^2(t), \tag{5}$$

where λ_f, λ_r represent the effort cost coefficients for the farmer and the retailer to reduce carbon emissions, respectively.

Table 1 provides the notations used in this paper.

Table 1. Major notations

Notations	Explanation
t	Time period
$H(t), H(0)$	Carbon emission reduction level of agricultural products at time t, and initial carbon emission reduction level, $H(0) \geq 0$.
$E_f(t), E_r(t)$	The emission reduction and carbon sequestration efforts of the farmer at time t, the low-carbon promotion efforts $E_r(t)$ of the retailer at time t.
λ_f, λ_r	Effort cost coefficients of the farmer and the retailer to reduce carbon emissions, $\lambda_f, \lambda_r > 0$.
μ_f, μ_r	The impact coefficient of the farmer's and the retailer's carbon emission reduction efforts on the level of carbon emission reduction, $\mu_f, \mu_r > 0$.
ω	The impact coefficient of the farmer's efforts to reduce carbon emissions on the carbon emission level, $\omega > 0$.
p_f, p_r, p, p_c	Marginal profits of the farmer and the retailer, unit agricultural product price, unit carbon sink price, $p_f, p_r, p, p_c \geq 0$.
$Q(t), D(t)$	Supply function for the farmer at time t, and demand function for the retailer at time t.
θ	Consumers' low-carbon preferences, $\theta > 0$.
a, b	The influence of agricultural product prices on product supply and demand, $a > 0$, $b > 0$.
$F(t)$	The total carbon sink at time t.
δ	Attenuation coefficient of carbon emission reduction level, $\delta > 0$.
ρ	profit discount rate, $\rho > 0$.

3   Analysis for Differential Game Equilibrium

Building on the assumptions mentioned earlier, we have developed three differential game models: decentralized, Stackelberg, and centralized. These game models are used to analyze the equilibrium strategies of the farmer and the retailer across three different cooperative scenarios in this section. For simplicity, the parameter t will be omitted in the following sections.

3.1   Decentralized Decision-Making Model

Under the decentralized decision-making model, indicated by the superscript GD, both the farmer f and the retailer r aim to maximize their respective profits over an infinite time horizon, with a profit discount rate assumed to be ρ. Thus, the objective functions for the farmer and the retailer are formulated as (6) and (7).

$$\max J_f^{GD} = \int_0^\infty e^{-\rho t} \left[p_f Q + p_c(1 + \omega E_f)Q - \frac{1}{2}\lambda_f E_f^2 \right] dt, \tag{6}$$

$$\max J_r^{GD} = \int_0^\infty e^{-\rho t} \left[p_r D - \frac{1}{2}\lambda_r E_r^2 \right] dt. \tag{7}$$

Proposition 1 propose the equilibrium strategies for both parties involved in the decentralized decision-making model. The complete proof of this proposition can be found in the full version, due to the limit of the space.

Proposition 1. *Under the decentralized decision-making model, the equilibrium carbon emission reduction efforts of the farmer and the retailer are*

$$E_f^{GD*} = \frac{p_c \theta \omega k_1 H + \mu_f (2A^{GD} H + B^{GD})}{\lambda_f}, \quad E_r^{GD*} = \frac{\mu_r M^{GD}}{\lambda_r},$$

the equilibrium level of carbon emission reduction is

$$H_d^{GD} = \frac{\lambda_r \mu_f B^{GD} + \lambda_f \mu_r M^{GD}}{\lambda_f \lambda_r \delta - 2A^{GD} \mu_s \lambda_r - p_c \omega k_1 \mu_f \lambda_r},$$

the optimal profits of the farmer and the retailer are

$$V_f^{GD*} = A^{GD} H^2 + B^{GD} H + C^{GD}, \quad V_r^{GD*} = M^{GD} H + N^{GD},$$

where $A^{GD}, B^{GD}, C^{GD}, M^{GD}, N^{GD}, T^{GD}$ *are shown in* (8)

$$A^{GD} = \frac{2\rho\lambda_f + 4\lambda_f \delta - 4\mu_f p_c \omega k_1 - \sqrt{\triangle^{GD}}}{8\mu_f^2},$$

$$M^{GD} = \frac{p_m k_2 \lambda_s}{p_c \omega k_1 \mu_s + 2A^{GD} \mu_s^2 + \lambda_s \rho - \delta \lambda_s},$$

$$B^{GD} = -\frac{\left(2\lambda_r^2 A^{GD} M^{GD} + k_1 \lambda_r p_c + k_1 \lambda_r p_f\right) \lambda_f}{\lambda_r \left(p_c \omega k_1 \mu_f + 2A\,\mu_f{}^2 - \lambda_f \rho - \delta \lambda_f\right)}, \qquad (8)$$

$$C^{GD} = \frac{B^{GD} \left(\lambda_s^2 \lambda_m B^{GD} + 2M^{GD} \lambda_m^2 \lambda_s\right)}{2\lambda_m \lambda_s \rho},$$

$$N^{GD} = \frac{M^{GD} \left(2\lambda_m \mu_s^2 B^{GD} + M^{GD} \mu_m^2 \lambda_s\right)}{2\lambda_m \lambda_s \rho},$$

with $\triangle^{GD} = (4\mu_f p_c \omega k_1 - 4\lambda_f \delta - 2\rho\lambda_f)^2 - (4\mu_f p_c \omega k_1)^2 > 0$, $k_1 = (Q_0 + ap)\theta$, *and* $k_2 = (D_0 - bp)\theta$.

3.2 Stackelberg Game Model

Under this model, the retailer provides subsidies to the farmer to encourage efforts towards reducing carbon emissions. In the first stage, the retailer acts as a leader, determining their efforts to low-carbon promotion E_r and the subsidy ratio x_f for the farmer. Then, in the second stage, acting as a follower, the farmer makes her optimal responses based on the decisions made by the retailer in the first stage. Thus, the objective functions for farmer and the retailer are formulated as follows:

$$\max J_r^{GS} = \int_0^\infty e^{-\rho t} \left[p_r D - \frac{1}{2} \lambda_r E_r^2 - \frac{1}{2} x_f \lambda_f E_f^2 \right] dt, \tag{9}$$

$$\max J_f^{GS} = \int_0^\infty e^{-\rho t} \left[p_f Q + p_c(1 + \omega E_f)Q - (1 - x_f)\frac{1}{2}\lambda_f E_f^2 \right] dt. \tag{10}$$

Similarly, the equilibrium strategies for both parties under the Stackelberg game model are provided in Proposition 2, whose proof is also in the full version.

Proposition 2. *Under the Stackelberg game model, the equilibrium carbon emission reduction efforts of the farmer and the retailer are*

$$E_f^{GS*} = \frac{p_c \omega k_1 H + \mu_f(2A^{GS}H + B^{GS})}{\lambda_f}, \quad E_r^{GS*} = \frac{\mu_r(2M^{GS}H + N^{GS})}{\lambda_r},$$

the equilibrium level of carbon emission reduction is

$$H_d^{GS} = \frac{2(\lambda_f \rho \mu_r + \lambda_r \mu_f)N^{GS} + \lambda_r \mu_f B^{GS}}{4\lambda_r \mu_f M^{GS} - \rho \lambda_r p_c \omega k_1 - 4\lambda_f \mu_r \rho M^{GS} - 2\lambda_r \mu_f A^{GS} + 2\lambda_r \lambda_f \rho \delta},$$

the equilibrium subsidy ratio is

$$x_f^{GS*} = \frac{(4M - 2A)H + 2N - B - p_c \omega k_1 H/\mu_f}{(4M + 2A)H + 2N + B + p_c \omega k_1 H/\mu_f},$$

The optimal profits for the farmer and the retailer are

$$V_f^{GS*} = A^{GS}H^2 + B^{GS}H + C^{GS}, \quad V_r^{GS*} = M^{GS}H^2 + N^{GS}H + F^{GS},$$

where $A^{GS}, B^{GS}, C^{GS}, M^{GS}, N^{GS}, T^{GS}$ *are shown in* (11)

$$A^{GS} = \frac{2\lambda_r \lambda_f \rho \delta - p_c \omega \ k_1 \lambda_r \rho \mu_f - \lambda_r \lambda_f \rho^3 - 2\mu_f^2 \lambda_r M^{GS} - 4\mu_r^2 \lambda_f M^{GS} - \sqrt{\triangle^{GS1}}}{2\mu_f^2 \lambda_r},$$

$$M^{GS} = \frac{2\lambda_f \lambda_r \rho \delta - p_c \omega k_1 \rho \lambda_r \mu_f - \lambda_f \lambda_r \rho^3 - 2A^{GS} \lambda_r \mu_f^2 - \sqrt{\triangle^{GS2}}}{4(\lambda_r \mu_f^2 + \lambda_f \mu_r^2)},$$

$$B^{GS} = \frac{\lambda_f \lambda_r((p_c + p_f)k_1 + p_r k_2)}{\lambda_f \lambda_r(\rho - \delta) - p_c \omega \lambda_r \mu_f k_1 - 2A^{GC}(\lambda_r \mu_f^2 + \lambda_f \mu_r^2)},$$

$$C^{GS} = \frac{\mu_f^2 \lambda_r B^{GS2} + (4\mu_r^2 \lambda_f + 2\mu_f^2 \lambda_r)B^{GS} N^{GS}}{4\rho^3 \lambda_f \lambda_r}, \tag{11}$$

$$N^{GS} = \frac{2\lambda_r \mu_f^2 B^{GS}(A^{GS} + 2M^{GS} - 4\lambda_r \lambda_f p_r \rho^2 k_2 - p_c \omega k_1 \lambda_r \mu_f \rho B^{GS})}{4\rho^3 \lambda_f \lambda_r - 2p_c \omega k_1 \mu_f \rho \lambda_r + 4\lambda_f \lambda_r \delta \rho + 4\mu_f^2 \lambda_r(A^{GS} + 2M^{GS}) + 8\mu_r^2 \lambda_f M^{GS}},$$

$$F^{GS} = \frac{\mu_f^2 \lambda_r(B^{GS2} + 4N^{GS} B^{GS}) + 4(\lambda_r \mu_f^2 + \lambda_f \mu_r^2)N^{GS2}}{8\rho^3 \lambda_r \lambda_f},$$

with $k_1 = (Q_0 + ap)\theta$, $k_2 = (D_0 - bp)\theta$,

$$\triangle^{GS1} = 2(\mu_f \omega k_1 p_c \rho)^2 + 2k_1 p_c \omega \lambda_f \mu_f \rho^2(\rho^2 - 2\delta) + \lambda_f \rho^3(\lambda_f \rho^3 + 4M\mu_f^2 - 4\lambda_f \rho \delta)$$
$$+ 4M^2 \mu_f^4 - 8M\lambda_f \rho \delta \mu_f^2 + 4\lambda_f^2 + 4(\lambda_f \rho \delta)^2 + 8M\mu_r^2 \lambda_f(\lambda_r \rho \omega p_c k_1 + \lambda_f \lambda_r \rho^3$$
$$+ 2M\mu_f^2 \lambda_r + 2M\mu_r^2 \lambda_f - 2\lambda_r \lambda_f \rho \delta) > 0, and$$

$$\triangle^{GS2} = 2\lambda_r(k_2 p_c \rho \omega)^2(\lambda_r \mu_f^2 + \lambda_f) + 2k_2 \lambda_r \lambda_f p_c \rho \omega(\lambda_r \rho^3 \mu_f - 2A\mu_r^2 - 2\lambda_r \rho \delta)$$
$$+ 4\lambda_r \lambda_f \rho^2(\rho^3 + \rho^2 \delta + \delta^2) + 4A\lambda_r \lambda_f \mu_f^2(\lambda_r \rho^3 - A\mu_r^2 - 2\lambda_r \rho \delta) > 0.$$

3.3   Centralized Decision-Making-Making Model

Under this model, the cooperation among the farmer and the retailer are aimed at maximizing the total profits of the entire system. Unlike the previous two models, the objective function in this case is derived by discounting the combined profits of the two participants, with a profit discount rate assumed to be ρ. Thus, we propose the objective function in the centralized decision-making model.

$$\max J_{fr}^{GC} = \int_0^\infty e^{-\rho t}\left[p_f Q + p_r D + p_c(1+\omega E_f)Q - \frac{1}{2}\lambda_f E_f^2 - \frac{1}{2}\lambda_r E_r^2\right]dt. \tag{12}$$

Under the centralized decision-making model, the equilibrium strategies of the corresponding differential game are proposed in the following Proposition. We also move its proof to the full version.

Proposition 3. *Under the centralized decision-making model, the equilibrium carbon emission reduction efforts of the farmer and the retailer are*

$$E_f^{GC*} = \frac{\mu_f(2A^{GC}H + B^{GC})}{\lambda s}, \quad E_r^{GC*} = \frac{p_c\theta\omega\theta(D_0 - bp)H + \mu_r(2A^{GC}H + B^{GC})}{\lambda_r},$$

the equilibrium level of carbon emission reduction is

$$H_d^{GC} = \frac{(\lambda_f\mu_r + \lambda_r\mu_f)B^{GC}}{\lambda_f\lambda_r\delta - \lambda_r p_c\omega k_1 - 2(\lambda_f\mu_r + \lambda_r\mu_f)A^{GC}},$$

the optimal profit of the supply chain is

$$V_{fr}^{GC*} = A^{GC}H^2 + B^{GC}H + C^{GC},$$

where A^{GC}, B^{GC}, C^{GC} *are shown in (13)*

$$A^{GC} = \frac{2\delta\lambda_f - \rho\lambda_f - 2\lambda_r\mu_f p_c\omega k_1 - \sqrt{\triangle^{GC}}}{4\lambda_r\mu_f^2 + 4\lambda_f\mu_r^2},$$

$$B^{GC} = \frac{\lambda_f\lambda_r((p_c + p_f)k_1 + p_r k_2)}{\lambda_f\lambda_r(\rho - \delta) - p_c\omega\lambda_r\mu_f k_1 - 2A^{GC}(\lambda_r\mu_f^2 + \lambda_f\mu_r^2)},$$

$$C^{GC} = \frac{B^{GC^2}(\lambda_r\mu_f^2 + \lambda_f\mu_r^2)}{2\lambda_r\lambda_f\rho}, \tag{13}$$

with $\triangle^{GC} = 4\lambda_r(p_c\omega k_1)^2(2\lambda_r\mu_f^2 + \lambda_f\mu_r^2 + \lambda_r^2\lambda_f(\rho - 2\delta)(4p_c\omega k_1\mu_f + \rho - 2\delta))$, $k_1 = (Q_0 + ap)\theta$, and $k_2 = (D_0 - bp)\theta$.

4   Numerical Simulation and Analysis

Referring to [4,5], the relevant parameter in our numerical experiments are set as follows: $\lambda_f = 500, \lambda_r = 200, \mu_f = 1.5, \mu_r = 0.5, \omega = 0.4, p_f = 5, p_r = 10, p = 25, p_c = 0.5, a = 3, b = 2, \delta = 1, \rho = 0.7, \theta = 0.8, Q_0 = 300, D_0 = 250$. The outcomes of the decision model are illustrated using distinct colors to distinguish among different cooperation modes. Specifically, red represents the decentralized decision-making model, green denotes the Stackelberg game model, and blue indicates the centralized game model. It's important to note that solid lines symbolize scenarios with carbon sink trading, whereas dashed lines depict scenarios without carbon sink trading.

4.1   Impact of Time t and Parameter ω on the Level of Carbon Emission Reduction and Total Profit

This section discusses the changes in the level of carbon emission reduction and total supply chain profits over time, considering three types of cooperation models.

Figure 2a depicts the carbon emission reduction level trajectories under three types of cooperation models, which increase over time t and stabilize as t approaches infinity. Clearly, the centralized decision-making model achieves the highest level of carbon emission reduction among the three models, followed by the Stackelberg game model, while the decentralized decision-making model exhibits the lowest level of carbon emission reduction.

Figure 2b depicts the changes of the total profit in the supply chain over time, under three types of cooperation models. We can clearly observe that the decentralized decision-making model shows the lowest total profit in the supply chain. While the Stackelberg game model initially outperforms the centralized decision model in terms of profit, over time, the centralized model gradually reveals more substantial advantages and significantly exceeds the other two models.

Specifically, the centralized model consolidates carbon emission reduction resources from stakeholders like the farmer and the retailer to optimize resource allocation. In contrast, the decentralized model, lacking unified coordination, risks redundant investments and resource wastage. In the Stackelberg model, the retailer set emission standards and targets, guiding the farmer and driving the supply chain towards a low-carbon transformation. Additionally, scenarios involving carbon sink trading outperform those without carbon sink trading under three types of cooperation models, indicating that appropriate incentive mechanisms can improve both the carbon reduction efficiency and the profits of supply chain participants.

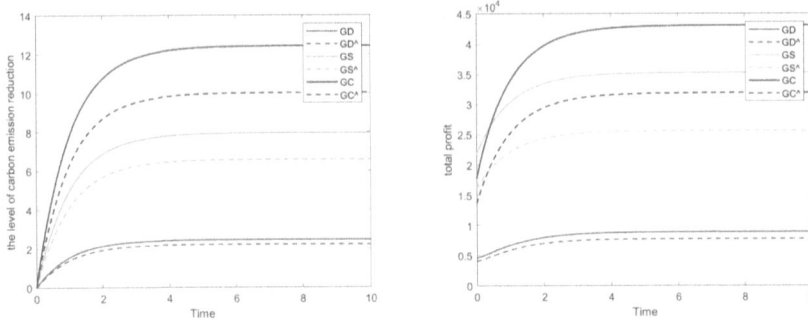

(a) The level of carbon emission reduction under three modes

(b) Total profit under three modes

Fig. 2. The level of carbon emission reduction and total supply chain profits

4.2 Impact of Time t and Parameter ω on the Farmer's and the Retailer's Profits

This section discusses the changes in the profits of the farmer and the retailer over time, under the decentralized decision model and the Stackelberg game model.

Figure 3a depicts the changes of the farmer's profits over time, under two models. And Fig. 3b depicts the changes of the retailer's profits over time, under two models. From the two figures, it is evident that under the Stackelberg model, both the farmer and the retailer achieve higher profits compared to those in the decentralized model.

In the Stackelberg model, the retailer provides partial cost subsidies to the farmer, thereby encouraging them to adopt carbon reduction measures. On the other hand, in the decentralized model, the farmer bears greater costs for carbon reduction. Additionally, the absence of effective communication between the

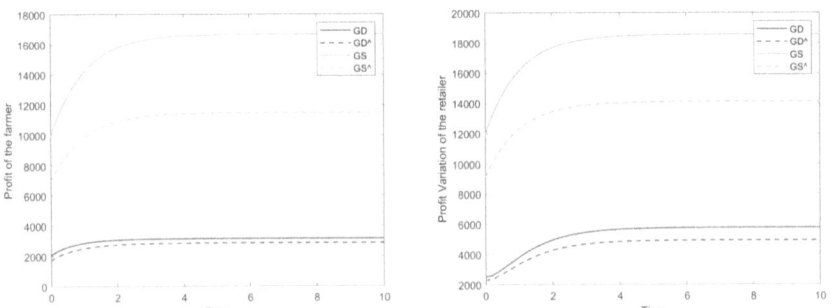

(a) Farmer's profit under different modes (b) Retailer's profit under different modes

Fig. 3. Profits of farmer and retailer under decentralized and Stackelberg model.

farmer and the retailer leads to inadequate access to accurate information, which negatively impacts the effectiveness of carbon reduction. Similar to the previous sections, under the Stackelberg and decentralized models, scenarios involving carbon trading result in significantly higher profits for the profits of both the farmer and the retailer.

4.3   Impact of λ and μ

This section conducts a sensitivity analysis on the emission reduction and carbon sequestration efforts of the farmer $E_f(t)$ and the low-carbon promotion efforts of the retailer E_r, considering three types of cooperation models.

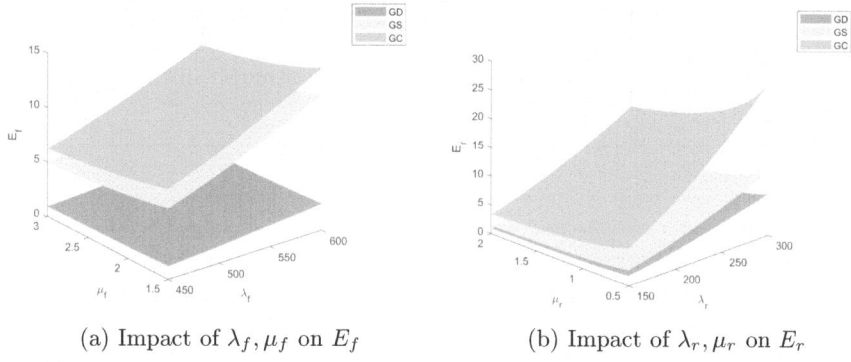

(a) Impact of λ_f, μ_f on E_f (b) Impact of λ_r, μ_r on E_r

Fig. 4. Sensitivity analysis on the efforts of the farmer and the retailer

Figure 4a depicts the impact of λ_f and μ_f on E_f, under three types of cooperation models. Figure 4a shows that as the λ_f and μ_f increase, the emission reduction and carbon sequestration efforts of the farmer E_f also increases. Furthermore, E_f is the highest in the centralized model, followed by the Stackelberg model, and the lowest in the decentralized model. Therefore, appropriately adjusting the relevant coefficients can incentivize both the farmer and the retailer to increase their efforts to reduce carbon emission.

Figure 4b depicts the impact of λ_r and μ_r on E_r, under three types of cooperation models. The trend of Fig. 4b is similar to that of Fig. 4a. Since the effectiveness of carbon reduction in the supply chain is significantly influenced by the efforts of the farmer and the retailer to reduce carbon emission, there is a strong correlation between the increase in these efforts and the enhanced levels of carbon reduction across three different cooperative models (as illustrated in Fig. 2a).

4.4   Impact of p_c

This section discusses the impact of unit carbon sink price p_c on the emission reduction efforts E_f and profit of the farmer, considering three types of cooperation models.

Figure 5a depicts the impact of carbon sink price (p_c) on the farmer's carbon emission reduction efforts(E_f). In Fig. 5a, the most significant variations are observed under the decentralized decision model. Under this model, when p_c is small, the changes in E_f are directly proportional to p_c. This indicates that an increase in carbon sink price can incentivize the farmer's carbon emission reduction efforts. However, as p_c continues to increase and reaches a peak, further increases in p_c start to inhibit the growth of E_f. This change can be attributed to two main reasons:

– Substantial increase in carbon reduction costs due to technological limitations. And the profits of farmers also show a downward trend, when the price of carbon sinks reaches its peak (in Fig. 5b).
– The farmer exhibit risk aversion behavior, which may include concerns about a potential future decline in carbon prices, leading them to prioritize maximizing their current profits over increasing their investments in carbon emission reduction.

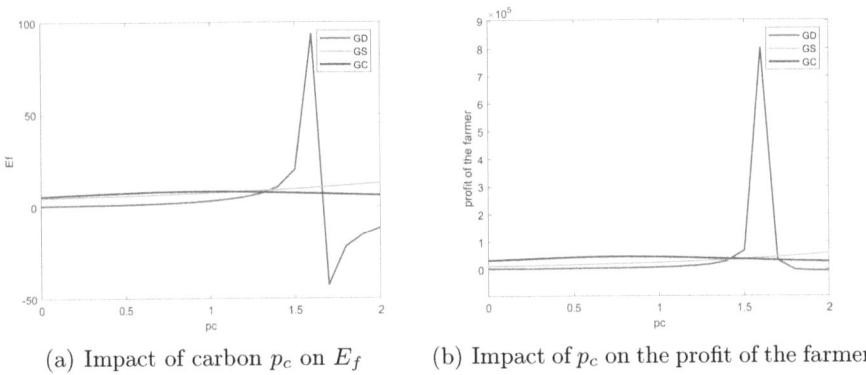

(a) Impact of carbon p_c on E_f (b) Impact of p_c on the profit of the farmer

Fig. 5. Sensitivity analysis on the efforts and profit of the farmer

Figure 5b depicts the impact of carbon sink price on the farmers' profit, and its trend is basically consistent with Fig. 5a, indicating that appropriately increasing carbon sink prices can increase farmer's profit.

5   Conclusion

In this paper, we explore the cooperative models between the farmer and the retailer within the context of agricultural carbon sink trading, as well as the optimal decisions on the efforts to reduce carbon emission for both parties under different cooperative models. In contrast to previous research that primarily focused on carbon emissions during processing and transportation in low-carbon supply chains, this paper mainly investigates carbon emissions during the production of agricultural products and innovatively incorporate carbon sink trading into the agricultural supply chain. Specifically, we delve into three distinct cooperative frameworks: the decentralized, the Stackelberg, and the centralized models, each accompanied by a corresponding differential game model. The Hamilton-Jacobi-Bellman equation is utilized to investigate the equilibrium decisions of each participant under these three cooperative models, respectively. The main conclusions are as follows:

- In the long term, the centralized model excels in all aspects, including the carbon emission reduction efforts of farmers and retailers, the carbon emission reduction level of the agricultural supply chain, and the overall profits of the supply chain, followed by the Stackelberg model, with the decentralized model showing the weakest performance. This indicates that the farmer and the retailer should collaborate to reduce carbon emissions in agricultural products. Specifically, in the initial stages of carbon reduction, supply chain members can adopt the Stackelberg model, where the retailers provides the farmer with essential subsidies for carbon reduction costs. As the carbon emission reduction efforts progress, both participants should strengthen their cooperation and shift to a centralized cooperative model to jointly pursue deeper carbon reduction goals.
- Under the three types of cooperation models, performance with carbon sink trading consistently exceeds that without carbon sink trading. Specifically, In the context of carbon trading, both the farmer and the retailer achieve higher profits, and the level of carbon emission reduction is also improved. Additionally, an appropriately set carbon sink price significantly enhances the efforts of the farmer to reduce carbon emission. Therefore, the government should adopt carbon sink trading as an effective incentive mechanism to promote the implementation of carbon emission reduction throughout the agricultural supply chain.

In conclusion, while this study has provided valuable insights into the effects of carbon emission reduction strategies in agricultural product supply chain, there are several avenues for further research. Carbon sink trading is an important policy tool in the carbon market, where there are sellers and buyers of carbon sinks. Farmers can earn additional income by selling carbon sinks, while companies in need can purchase these sinks to offset their excess carbon emissions. This study is limited to the scenario where only the farmer sells carbon sinks. Therefore, future research could further explore the involvement of companies in the carbon market and assess their potential impact on the sustainability of the agricultural product supply chain. Moreover, carbon trading policies merit consideration during the processing and transportation stages of the agricultural supply chain. Future research could elucidate the synergistic benefits of carbon trading and sink trading on supply chain efficiency and carbon emission reduction.

References

1. Wan, B., Tian, L., Fu, M., et al.: Green development growth momentum under carbon neutrality scenario. J. Clean. Prod. **316**, 128327 (2021)
2. Xu, J., Chen, Y., Bai, Q.: A two-echelon sustainable supply chain coordination under cap-and-trade regulation. J. Clean. Prod. **135**, 42–56 (2016)
3. Xu, X., He, P., Xu, H., et al.: Supply chain coordination with green technology under cap-and-trade regulation. Int. J. Prod. Econ. **183**, 433–442 (2017)
4. Zhang, Y., Chai, Y., Ma, L.: Research on multi-echelon inventory optimization for fresh products in supply chains. Sustainability **13**(11), 6309 (2021)
5. Cai, J., Jiang, F.: Decision models of pricing and carbon emission reduction for low-carbon supply chain under cap-and-trade regulation. Int. J. Prod. Econ. **264**, 108964 (2023)
6. Wang, F., Harindintwali, J.D., Yuan, Z., et al.: Technologies and perspectives for achieving carbon neutrality. Innovation **2**(4) (2021)
7. Ke, S., Zhang, Z., Wang, Y.: China's forest carbon sinks and mitigation potential from carbon sequestration trading perspective. Ecol. Ind. **148**, 110054 (2023)
8. Wang, Z., Wang, C.: How carbon offsetting scheme impacts the duopoly output in production and abatement: analysis in the context of carbon cap-and-trade. J. Clean. Prod. **103**, 715–723 (2015)
9. Yu, Y., Xiao, T.: Pricing and cold-chain service level decisions in a fresh agri-products supply chain with logistics outsourcing. Comput. Ind. Eng. **111**, 56–66 (2017)
10. Yan, B., Wu, X., Ye, B., et al.: Three-level supply chain coordination of fresh agricultural products in the Internet of Things. Ind. Manag. Data Syst. **117**(9), 1842–1865 (2017)
11. Song, Z., He, S.: Contract coordination of new fresh produce three-layer supply chain. Ind. Manag. Data Syst. **119**(1), 148–169 (2019)
12. Ma, X., Wang, S., Islam, S.M.N., et al.: Coordinating a three-echelon fresh agricultural products supply chain considering freshness-keeping effort with asymmetric information. Appl. Math. Model. **67**, 337–356 (2019)
13. Zu, Y., Chen, L., Fan, Y.: Research on low-carbon strategies in supply chain with environmental regulations based on differential game. J. Clean. Prod. **177**, 527–546 (2018)
14. Wang, Y., Xu, X., Zhu, Q.: Carbon emission reduction decisions of supply chain members under cap-and-trade regulations: A differential game analysis. Comput. Ind. Eng. **162**, 107711 (2021)
15. Xia, L., Guo, T., Qin, J., et al.: Carbon emission reduction and pricing policies of a supply chain considering reciprocal preferences in cap-and-trade system. Ann. Oper. Res. **268**, 149–175 (2018)
16. Poyago-Theotoky, J.A.: The organization of R&D and environmental policy. J. Econ. Behav. Organ. **62**(1), 63–75 (2007)
17. Ji, J., Zhang, Z., Yang, L.: Carbon emission reduction decisions in the retail-/dual-channel supply chain with consumers' preference. J. Clean. Prod. **141**, 852–867 (2017)
18. Yang, Y., Yao, G.: Fresh keeping decision and coordination of fresh agricultural products supply chain under carbon cap-and-trade. PLoS ONE **18**(4), e0283872 (2023)
19. Wu, D., Yang, Y.: The low-carbon supply chain coordination problem with consumers' low-carbon preference. Sustainability **12**(9), 3591 (2020)

Active Learning Supported Iterative Combinatorial Auctions

Benjamin Estermann[1], Stefan Kramer[1], Roger Wattenhofer[1],
and Kanye Ye Wang[2]($\boxtimes$)

[1] ETH Zurich, Zurich, Switzerland
{estermann,wattenhofer}@ethz.ch, stefan.kramer@inf.ethz.ch
[2] University of Macau, Macao, China
wangye@um.edu.mo

Abstract. In deep learning-based iterative combinatorial auctions (DL-ICA), bidders do not have to report valuations for all bundles up front. Instead, DL-ICA iteratively asks bidders to report their value for specific bundles and determines the allocation of the items using a winner determination problem. In the process, bidder profiles are modeled using neural networks. However, due to the relatively small number of reported bundles, DL-ICA may not always realize the optimal winner allocation, resulting in reduced economic efficiency. In this work, we improve the economic efficiency, i.e., the social welfare, of DL-ICA by optimizing the underlying machine learning-based elicitation algorithm. To this end, we introduce two different active learning-based initial sampling strategies, called GALI and GALO. GALI ensures an optimal coverage of the entire bundle space during sampling, while GALO tries to find bundles with a high diversity of the bidders' value estimated by the underlying neural network. By doing so, our work extends the scope of active learning, which was previously constrained to small pool sizes. We show how linear programs can be used for active learning to handle pool sizes larger than 10^{30} samples. We prove the correctness of our approach theoretically and verify its performance experimentally.

Keywords: Active Learning · Combinatorial Auctions · Deep Learning

1 Introduction

A combinatorial auction (CA) is a type of auction in which bidders are allowed to place bids not only on individual items but also on bundles of items [14]. CAs allow bidders to fully express their preferences. As a result, bidders will not suffer from the exposure problem [7], where a bidder tries to acquire certain collections of items, but then ends up with only incomplete subsets of those collections. Combinatorial auctions are widely used in practice, for example in the sale of TV advertising slots [9] and in government spectrum auctions [5]. On the downside, as the number of items in the CA increases, the CA's bundle

© The Author(s), under exclusive license to Springer Nature Singapore Pte Ltd. 2025
B. Li et al. (Eds.): IJTCS-FAW 2024, LNCS 14752, pp. 301–316, 2025.
https://doi.org/10.1007/978-981-97-7752-5_22

space grows exponentially. This makes it challenging for bidders to report their full value functions. Instead, algorithms have to be used to achieve an efficient allocation without asking the bidders to value all possible combinations of items. Iterative combinatorial auctions interact with the bidders in rounds. In each round, the algorithm asks the bidders about a limited number of bundles. Deep learning techniques have been used because they provide a better estimate of the bidders' valuation of the unreported bundles [19].

Previous studies [19] have not considered the effects of sampling strategies for eliciting bidders' valuations. This leads to inefficient allocations where the auction outcome is unlikely to reach the social optimum. To improve the efficiency of deep learning-based iterative combinatorial auctions (DL-ICA), we propose to incorporate active learning strategies when determining the bundles that bidders are initially asked to report their willingness to pay for. We implement two different versions of pool-based active learning for regression (ALR), namely greedy active learning on input values (GALI) and greedy active learning on output values (GALO). We propose a linear programming method that enables GALI and GALO to handle sample pools of exponential size. We prove that our formulation is equivalent to the original formulation of GALI and GALO in [20].

Finally, we verify the performance of our algorithms in real-world CA models with different instances in various domains, including the Global Synergy Value Model (GSVM) [8], the Local Synergy Value Model (LSVM) [15], and the Multi-Region Value Model (MRVM) [18]. Across all models, our algorithms achieve a better economic efficiency than the original DL-ICA algorithm, while generally having a lower average number of queries to bidders.

In summary, we introduce ALR to DL-ICA and propose a method for dealing with exponential pool sizes. We formally prove the correctness of our method and experimentally verify its performance on several auction models. Thus, our research provides a new perspective on DL-ICA. We demonstrate the importance of sampling strategies for accurately representing the bidders' valuation function. This has implications for the mechanism design of DL-ICA, which contributes to improving the social welfare of auction allocations.

2   Background and Related Work

2.1   Machine Learning Powered CAs

Machine learning algorithms have been used in the mechanism design of CAs for almost two decades. The majority of previous work can be classified in two ways: (i) generating CA mechanisms with machine learning algorithms, and (ii) reducing the computational complexity within the auction process with machine learning methods. In (i), researchers utilized the automated mechanism design paradigm to generate CA mechanism directly with machine learning algorithms [6,10]. In (ii), previous studies focused on improving the computational efficiency during the auction process [3,11]. In particular, Bayesian learning techniques have been applied in CAs to converge faster [4].

Recently, machine learning algorithms have been integrated into CAs, especially as iterative combinatorial auctions (ICAs) [13]. ICAs do not require bidders to report their full valuation functions on all bundles. Instead, ICAs only query a limited number of random bundles at the beginning of the auctions and interact with the bidders iteratively to ask their value of specific bundles in multiple rounds until they compute the final allocation. Previous work utilizes machine learning techniques, such as kernelized support vector regressions (SVRs), to learn the valuation functions of bidders on different bundles of items [2,4], improving traditional ICAs such as combinatorial clock auctions (CCAs) [1]. Weissteiner and Seuken [19] introduce deep neural networks (DNNs) into the elicitation algorithm in ICAs, which realizes a better computational scalability and enables ICAs in large auction domains. In this study, we investigate the mechanism design of DL-ICA and reveal the importance of the sampling strategy preceding the iterative phases.

2.2 Active Learning

Active learning offers a strategic approach to enhancing machine learning models by allowing them to selectively query the most informative data points for training, addressing the challenge of costly data acquisition, especially in contexts like iterative combinatorial auctions (ICAs) where bidder valuations are crucial yet expensive to obtain [16,17]. The goal is to minimize the number of queries to bidders, thus reducing their burden while maintaining or improving the model's performance. Inspired by greedy passive sampling, an advanced active learning regression method (ALR) has been shown to outperform conventional ALR strategies by maximizing diversity in both the sample and model output spaces [20,21].

Incorporating active learning into DL-ICA significantly enhances social welfare outcomes of auction allocations. This section outlines the foundation of two ALR approaches adapted for our study, following the methodology and notation established by [20].

Greedy Active Learning on Input Values (GALI): Originally proposed by [21], GALI values tries to achieve a good diversity in the input space for the sampled data points. The data point we want to sample next should always lie as far as possible from the already labeled points. Let H be the set of all data points, l be the dimension of our input space, k the number of already labeled data points and $S = \{s_1, ..., s_k\}$ the set of these labeled data points, with $s_i = (s_{i,1}, ..., s_{i,l}) \in H$. For every $x \in H \setminus S$ we then compute the distance to the closest point in S as $d_x^{in} = \min_{i=1,...,k} \|s_i - x\|$. A data point x^* with the largest distance to its closest neighbor in S will be the next one to label.

$$x^* \in \arg\max_{x \in H \setminus S} \quad d_x^{in} \tag{1}$$

After being labeled, the point is added to S and the process is repeated until we have reached our desired number of samples. Notice that this approach is different than the other active learning methods as it does not depend on any learning algorithm.

Greedy Active Learning on Output Values (GALO): Originally proposed by [20], GALO aims to achieve diversity in the model output space. In other words, the next data point to be sampled should lie the furthest away from the already labeled points in the model output space. Unlike GALI, this approach depends on a learned model $\mathcal{M}$. This model $\mathcal{M}$ needs to be trained with k samples before GALO is applied, so that it is able to provide a reasonable estimate of the labels. Let then $S = \{(s_1, y_1), ..., (s_k, y_k)\}$ be the set containing the pairs of the already labeled data points s_i and their label y_i. Let further f denote the function learned by the machine learning model. The other notations stay the same as before. We again compute for every data point $x \in H \setminus S$ the distance to its closest point in S as $d_x^{out} = \min_{i=1,...,k} |y_i - f(x)|$. As before (1), we choose a point with the largest distance to be sampled next.

$$x^* \in \arg\max_{x \in H \setminus S} d_x^{out} \tag{2}$$

3   Methods

This section extends the Active Learning for Revealed Preferences (ALR) methods to improve the Data-Driven Iterative Combinatorial Auctions (DL-ICA) approach, particularly in refining the initial bundle sampling strategy for bidder queries. By optimizing the sampling process, we aim to enhance DL-ICA's economic efficiency and ease bidder participation by reducing the number of bundle queries. Here, we use "already sampled" and "queried" interchangeably with "labeled" to indicate known data point labels.

3.1   Greedy Active Learning on Input Values (GALI)

For GALI, inspired by [20], we compute the distance from each unlabeled bundle to its nearest labeled counterpart to identify the next bundle for sampling. Given the exponential growth of our bundle space with item count, direct sampling becomes computationally prohibitive. We introduce an efficient approach using a limited set of integer linear programs (ILP) solvable via an ILP-solver, with the ILP count matching the labeled bundle count. Given our mechanism aims for minimal bundle queries, the labeled set remains small. We'll redefine GALI as an optimization issue, highlight its computational challenges, and detail our ILP-based solution.

Reformulation of GALI

Definition 1. *(closest to ... in set ...). Let C, D be sets with $C \subset D$. Then $d \in D$ is closest to $c' \in C$ in set C if and only if $\forall c \in C : \|d - c\| \geq \|d - c'\|$. Notice, that d can be closest to more than one element in C.*

Let m be the number of items, $\mathcal{X} = \{0,1\}^m$ the bundle space and $S \subset \mathcal{X}$ the set of already sampled bundles. Note that due to the form of the items in $\mathcal{X}$, the Euclidian norm $\|\cdot\|$ between to items is simply the symmetric difference. To overcome the obstacle of exponential iterations, we reformulate the rule for the determination of the next sample. For all $s \in S$ we denote $H_s \subseteq \mathcal{X} \backslash S$ as the set of not already sampled bundles x, for which x is closest to s in set S. The elements of H_s, which have maximal distance to s, build up the set $\arg\max_{x \in H_s} \|x - s\|$. Let further $dist_s = \max_{x \in H_s} \|x - s\|$ be the maximal distance between s and a bundle in H_s.

The next bundle to query is then:

$$x^* \in \arg\max_{x \in H_{s^*}} \|x - s^*\|$$

$$\text{s.t.} \quad s^* \in \arg\max_{s \in S} dist_s \tag{3}$$

Definition (3) is in fact equivalent to the old one in (1).

As already mentioned at the beginning of this section, B_{new} cannot be computed efficiently in a straightforward way. The problem is the constraint $x \in H_{s^*}$, demanding that H_{s^*} has to be known. Building this set requires computing for every $x \in \mathcal{X} \setminus S$ its distance to each $s \in S$, so that we can determine whether x is closest to s^*. As the set $\mathcal{X} \setminus S$ is of exponential size in the number of items, deriving H_{s^*} takes exponential time.

Implementation with ILPs. We now present a way to compute (3) more efficiently. The idea is to find a $x^* \in \arg\max_{x \in H_s} \|x - s\|$ for every already queried bundle $s \in S$ without having to build H_s. This can be done by solving the following small optimization problem for all $s \in S$.

$$\arg\max_{x \in \mathcal{X}} \|x - s\| \quad (\mathcal{X} = \{0,1\}^m)$$

$$\text{s.t.} \quad \|x - s\| \leq \|x - s'\| \quad \forall s' \in S.\ s' \neq s \tag{4}$$

In this form, the Optimization Problem (4) is nonlinear due to the norms and cannot be solved by an ILP-solver. Therefore, we reformulate it into an integer linear program (ILP).

$$\arg\max_{x \in \mathcal{X}} \sum_{j=1}^{m} x_j + s_j - 2x_j s_j$$

$$\text{s.t.} \quad \forall s' \in S.\ s' \neq s$$

$$\sum_{j=1}^{m} x_j + s_j - 2x_j s_j \leq \sum_{j=1}^{m} x_j + s_j' - 2x_j s_j' \tag{5}$$

$$x, s, s' \in \{0,1\}^m$$

The following Lemma will help prove the equivalence of the ILP (5) and the Optimization Problem (4) in Theorem 1.

Lemma 1. *Let $x, y \in \mathcal{X} = \{0,1\}^m$ be two bundles of size m. Then, $\|x - y\|^2 = \sum_{j=1}^{m} x_j + y_j - 2x_j y_j$.*

Proof.

$$\|x - y\|^2 = \sum_{j=1}^{m} (x_j - y_j)^2$$

$$= \sum_{j=1}^{m} x_j^2 + y_j^2 - 2x_j y_j$$

$$= \sum_{j=1}^{m} x_j + y_j - 2x_j y_j \qquad (\text{since } 0^2 = 0 \text{ and } 1^2 = 1)$$

Theorem 1. *The Integer Linear Program (5) is equivalent to Optimization Problem (4).*

Proof. As the general square function $f(a) = a^2$ is strongly monotone, one can square all the norms in (4) while preserving equivalence. Next, by using the result of Lemma 1, all the squared norms $\|x - s\|^2 \forall s \in S$ can be replaced with the sum $\sum_{j=1}^{m} x_j + s_j - 2x_j s_j$, which results in the ILP (5).

For an existing bundle $s \in S$, $x^* \in \arg\max_{x \in H_s} \|x - s\|$ can be computed by solving the presented ILP (5). In order to find the next sample to query, one has to solve the ILP for every bundle $s \in S$. The resulting bundle of the ILP is then used to compute the maximal distance $dist_s$ of its corresponding bundle s. Out of all the computed bundles, the one that was found by the ILP for the bundle $s \in S$ with the largest $dist_s$ is chosen to be our next sample.

The utilization of ILP (5) to efficiently generate the initial set of bundles that bidders are queried about in ICAs. It takes as an input the number of initial bids c_0 and the number of items m and produces for each bidder the set S of bundles with corresponding valuations Y. It should be noted that for this sampling technique, the bids of the bundles in S are not necessary for the decision of the next sample. Hence, the bidder could also be asked to submit all his bids in one go, instead of being queried every iteration. For this reason, such a sampling strategy is also called *passive sampling*. Based on the results of [12], GALI exhibits a pseudopolynomial runtime in the number of items because all possible solutions are in $\{0,1\}^m$ and the dominant coefficient of the constraint coefficient vector is bounded by the number of items.

3.2 Greedy Active Learning on Output Values (GALO)

Similar to GALI, the problem of exponential iterations occurs in GALO as well. The algorithm proposed by [20] suggests to evaluate the model output for each bundle in the pool, which for our case is of exponential size in the number of items. We can reformulate the way how GALO selects the next point, (2), as an optimization problem. Later on, we use the newly stated optimization problem to derive an ILP-based implementation that is efficiently solvable.

Reformulation of GALO. Let $S \subset \mathcal{X} = \{0,1\}^m$ be the set of already labeled bundles. The labels of the bundles are saved in $Y = \{y_s \mid s \in S$ and y_s is the bidders reported value for $s\}$. Additionally, we have the estimated value function of bidder i $\tilde{v}_i : \mathcal{X} \longrightarrow \mathbb{R}$, which was learned by a machine learning regression model using S together with the labels Y as its training set. We do not specify closer how this model looks like in this section as it could be any regression model. For every y in Y we introduce a set $H_y = \{x \mid \tilde{v}_i(x)$ is closest to y in $Y\}$. Note that the distance metric $|\cdot|$ in label space simply denotes the absolute difference between two real numbers. The idea of the reformulation remains similar. Instead of iterating over the queried bundles S, we do so over their labels Y. For every $y \in Y$, we compute a bundle $x^* = \arg\max_{x \in H_y} |y - \tilde{v}_i(x)|$ for which its estimated value $\tilde{v}_i(x)$ is closest to y in Y, but their distance is maximized. Further, we have $dist_y = \max_{x \in H_y} |y - \tilde{v}_i(x)| = |y - \tilde{v}_i(x^*)|$. The next bundle to query is then

$$x^* = \arg\max_{x \in H_{y^*}} |y^* - \tilde{v}_i(x)|$$

$$\text{with } y^* \in \arg\max_{y \in Y} dist_y \tag{6}$$

The two definitions (6) and (2) are in fact equivalent. As with GALI, the Optimization Problem (6) cannot directly be computed in polynomial time. The problem is the constraint $x \in H_{y^*}$, which demands that H_{y^*} has to be known. Unfortunately, building this set requires computing for every $x \in \mathcal{X} \setminus S$ and $y \in Y$ the distance $|y - \tilde{v}_i(x)|$. As $\mathcal{X} \setminus S$ is of exponential size, deriving H_{y^*} takes exponential time.

Implementation with ILPs. We show how we can again can use ILPs to compute H_{y^*}. In order to solve the presented Optimization Problem (6), we have to compute $x^* \in \arg\max_{x \in H_y} |y - \tilde{v}_i(x)|$ for every label $y \in Y$ without having to build H_y. This can be done by solving the following optimization problem for every $y \in Y$.

$$\arg\max_{x \in \mathcal{X}} |y - \tilde{v}_i(x)|$$

$$\text{s.t.} \quad |y - \tilde{v}_i(x)| \le |y' - \tilde{v}_i(x)| \qquad \forall y' \in Y. y' \neq y \tag{7}$$

The Optimization Problem (7) is not directly of a linear kind due to the absolute difference and the value function $\tilde{v}_i(x)$. In the following sections we derive a linear formulation depending on the chosen machine learning model for the value function $\tilde{v}_i(x)$. We are going to present solutions for two different models, a linear model and a fully-connected multi-layer neural network.

Solution for Linear Model. We first treat the case where the machine learning model is a simple linear regression model and call this approach **GALO (linear)**.

In the linear model the estimated value function is $\tilde{v}_i(x) = w^T x + w_0$, where $w = (w_1, ..., w_m) \in \mathbb{R}^m$ and $w_0 \in \mathbb{R}$. The function is obviously linear. The nonlinearity is only caused by the absolute difference between the real label and the model output. Let C be a large constant. The following ILP erases the nonlinearity while staying equivalent to (7).

$$\underset{x \in \mathcal{X}}{\arg\max} \ r$$

s.t.

$$y - \tilde{v}_i(x) \le r$$
$$\tilde{v}_i(x) - y \le r \tag{8}$$
$$\left.\begin{array}{l} y' - \tilde{v}_i(x) + C \cdot b_{y'} \ge r \\ \tilde{v}_i(x) - y' + C \cdot (1 - b_{y'}) \ge r \end{array}\right\} \forall y' \in Y$$
$$b_{y'} \in \{0,1\} \qquad \forall y' \in Y$$
$$r \in \mathbb{R}$$

Before we can proof Theorem 2, we introduce Lemma 2 which will be helpful for the equivalence proof.

Lemma 2. *Let $b_{y'} \in \{0,1\}$. Let further C be a large enough constant. Then it holds that $|y' - \tilde{v}_i(x)| \ge r$, if and only if $y' - \tilde{v}_i(x) + C \cdot b_{y'} \ge r$ and $\tilde{v}_i(x) - y' + C \cdot (1 - b_{y'}) \ge r$.*

Proof. We make a case distinction.

- $y' \ge \tilde{v}_i(x)$: We choose $b_{y'} = 0$ and $C \ge y' - \tilde{v}_i(x) + r$.

$$|y' - \tilde{v}_i(x)| \ge r$$
$$\iff y' - \tilde{v}_i(x) \ge r$$
$$\iff y' - \tilde{v}_i(x) + (y' - \tilde{v}_i(x) + r) \cdot 0$$
$$\ge r \ \wedge \ \tilde{v}_i(x) - y' + (y' - \tilde{v}_i(x) + r) \cdot 1 \ge r$$
$$\iff y' - \tilde{v}_i(x) + C \cdot b_{y'}$$
$$\ge r \ \wedge \ \tilde{v}_i(x) - y' + C \cdot (1 - b_{y'}) \ge r$$

- $\tilde{v}_i(x) > y'$: We choose $b_{y'} = 1$ and $C \ge \tilde{v}_i(x) - y' + r$.

$$|y' - \tilde{v}_i(x)| \ge r$$
$$\iff \tilde{v}_i(x) - y' \ge r$$
$$\iff y' - \tilde{v}_i(x) + (\tilde{v}_i(x) - y' + r) \cdot 1$$
$$\ge r \ \wedge \ \tilde{v}_i(x) - y' + (\tilde{v}_i(x) - y' + r) \cdot 0 \ge r$$
$$\iff y' - \tilde{v}_i(x) + C \cdot b_{y'}$$
$$\ge r \ \wedge \ \tilde{v}_i(x) - y' + C \cdot (1 - b_{y'}) \ge r$$

Let us now state the equivalence of the two presented formulations.

Theorem 2. *The Integer Linear Program (8) is equivalent to the Optimization Problem (7).*

Proof. The proof is divided into two steps. In the first part we show the equivalence of the constraints:

$$|y - \tilde{v}_i(x)| \leq |y' - \tilde{v}_i(x)| \qquad (\forall y' \in Y)$$

$$\Longleftrightarrow |y - \tilde{v}_i(x)| \leq r \leq |y' - \tilde{v}_i(x)| \land r \in \mathbb{R} \qquad (\forall y' \in Y)$$

$$\Longleftrightarrow \begin{array}{l} y - \tilde{v}_i(x) \leq r \land \tilde{v}_i(x) - y \leq r \\ |y' - \tilde{v}_i(x)| \geq r \qquad (\forall y' \in Y) \\ r \in \mathbb{R} \end{array}$$

$$\xLeftrightarrow{\text{Lemma 2}} \left. \begin{array}{l} y - \tilde{v}_i(x) \leq r \land \tilde{v}_i(x) - y \leq r \\ y' - \tilde{v}_i(x) + C \cdot b_{y'} \geq r \\ \tilde{v}_i(x) - y' + C \cdot (1 - b_{y'}) \geq r \\ b_{y'} \in \{0, 1\} \\ r \in \mathbb{R} \end{array} \right\} \qquad (\forall y' \in Y)$$

Since the constraint of the linear program and the original optimization problem can be mutually derived from each other, the sets of feasible solutions of the two approaches are equal. Now we get to the second part. We'd like to show that $r = |y - \tilde{v}_i(x)|$ for any feasible x.

Let x be feasible. Hence, there exists $r \in \mathbb{R}$ satisfying the constraints

$$y - \tilde{v}_i(x) \leq r$$
$$\tilde{v}_i(x) - y \leq r$$
$$y - \tilde{v}_i(x) + C \cdot b_y \geq r$$
$$\tilde{v}_i(x) - y + C \cdot (1 - b_y) \geq r$$

The first two constraints imply $|y - \tilde{v}_i(x)| \leq r$, the last two $|y - \tilde{v}_i(x)| \geq r$ (through Lemma 2). It follows $|y - \tilde{v}_i(x)| = r$.

We have now proven that the sets of feasible solutions are equal for both, the ILP (8) and the Optimization Problem (7), and that for all these feasible solutions $|y - \tilde{v}_i(x)| = r$. Therefore, maximizing r and maximizing $|y - \tilde{v}_i(x)|$ produces the same results.

With the help of the ILP, we can introduce the algorithm which generates the initial bundle-value pairs for a bidder. The input k defines how many starting bundles are needed. Every round we call the machine learning algorithm $\mathcal{M}$, which takes a training set and their labels as inputs, trains a model on this data and returns the trained model. In this case, the model would be a linear one. Using the ILP (8) we can then determine the next bundle to sample.

Algorithm 1: Generate Initial Bids based on GALO

Input c_0, k

Parameter Machine learning algorithm $\mathcal{M}$

generate k initial bids based on uniform sampling or using Algorithm ?? $\longrightarrow S, Y$;

$count \leftarrow 0$;

while count $\leq c_0 - k$ **do**

    $\tilde{v}_i \leftarrow \mathcal{M}(S, Y)$;

    for all $y \in Y$ **do**

        compute $x_y \in \arg\max_{x \in H_y} |y - \tilde{v}_i(x)|$ (ILP (8));

        $dist_y \leftarrow |y - \tilde{v}_i(x_y)|$;

    end for

    choose $y^* \in \arg\max_{y \in Y} dist_y$;

    $S \leftarrow S \cup \{x_{y^*}\}$;

    ask bidder to submit value for bundle x_{y^*}, add value to Y;

    $count \leftarrow count + 1$

end while

Output (S, Y)

Solution for Neural Networks We present how we can efficiently compute GALO for the neural networks used in DL-ICA and call this approach **GALO (NN)**. More precisely we consider MLPs with ReLU activation functions. Since the ReLU function is nonlinear, our neural networks are as well. Consequently, the learned estimated value function for bidder i, $\tilde{v}_i$, cannot directly be used inside a linear program. Let d_k be the dimension, b^k the bias of layer k and W^k the matrix used to compute the output values of the layer $k + 1$ based on the output values of layer k. Further, o^{k+1} denotes the output of layer $k + 1$ which is computed as the element-wise maximum of 0 and $W^k \cdot o^k + b^k$ due to the ReLU activation function. We can convert the neural network into a set of linear constraints by means of a technique first introduced in [19]. For any layer k (besides the input layer), we introduce three new vectors y^k, z^k, q^k, all of size d_k, together with three linear constraints.

$$z^k - s^k = W^{k-1} \cdot o^{k-1} + b^{k-1} \tag{9}$$

$$0 \leq z^k \leq y^k \cdot L \tag{10}$$

$$0 \leq s^k \leq (1 - y^k) \cdot L \tag{11}$$

Following [19], every element of this polytope satisfies $z^k = o^k$. Let $K + 1$ be the number of layers, input and output layer included. The output of the neural network corresponds to the output of the output layer and is therefore equal to o^K (indexing starts at 0). Hence, we can write $\tilde{v}_i(x) = z^K$ if the following constraints are satisfied:

$$z^0 = x$$

$$\left.\begin{array}{l} z^k - q^k = W^{k-1} \cdot z^{k-1} + b^{k-1} \\ 0 \leq z^k \leq y^k \cdot L \\ 0 \leq q^k \leq (1 - y^k) \cdot L \\ y^k \in \{0,1\}^{d_k} \end{array}\right\} \forall k \in \{1,...,K\}$$

Some remarks on the constraints: L is a big constant, such that for $\forall k \in \{1,...,K\}$ and for every element $z \in z^k$, $z < L$ and for every element $q \in q^k$, $q < L$. Further, z^k and q^k are vectors in the real number space. We omit the proof and refer to the proof of Theorem 1 in the paper of [19].

To solve the optimization problem, we use the ILP (8) from the linear model defined in the previous section and replace the now nonlinear expression $\tilde{v}_i(x)$ with the just derived linear representation.

All together the resulting ILP for GALO (NN) looks as follows:

$$\arg\max_{x \in \mathcal{X}} r$$

s.t.

$$y - z^K \leq r$$

$$z^K - y \leq r$$

$$\left.\begin{array}{l} y' - z^K + C \cdot b_{y'} \geq r \\ z^K - y' + C \cdot (1 - b_{y'}) \geq r \end{array}\right\} \forall y' \in Y$$

$$b_{y'} \in \{0,1\} \forall y' \in Y$$

$$r \in \mathbb{R}$$

$$z^0 = x$$

$$\left.\begin{array}{l} z^k - q^k = W^{k-1} \cdot z^{k-1} + b^{k-1} \\ 0 \leq z^k \leq y^k \cdot L \\ 0 \leq q^k \leq (1 - y^k) \cdot L \\ y^k \in \{0,1\}^{d_k} \end{array}\right\} \forall k \in \{1,...,K\}$$

(12)

We have shown that the ILP for the linear model defined in (8) is equivalent to the Optimization Problem (7). The additional constraints in (12) only ensure that $\tilde{v}_i(x)$ is expressed in a linear way. They do not affect the set of feasible solutions in any way. Hence, the ILP (12) is also equivalent to the Optimization Problem (7).

Finally, to generate the initial bids for bidder i, we can adapt Algorithm 1 to a neural network by solving ILP (12) instead of ILP (8) in line 6.

4   Results

Our experiments were conducted using the Spectrum Auction Test Suite (SATS) version 0.6.4 [18]. SATS enables the generation of diverse Combinatorial Auction (CA) instances across various domains, granting access to bidders' complete value functions and the optimum allocation a that maximizes social welfare. This facilitates the efficiency calculation $\frac{V(a)}{V(a)}$ of any allocation a. Within SATS, we focused on the LSVM, GSVM, and MRVM auction models (Appendix ??), representing spectrum auctions for electromagnetic frequency band licenses. We executed 51 instances each for LSVM and GSVM, and due to MRVM's higher computational demands, we conducted 21 instances.

Table 1. Comparison of uniform sampling (UF) and greedy active learning on input values (GALI) in the LSVM model. The results "Average # Queries" and "Max # Queries" are measured per bidder. "Optimal Alloc. in %" means the fraction that the algorithm finds the optimal allocation in all of the instances. "Efficiency in %" means the average ratio between the *reported* social welfare of the allocation found by our algorithm and the optimal *reported* social welfare. The value inside the parenthesis denotes the standard deviation of the efficiency.

Model	Sampling Technique	Average #Queries	Average #Iterations	Optimal Alloc. in %	Average Runtime	Efficiency in %
GSVM	UF	39.7	4.6	27.45	15 min	**97.95** (0.32)
	GALI	37.2	3.9	60.78	17 min	**99.18** (0.20)
	GALO (NN)	37.95	4.2	62.00	71 min	**99.53** (0.13)
LSVM	UF	50.9	5.0	13.73	33 min	**96.80** (0.41)
	GALI	47.5	4.0	11.76	27 min	**97.55** (0.32)
	GALO (NN)	45.69	3.3	18.00	70 min	**97.45** (0.33)
MRVM	UF	250.3	61.5	0.0	422 min	**96.20** (0.18)
	GALI	252.5	61.5	0.0	455 min	**96.42** (0.20)
	GALO (NN)	312.5	62.8	0.0	760 min	**96.17** (0.13)

Our results, summarized in Table 1, demonstrate that both GALI and GALO (NN) significantly outperform the uniform sampling (UF) baseline by achieving higher efficiency with less variation. Notably, in the GSVM and LSVM models, the average number of samples required decreased. The MRVM model, with its 98 items, proved to be the most challenging, where no instance achieved the optimal allocation. Although GALO (NN)'s performance slightly lagged behind UF in the MRVM model, it still secured better worst-case efficiency. For instance, UF's lowest efficiency across all attempts was 94.40%, with the 25th percentile at 95.66%, whereas GALO (NN) reached 94.70% and 95.92%, respectively.

In combinatorial auctions, even minor improvements in allocation efficiency can significantly affect the outcome due to the high value of transactions involved. These efficiency improvements lead to meaningful financial benefits for all parties involved and positively influence market dynamics. Furthermore, enhanced efficiency boosts trust and engagement in the auction process, creating a ripple effect that enhances market competitiveness and spurs innovation in bidding strategies over time.

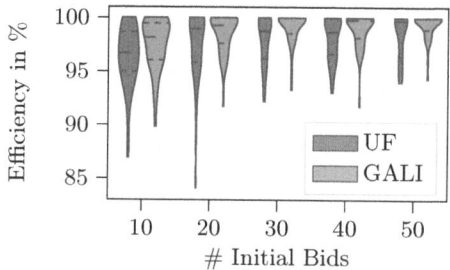

Fig. 1. Violin Plots of efficiency for different numbers of initial bids using uniform sampling (UF) and greedy active learning on input values (GALI) in **GSVM**. The dotted lines inside the violins represent the quartiles. Note that quartile lines are not visible if they denote 100%.

GALI has an average runtime similar to uniform sampling and both can be precomputed before any evaluation is requested from the bidders. The runtime of GALO is of the same magnitude as the uniform sampling approach, scaling well even for auctions with a pool size of 2^{98} samples. In real-world scenarios, large-scale auctions may take multiple days or weeks. Therefore, runtimes in the range of hours mean that it is feasible to apply both GALI and GALO to such auctions. Further, GALI and GALO can contribute to a lower number of total valuations requested from the bidders, leading to a decrease in the total execution time of a real-world auction.

4.1 Robustness to Number of Initial Queries

We examined the impact of varying the number of initial queries on allocation efficiency, using 10 to 50 initial bundles (c_0) and a constant c_e of 10, across 51 auction instances. Our findings, illustrated in Violin Plots for GSVM (Fig. 1), indicate GALI's allocations are more consistent and efficient than those from Uniform Sampling (UF), particularly as initial query numbers decrease. On average, GALI requires 16% fewer queries for GSVM and 18% fewer for LSVM than UF. Notably, GALI outperforms UF even in the least efficient outcomes, with a minimal efficiency of 91.7% for GSVM with 20 initial queries, compared to UF's 84.11%. These results highlight GALI's practical advantage in delivering consistent results with fewer queries.

4.2 Ablations on GALO

We analyze the individual building blocks of GALO in more detail and perform an ablation with two different variants. GALO (NN) is our baseline, where GALO is performed on the basis of a neural network as described in Sect. 3.2. For GALO (linear), we use a simple linear regression model as described in Sect. 3.2 instead of the neural network. For the second one, GALI + UF, we sample the first k bundles with GALI (similar to GALO, see Algorithm 1). We then sample the remaining bundles according to a uniform distribution.

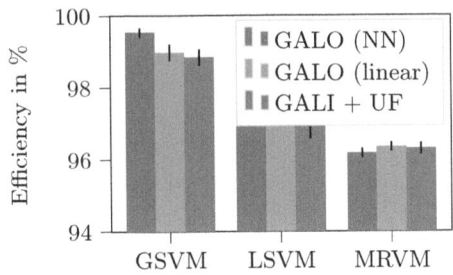

Fig. 2. Ablation study on GALO. GALO (NN) is based on a neural network whereas GALO (linear) uses a simple linear regression model. For GALI + UF, the first k bundles are sampled with GALI (as in GALO), but then uniform sampling (UF) is used for the remaining bundles instead of GALO.

We summarize results in Fig. 2. As can be seen for GSVM and LSVM, GALO (NN) clearly performs the best. This gives us two insights. First, sampling bundles to increase diversity in the output of a neural networks provides a benefit over a simpler linear regression model. Second, GALO (NN) yields a benefit that is not captured by either GALI or uniform sampling. Results for MRVM are slightly different because it contains much more items and is more complex. We hypothesize that due to the large complexity of MRVM, the chosen neural network architecture is not capable enough to learn a useful representation with the first k samples. As a piece of evidence, we find that the number of iterative phases and the number of samples queried in the iterative phase for MRVM are extremely high, with many bidders reaching the upper bound of queries set to terminate an iteration. This already greatly reduces the effect of the initial sampling strategy. Furthermore, the results indicate that for an auction with a large bundle space and complex bidder profiles, broader exploration of the input space is necessary. This is supported by the stronger performance of GALI, which reached an efficiency of 96.42%, compared to GALO (NN)'s 96.17%. For future research, we plan to deviate from the query sizes used in [19] in order to further optimize the performance of GALI and GALO.

5   Conclusion

We introduced active learning to deep learning based iterative combinatorial auctions. We demonstrated how to solve the exponential pool size problem using linear programming. Moreover, we proved the correctness of the linear programming approach. Our results show that active learning yields a significant increase in the efficiency of the resulting allocation, even with fewer queries than the baseline. Furthermore, we have shown that active learning is more robust to different initial sample sizes. Together, these features give our approach significant practical appeal.

References

1. Ausubel, L.M., Milgrom, P., et al.: The lovely but lonely vickrey auction. Comb. Auctions **17**, 22–26 (2006). https://www.researchgate.net/profile/Paul_Milgrom/publication/247926036_The_Lovely_but_Lonely_Vickrey_Auction/links/54bdcfe10cf27c8f2814ce6e/The-Lovely-but-Lonely-Vickrey-Auction.pdf
2. Beyeler, M., Brero, G., Lubin, B., Seuken, S.: imlca: machine learning-powered iterative combinatorial auctions with interval bidding. In: Proceedings of the 22nd ACM Conference on Economics and Computation, p. 136 (2021)
3. Blum, A., Jackson, J., Sandholm, T., Zinkevich, M.: Preference elicitation and query learning. J. Mach. Learn. Res. **5**(Jun), 649–667 (2004)
4. Brero, G., Lubin, B., Seuken, S.: Combinatorial auctions via machine learning-based preference elicitation. In: IJCAI, pp. 128–136 (2018)
5. Cramton, P.: Spectrum auction design. Rev. Ind. Organiz. **42**(2), 161–190 (2013). https://EconPapers.repec.org/RePEc:kap:revind:v:42:y:2013:i:2:p:161-190
6. Dütting, P., Feng, Z., Narasimhan, H., Parkes, D., Ravindranath, S.S.: Optimal auctions through deep learning. In: International Conference on Machine Learning, pp. 1706–1715. PMLR (2019)
7. Englmaier, F., Guillén, P., Llorente, L., Onderstal, S., Sausgruber, R.: The chopstick auction: a study of the exposure problem in multi-unit auctions. Int. J. Ind. Organ. **27**(2), 286–291 (2009). https://doi.org/10.1016/j.ijindorg.2008.09.001. https://www.sciencedirect.com/science/article/pii/S0167718708000945
8. Goeree, J.K., Holt, C.A.: Hierarchical package bidding: a paper & pencil combinatorial auction. Games Econ. Behav. **70**(1), 146–169 (2010). https://doi.org/10.1016/j.geb.2008.02.013. https://www.sciencedirect.com/science/article/pii/S0899825608000626, Special Issue In Honor of Ehud Kalai
9. Goetzendorff, A., Bichler, M., Shabalin, P., Day, R.: Compact bid languages and core pricing in large multi-item auctions. Manag. Sci. **61**, 150316095449006 (2015). https://doi.org/10.1287/mnsc.2014.2076
10. Golowich, N., Narasimhan, H., Parkes, D.C.: Deep learning for multi-facility location mechanism design. In: IJCAI, pp. 261–267 (2018)
11. Lahaie, S.M., Parkes, D.C.: Applying learning algorithms to preference elicitation. In: Proceedings of the 5th ACM Conference on Electronic Commerce, pp. 180–188 (2004)
12. Papadimitriou, C.H.: On the complexity of integer programming. J. ACM (JACM) **28**(4), 765–768 (1981)
13. Parkes, D.C.: Iterative Combinatorial Auctions. MIT Press (2006)
14. Rassenti, S.J., Smith, V.L., Bulfin, R.L.: A combinatorial auction mechanism for airport time slot allocation. Bell J. Econ. 402–417 (1982)
15. Scheffel, T., Ziegler, G., Bichler, M.: On the impact of package selection in combinatorial auctions: an experimental study in the context of spectrum auction design. Exp. Econ. **15**(4), 667–692 (2012). https://doi.org/10.1007/s10683-012-9321-0. https://ideas.repec.org/a/kap/expeco/v15y2012i4p667-692.html

16. Settles, B.: Active learning. Synth. Lect. Artif. Intell. Mach. Learn. **6**(1), 1–114 (2012)
17. Sugiyama, M., Nakajima, S.: Pool-based active learning in approximate linear regression. Mach. Learn. **75**(3), 249–274 (2009)
18. Weiss, M., Lubin, B., Seuken, S.: Sats: a universal spectrum auction test suite. In: Proceedings of the 16th Conference on Autonomous Agents and MultiAgent Systems, AAMAS 2017, pp. 51–59. International Foundation for Autonomous Agents and Multiagent Systems, Richland, SC (2017)
19. Weissteiner, J., Seuken, S.: Deep learning–powered iterative combinatorial auctions. In: The Thirty-Fourth AAAI Conference on Artificial Intelligence (AAAI 2020), New York City, United States of America, pp. 2284–2293 (2020). https://doi.org/10.1609/aaai.v34i02.5606. https://ojs.aaai.org//index.php/AAAI/article/view/5606
20. Wu, D., Lin, C.T., Huang, J.: Active learning for regression using greedy sampling (2018). https://doi.org/10.48550/ARXIV.1808.04245
21. Yu, H., Kim, S.: Passive sampling for regression. In: 2010 IEEE International Conference on Data Mining, pp. 1151–1156. IEEE (2010)

Locating Two Facilities on a Square with a Minimum Distance Requirement

Weian Li[1] and Yu Zhou[2]([✉])

[1] School of Software, Shandong University, Jinan, China
`weian.li@sdu.edu.cn`
[2] The Hong Kong Polytechnic University, Hung Hom, Hong Kong, China
`yu-comp.zhou@polyu.edu.hk`

Abstract. Classic works on facility location problems have been focused on the basic model where facilities and agents are distributed on one line. In this work, we study a new model where one facility or two facilities with a minimum distance requirement are to be located on a square (e.g., a plaza) to serve the agents who are distributed on a line (e.g., a street) that crosses the square. The actual positions of the agents are their private information, and our goal is to design strategyproof mechanisms that decide the locations to build the facilities such that the agents are incentivized to report their true positions and the social welfare is (approximately) maximized. We study different settings, where the facilities can be favorable or obnoxious and the distance metrics can be Manhattan or Euclidean. Interestingly, for Manhattan distances, all of our mechanisms achieve the optimal social welfare. For Euclidean distances, however, the optimal algorithms are not strategyproof. Accordingly, for each setting with Euclidean distances, we design strategyproof mechanisms that guarantee constant approximations of the optimal social welfare.

1 Introduction

Since the work by [22], the field of approximate mechanism design without money has become unprecedentedly flourishing in the recent decade in the fields of multi-agent systems and algorithmic game theory. One popular application scenario is the facility location problem, where facilities are to be built on a line to serve a number of agents who are also distributed on this line, and the goal is to maximize the social welfare. However, the agents' locations are their private information and they may misreport to manipulate the output locations of the facilities to improve their own utilities. On one hand, the mechanism designer wants to design *strategyproof* mechanisms, which informally means the agents' utilities are maximized by telling the truth. On the other hand, due to ethical and legal considerations, there are many situations, such as political election and kidney exchange, when monetary transfers between the agents and the mechanism designer are not allowed [21]. Unfortunately, without monetary transfers, only

---

The authors are listed in an alphabetical order.

© The Author(s), under exclusive license to Springer Nature Singapore Pte Ltd. 2025
B. Li et al. (Eds.): IJTCS-FAW 2024, LNCS 14752, pp. 317–333, 2025.
https://doi.org/10.1007/978-981-97-7752-5_23

for very simple and specific settings, the optimal algorithms are strategyproof [14,22,23]. Accordingly, investigating the extent to which strategyproof mechanisms can approximate the optimal social welfare establishes a large research agenda.

Motivated by real-world applications, in this work, we study the case when the space of agents is different from that of the facilities. Let us imagine a scenario when the government plans to open new retail shops in the town square. The locations of the shops can only be somewhere in the surrounding buildings. To attract more customers and provide better service, the government wants to minimize the sum of total distance between new locations and residents. We assume the residents are living on a street crossing the square, consisting of west and east parts. To better serve the residents, it is desired that the two locations are relatively far from each other (by satisfying a minimum distance requirement). There are other similar problems, such as deciding the entrances of sports stadium, school campus and park, where the locations of these entrances are not on a line and nearby athletes, students and residents would like the entrances to be far away from each other. These problems cannot be described by the classic one-dimensional facility location game model. In this work, we model the above problems as a constrained facility location game.

The minimum distance requirement is first introduced to facility location games in [8], following which we extend the space of facilities from a real line to a square. The situation when agents are located in a line and the facility can be placed on a Euclidean plane is studied in [16], but they focus on the single facility case and the goal is to characterize the conditions under which socially optimal algorithms are strategyproof. In this work, we study the case of two facilities and design strategyproof mechanisms with constant approximations. We study various settings, where the facilities can be favorable or obnoxious, and the agents' distances can be Manhattan or Euclidean. Interestingly, for Manhattan distances, all of our mechanisms achieve the optimal social welfare. For Euclidean distances, we prove that there is no strategyproof mechanism that achieves the optimal social welfare when facilities are obnoxious, and then complement this part by mechanisms with constant approximation ratios. In the following, we briefly state our results.

1.1   Main Results

In this paper, we mainly focus on two-facility location games. The model is parameterized by a minimum distance requirement $d \geq 0$, so that the distance between the two facilities needs to be at least d. The facility can be favorable or obnoxious. A favorable facility means that all agents want to be close to it, and thus agents' *costs* are defined as their distances to the location of the facility. In contrast, an obnoxious facility means that agents want to be far away, and thus agents' *utilities* are defined as their distances to the location of the facility. We discuss two distance metrics, *Manhattan* and *Euclidean*. For each case, we design a strategyproof mechanism to (approximately) minimize (or maximize) the *social cost* (or *social welfare*), i.e., the total cost (or utility) of all agents.

We start this part with two homogeneous facilities of the same type, such as two identical supermarkets or two identical landfills, where every agent's cost or utility is determined by her distance to the closer facility. We first observe that when $d \leq 2$, we can make use of the single facility case (shown in Table 3) and obtain similar results. When $d > 2$, however, the problem becomes more complicated. Our results are summarized in Table 1.

Table 1. Approximation ratio of two homogeneous facilities when $2 < d < 2\sqrt{2}$.

Distances	Favorable	Obnoxious
Manhattan	1	1
	(Theorem 1)	(Theorem 4)
Euclidean	$(1, 2]$	$(1, 2.42]$
	(Theorem 2 & 3)	(Theorem 5 & 6)

Next, we study two heterogeneous facilities of different types, such as one supermarket and one bus station or one landfill and one chemical plant, where every agent's cost or utility is determined by her total distance to both facilities. Our results are summarized in Table 2. For heterogeneous facilities, the problem becomes even trickier. This is partly because the optimal algorithmic solution is highly sensitive to the distance requirement d, which cannot be simply separated by $d = 2$, and it is not easy to utilize the results for single-facility case, even for small d.

Table 2. Approximation ratio of two heterogeneous facilities

Distances	Favorable	Obnoxious
Manhattan	1	1
	(Theorem 7)	(Theorem 9)
Euclidean	$[1, 4]$	$(1, 2.42]$
	(Theorem 8)	(Theorems 10 & 11)

In addition, in Sect. 5, we briefly discuss how to extend our results to several more general models. Regarding the distance metric, we consider the walking distance model. Regarding the facility space, we discuss the case when it also contains the inside area and when its shape is rectangle.

In the full version of this paper, we also investigate the single-facility location games in our model. The results are presented in Table 3, except for the case of Euclidean & Obnoxious, all the mechanisms are optimal. For Euclidean & Obnoxious, we prove that the optimal strategyproof mechanism has approximation ratio between 1.06 and 1.59.

Table 3. Approximation ratio of a single facility

Distances	Favorable	Obnoxious
Manhattan	1	1
	(Theorem 16)	(Theorem 18)
Euclidean	1	[1.06, 1.59]
	(Theorem 17)	(Theorem 19 & 20)

1.2 Related Works

The study of facility location games was initiated by [20]. Since [22], more works have been studying this game through the lens of approximate mechanism design without using payment. Most of the classic models are concentrated on the situation when the facilities can only be placed on a real line [1,11,14,17]. Real line is widely considered not only because it is geographically interesting but also because it captures many non-geographical settings, such as choosing the temperature for a room and selecting a committee to represent people with different political views. A more general setting is considered in [15], where the agents' cost is measured by a concave function of the distance. Beyond real lines, networks such as cycles [7,18] and trees [6,10,25] are also well studied domains. However, to the best of our knowledge, in all these works, the space for facilities coincides with that for agents. A recent survey on mechanism design for facility location games can be found in [4].

While most earlier works on facility location games assume that the agents want to be close to one of the facilities, the past decade witnessed the emergence of a plethora of results on different preference models. Some widely studied models include obnoxious facilities [6], heterogeneous facilities with dual preference [9,28], optional preferences [27] and fractional preference [12].

The minimum distance requirement has been studied from the algorithmic perspective. For example, dispersion problem (spreading a set of points within a metric space or network) has long been studied in computational geometry and facility location [2,3,5,13,19,24]. But all these works focus on designing heuristic, approximate or online algorithms, and incentives have been overlooked. To the best of our knowledge, [8,26] were the first to study facility location problems with minimum distance requirements. However, in their work, the facilities can only be placed on a real line.

2 Preliminaries

In this section, we formally introduce our model, which is illustrated in Fig. 1. There is a square located at the center $(0,0)$ of a Euclidean plane. The coordinates of the four corners are $(1,1), (-1,1), (-1,-1)$ and $(1,-1)$, respectively. On the left and right hand sides of the square, there is a street, $I = L \cup R$, consisting of left part $L = (-\infty, -1]$ and right part $R = [1, +\infty)$. The intersections of the street and the square are $l = (-1, 0)$ and $r = (1, 0)$. The facilities, represented by blue squares or diamonds in the figure, are restricted to be placed on the boundaries of the square, and the agents, represented by red circles, are on the street. We use S to represent the boundary of the square and $N = \{1, 2, \cdots, n\}$ to be the set of agents. Since the agents are all on the x-axis, for simplicity, we ignore the y-dimension and denote by $\mathbf{x} = (x_1, x_2, \cdots, x_n)$ the agents' true location profile. Without loss of generality, suppose that $x_1 \leq \cdots \leq x_n$, and there are n_1 agents on the left, denoted by N_L, and $n_2 = n - n_1$ agents on the right, denoted by N_R. In this paper, we always assume $n_1 \geq n_2$; otherwise, we can reverse x-axis. For any agents $i \in N_L$ and $j \in N_R$, let $l_i = -1 - x_i$ and $r_j = x_j - 1$, respectively, be their distances to the boundary of the square.

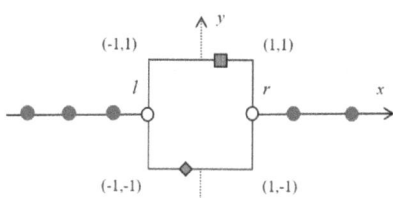

Fig. 1. Model of square facility location (Color figure online)

In this paper, we mainly investigate both single-facility and two-facility location games. Denote by $\mathbf{x}$ the agents' position profile, based on which, the designer needs to decide the locations to build the facilities. Specifically, for single-facility case, a mechanism f outputs a facility's location $(x, y) \in S$ based on the reported location profile $\mathbf{x}$, i.e., $(x, y) = f(\mathbf{x}) : I^n \rightarrow S$. Similarly, for the two-facility case, a mechanism f outputs two facilities' locations $\{f^A = (x^A, y^A), f^B = (x^B, y^B)\} = f(\mathbf{x}) : I^n \rightarrow S^2$. In this work, we only focus on deterministic mechanisms. For two-facility location games, the distance between the two facilities is required to be no less than $d \in [0, 2\sqrt{2})$, i.e.,

$$\sqrt{(x^A - x^B)^2 + (y^A - y^B)^2} \geq d.$$

Note that when $d = 2\sqrt{2}$, it is trivial because the location space only contains $\{(-1, -1), (1, 1)\}$ and $\{(-1, 1), (1, -1)\}$.

To capture agents' utilities, we consider two distance metrics, namely (1-Norm) *Manhattan* Distance and (2-Norm) *Euclidean* Distance. For single-facility location games, agent i's cost (or utility[1]) equals the distance to the facility:

$$c_i(f(\mathbf{x}), x_i) = d_i = |x - x_i| + |y| \quad \text{(Manhattan Distance), or}$$

$$c_i(f(\mathbf{x}), x_i) = d_i = \sqrt{(x - x_i)^2 + y^2} \quad \text{(Euclidean Distance).}$$

---

[1] Since cost and utility have the same value, we do not distinguish them in this paper and use c_i to represent them.

For the two-facility location games, denote by d_i^A and d_i^B the corresponding distances of agent i to facilities A and B. If the two facilities are *homogeneous*, the cost (or utility) of agent i is determined by her distance to the closer facility, i.e.,

$$c_i(f(\mathbf{x}), x_i) = \min\{d_i^A, d_i^B\}. \tag{1}$$

If two facilities are *heterogeneous*, the cost (utility) is determined by the sum of the distances, i.e.,

$$c_i(f(\mathbf{x}), x_i) = d_i^A + d_i^B. \tag{2}$$

Note that, for any agent, the distances to a location (x, y) and its symmetric location with respect to x-axis, $(x, -y)$, are equal. For simplicity, in the following, without loss of generality, it suffices to restrict the location space to be the upper part of the square ($y \geq 0$) for single-facility location games. In two-facility location games, we suppose that facility A, (x^A, y^A), is located at the bottom left of facility B's location (x^B, y^B).

We investigate two types of facilities: *favorable* and *obnoxious*. For favorable ones, agents hope to be as close to them as possible in order to minimize the cost. On the contrary, for obnoxious ones, agents want to keep far away from them to maximize the utility.

As the agents' true locations are their private information, they may misreport to manipulate the final facility locations in order to gain benefit. The designer wants to design strategyproof mechanisms to elicit the true location profile. Informally, a mechanism is strategyproof if no agent can benefit by misreporting her location.

Definition 1. *A mechanism f is strategyproof if for any $\mathbf{x} \in I^n$ and any agent i,*

$$c_i(f(\boldsymbol{x}), x_i) \leq c_i(f(x_i', \boldsymbol{x}_{-i}), x_i) \quad \text{for favorable facilities, and}$$
$$c_i(f(\boldsymbol{x}), x_i) \geq c_i(f(x_i', \boldsymbol{x}_{-i}), x_i) \quad \text{for obnoxious facilities.}$$

Since we always focus on strategyproof mechanisms in this work, we continue to use $\mathbf{x}$ to denote the reported location profile. The social cost (welfare) of a mechanism f with respect to profile $\mathbf{x}$ is defined as the sum of costs (utilities) of all n agents,

$$SC(f(\mathbf{x}), \mathbf{x}) = \sum_{i=1}^{n} c_i(f(\mathbf{x}), x_i). \tag{3}$$

Our objective is to design strategyproof mechanisms such that the social cost is close to the optimal algorithmic solution, $OPT(\mathbf{x})$, i.e., the minimum social cost for favorable facilities or maximum social welfare for obnoxious facilities. A strategyproof mechanism f has an approximation γ, if for any location profile $\mathbf{x} \in I^n$,

$$SC(f(\mathbf{x}), \mathbf{x}) \leq \gamma \cdot OPT(\mathbf{x}) \text{ for favorable facilities, or}$$
$$\gamma \cdot SC(f(\mathbf{x}), \mathbf{x}) \geq OPT(\mathbf{x}) \text{ for obnoxious facilities.}$$

3   Two Homogeneous Facilities

We elaborate the results for homogeneous facilities as shown in Table 1 in this section. Note that all missing proofs in this and the following sections can be found in the full version of this paper. We start with the simple case when $d \le 2$. As we will see, this case actually degenerates to the case of a single facility, and we only provide some high-level ideas in the following. Consider the following two mechanisms, where (x^A, y^A), (x^B, y^B), (x'^A, y'^A) and (x'^B, y'^B) are prefixed locations. To ensure feasibility, it is required that the distance between (x^A, y^A) and (x^B, y^B) (or (x'^A, y'^A) and (x'^B, y'^B)) is at least d.

Mechanism 1 (Two-Fixed). *Disregard the reports and output* (x^A, y^A) *and* (x^B, y^B).

Mechanism 2 (Two-Counting). *Output* (x^A, y^A) *and* (x^B, y^B) *if* $|N_L| \ge |N_R|$; *otherwise output* (x'^A, y'^A) *and* (x'^B, y'^B).

For homogeneous favorable facilities, regardless of distance metric, $(-1, 0)$ and $(1, 0)$ form an optimal algorithmic solution. Thus by setting $(x^A, y^A) = (-1, 0)$ and $(x^B, y^B) = (1, 0)$ in the Two-Fixed mechanism, we obtain the optimal mechanism.

Similarly, for homogeneous obnoxious facilities and $|N_L| \ge |N_R|$, if the distance metric is Manhattan, we can set $(x^A, y^A) = (1, -1)$, $(x^B, y^B) = (1, 1)$ and $(x'^A, y'^A) = (-1, -1)$, $(x'^B, y'^B) = (-1, 1)$ in the Two-Counting mechanism and obtain the optimal social welfare for each facility, which is also optimal as a whole. If the metric is Euclidean, $(1, -1)$ and $(1, 1)$ are not necessarily the optimal algorithmic positions. On one hand, we prove that no strategyproof mechanism can be better than 1.06-approximation (Theorem 19). On the other hand, by setting $(x^A, y^A) = (0, -1)$ and $(x^B, y^B) = (0, 1)$ in the Two-Fixed mechanism, we can obtain an approximation ratio of $(\sqrt{10}/2) \approx 1.59$ as guaranteed by Theorem 20. Symmetric analysis holds for the case of $|N_L| < |N_R|$.

Therefore, in the following two subsections, it suffices for us to focus on the case when $2 < d < 2\sqrt{2}$, where the problem becomes more complicated.

3.1   Favorable and Homogeneous Facilities

Note that when $2 < d < 2\sqrt{2}$, the two facilities cannot be located on the same side of the square. The optimal locations can be characterized by Proposition 1, shown in Fig. 2, whatever the type of distance is used.

Proposition 1. *For favorable facilities and* $2 < d < 2\sqrt{2}$, *the optimal algorithmic locations* (x^A, y^A) *and* (x^B, y^B) *satisfy* $x^A x^B = -1$, $y^A y^B \le 0$, *and the distance between them is exactly* d.

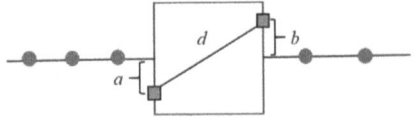

Fig. 2. Two-homogeneous-favorable-facility game

Manhattan Distance. Given this optimal layout, we define a and b (shown in Fig. 2) as the distance from two facilities to x-axis, respectively. The optimal locations $(-1, -a^*)$ and $(1, b^*)$ can be found by Programming (4):

$$\text{Min} \ \sum_{i=1}^{n} \min\{d_i^A, d_i^B\} = \sum_{i=1}^{n_1}(-1 - x_i) + \sum_{i=n_1+1}^{n}(x_i - 1) + n_1 a + n_2 b$$

$$= C + (n_1 - n_2)a + n_2(a + b)$$

$$\text{s.t.} \quad (a + b)^2 + 4 = d^2$$

$$0 \le a \le 1, \quad 0 \le b \le 1, \tag{4}$$

where $C = \sum_{i=1}^{n_1}(-1-x_i) + \sum_{i=n_1+1}^{n}(x_i-1)$ is a constant. Since $a+b = \sqrt{d^2 - 4}$ and $n_1 \ge n_2$, minimizing the social cost is equivalent to minimizing a within the feasible region. Accordingly, we can use the corresponding optimal locations to set up the Two-Counting mechanism as follows and show that it is the desirable mechanism.

Mechanism 3 (Two-Counting(F))[2]

- *Case 1. $2 < d \le \sqrt{5}$: Output $(-1, 0)$ and $(1, \sqrt{d^2 - 4})$ if $|N_L| \ge |N_R|$; otherwise output $(-1, -\sqrt{d^2 - 4})$ and $(1, 0)$;*
- *Case 2. $\sqrt{5} < d < 2\sqrt{2}$: Output $(-1, -(\sqrt{d^2 - 4} - 1))$ and $(1, 1)$ if $|N_L| \ge |N_R|$; otherwise output $(-1, -1)$ and $(1, \sqrt{d^2 - 4} - 1)$.*

Theorem 1. *Mechanism Two-Counting (F) is strategyproof and optimal for two-homogeneous-favorable-facility location games with Manhattan distance.*

Euclidean Distance. Unfortunately, for Euclidean distance, Two-Counting (F) is no longer optimal. Though the optimal algorithmic locations can still be output by the following Programming (5), they are not necessarily the same with the locations in Two-Counting (F). Thus the metrics make a huge difference in this case.

$$\text{Min} \ \sum_{i=1}^{n} \min\{d_i^A, d_i^B\} = \sum_{i=1}^{n_1}\sqrt{(-1 - x_i)^2 + a^2} + \sum_{i=n_1+1}^{n}\sqrt{(x_i - 1)^2 + b^2}$$

$$\text{S.t.} \quad (a + b)^2 + 4 = d^2$$

$$0 \le a \le 1, \quad 0 \le b \le 1, \tag{5}$$

---

[2] In this paper, we always use F to represent the mechanisms that are related to the favorable case. Similarly, O represents the mechanisms for the obnoxious case.

We first prove that there exists no optimal strategyproof mechanism and give a lower bound of approximation ratio which depends on the value of d.

Theorem 2. *There is no strategyproof mechanism which can achieve the optimal social cost for two-homogeneous-favorable-facility location games with Euclidean distance when $2 < d < 2\sqrt{2}$.*

Even though Mechanism Two-Counting (F) may not output the optimal locations, we show that it still achieves 2-approximation in this setting.

Theorem 3. *Mechanism Two-Counting (F) is strategyproof and achieves 2-approximation for two-homogeneous-favorable-facility location games with Euclidean distance when $d > 2$.*

3.2   Obnoxious and Homogeneous Facilities

Again, we first characterize the optimal algorithmic locations of obnoxious cases, given $2 < d < 2\sqrt{2}$, in Proposition 2.

Proposition 2. *When facilities are obnoxious and $2 < d < 2\sqrt{2}$, the optimal algorithmic locations (x^A, y^A) and (x^B, y^B) satisfy $y^A y^B = -1$, and the distance between them is exactly d (Fig. 3).*

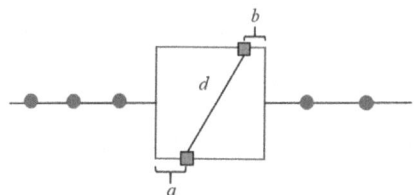

Fig. 3. Two-homogeneous-obnoxious-facility game

Manhattan Distance. We characterize the optimal locations for Manhattan distance. When $d > 2$, x^A cannot be the same with x^B. The social welfare on the x-axis can be written as

$$\sum_{i=1}^{n_1}(x^A - x_i) + \sum_{i=n_1+1}^{n}(x_i - x^B) = \sum_{i=1}^{n_1-n_2}(x^A - x_i)$$

$$+ \sum_{j=1}^{n_2}(x^A - x_{n_1-j+1} + x_{n_1+j} - x^B).$$

Since $x^A - x^B = -\sqrt{d^2 - 4}$ is a constant, the maximum is reached at $x^A = 1 - \sqrt{d^2 - 4}$ and $x^B = 1$. Accordingly, we can design the following variant of Two-Counting mechanism and show its strategyproofness and optimality.

Mechanism 4 (Two-Counting (O)). *Output* $(1 - \sqrt{d^2 - 4}, -1)$ *and* $(1, 1)$ *if* $|N_L| \geq |N_R|$; *otherwise output* $(-1, -1)$ *and* $(-1 + \sqrt{d^2 - 4}, 1)$.

Theorem 4. *Mechanism Two-Counting (O) is strategyproof and optimal for two-homogeneous-obnoxious-facility location games with Manhattan distance.*

Euclidean Distance. We first show that the optimal algorithm is not strategyproof in this setting. Using the similar construction and argument with the proof of Theorem 2, we can present the following theorem.

Theorem 5. *There is no strategyproof mechanism which can achieve the optimal social welfare for two-homogeneous-obnoxious-facility location games with Euclidean distance when $2 < d < 2\sqrt{2}$.*

Next, we design a mechanism that ensures a constant approximation ratio. As $2 < d < 2\sqrt{2}$, using the part of argument in Proposition 2, we claim that the optimal locations, i.e., the optimal values a^* and b^* can be decided by Programming (7).

$$\text{Max} \quad \sum_{i=1}^{n} \min\{d_i^A, d_i^B\} \tag{6}$$

$$= \sum_{i=1}^{n_1} \sqrt{(-1 + a - x_i)^2 + 1} + \sum_{i=n_1+1}^{n} \sqrt{(x_i - (1 - b))^2 + 1}$$

$$\text{S.t.} \quad (2 - b - a)^2 + 4 = d^2$$

$$0 \leq a \leq 2 - \sqrt{d^2 - 4}, \quad 0 \leq b \leq 2 - \sqrt{d^2 - 4}. \tag{7}$$

Though the optimal solution is not strategyproof, the following mechanism achieves a constant approximation.

Mechanism 5 (Two-Fixed (O)). *Disregard the reports and output* $(-1, -1)$ *and* $(1, 1)$.

Theorem 6. *Mechanism Two-Fixed (O) is a $(\sqrt{2} + 1)$-approximation strategyproof mechanism for two-homogeneous-obnoxious-facility location games with Euclidean distance.*

4   Two Heterogeneous Facilities

For two-homogeneous-facility location games, the problem becomes more complicated, as the agents' utilities are determined by the total distance between them and the two facilities. Even for some small d, the optimal locations are hard to characterize and analyze. In this section, different from two-homogeneous-facility location games, we may consider more than two cases based on the value of d.

4.1   Favorable and Heterogeneous Facilities

We first introduce the case of favorable facilities where strategic agents want to be close to them. Besides the mechanisms introduced in Sect. 3, we will also utilize one more involved mechanism, which is generalization of the counting mechanisms.

Mechanism 6 (Two-Counting-Parametric). *Given parameters* $c \geq 0$, $\tilde{y}^A$, $\tilde{y}^B$ *and* $\hat{y}^A$, $\hat{x}^B$, *output* $(-1, \tilde{y}^A)$ *and* $(-1, \tilde{y}^B)$ *if* $|N_L| - |N_R| \geq c$; *otherwise output* $(-1, \hat{y}^A)$ *and* $(\hat{x}^B, 1)$.

Intuitively, Mechanism Two-Counting-Parametric outputs locations by the difference between the number of agents on the left and right. If N_L exceeds N_R by at least a certain number, the locations are set both on the left side; otherwise the locations are set on the left and upper sides.

Manhattan Distance. We first simplify the social cost as

$$SC(f(\mathbf{x}), \mathbf{x}) = \sum_{j \in \{A,B\}} \sum_{i=1}^{n} |x^j - x_i| + n(|y^A| + |y^B|)$$

$$= C + \sum_{j \in \{A,B\}} \sum_{i=1}^{n_1 - n_2} (x^j - x_i) + n(|y^A| + |y^B|), \qquad (8)$$

where $C = 2 \sum_{k=1}^{n_2} (x_{n_1 + k} - x_{n_1 - k + 1})$ is a constant. In this setting, we consider four cases, $d \in [0,1], d \in (1,2], d \in (2, \sqrt{5}]$ and $d \in (\sqrt{5}, 2\sqrt{2})$. In each case, we will give a strategyproof mechanism which achieves the optimal social cost, combining them into Mechanism 7.

Mechanism 7. *The mechanism always outputs two locations by the following rules:*

- *If* $0 \leq d \leq 1$, *run Mechanism Two-Counting with* $(x^A, y^A) = (-1, 0)$, $(x^B, y^B) = (-1, d)$ *and* $(x'^A, y'^A) = (1, 0)$, $(x'^B, y'^B) = (1, d)$.
- *If* $1 < d \leq 2$, *run Mechanism Two-Counting-Parametric with constant* C_1.
- *If* $2 < d \leq 2\sqrt{2}$, *run Mechanism Two-Counting (F).*

Theorem 7. *For two-heterogeneous-favorable-facility location games with Manhattan distance, Mechanism 7 is a strategyproof and optimal mechanism.*

Euclidean Distance. In this case, we consider the general case directly. The optimal locations can be solved by Programming (9):

$$\text{Min} \ \sum_{i=1}^{n} (d_i^A + d_i^B)$$

$$\text{S.t.} \ (x^A - x^B)^2 + (y^A - y^B)^2 \geq d^2. \qquad (9)$$

With Euclidean distance, the optimal locations may vary depending on the true positions of agents and have no elegant property. However, we exploit Mechanism Two-Counting (HtF) in this setting and show that Mechanism Two-Counting (HtF) has an approximation ratio 4.

Mechanism 8 (Two-Counting (HtF)). *The mechanism always outputs two locations f_A and f_B by the following rules:*

- *If $0 \leq d \leq 2$, mechanism outputs $(x^A, y^A) = (-1, -d/2)$ and $(x^B, y^B) = (-1, d/2)$ if $|N_L| \geq |N_R|$; otherwise $(x^A, y^A) = (1, -d/2)$ and $(x^B, y^B) = (1, d/2)$;*
- *If $2 < d < 2\sqrt{2}$, disregard the reports and output $(x^A, y^A) = (-1, -\sqrt{d^2 - 4}/2)$ and $(x^B, y^B) = (1, \sqrt{d^2 - 4}/2)$.*

Theorem 8. *Mechanism Two-Counting (HtF) is a 4-approximation strategyproof mechanism for two-heterogeneous-favorable-facility location games with Euclidean distance.*

4.2 Obnoxious and Heterogeneous Facilities

Finally, we study the obnoxious case in two-heterogeneous-facility location games, where strategic agents want to maximize the total distance to two facilities.

Manhattan Distance. In this setting, we investigate two cases, $d \in [0, 2]$ and $d \in (2, 2\sqrt{2})$, and design a strategyproof and optimal mechanism in each setting.

Mechanism 9. *The mechanism always outputs two locations by the following rules:*

- *If $0 \leq d \leq 2$, run Mechanism Two-Counting with $(x^A, y^A) = (1, -1)$, $(x^B, y^B) = (1, 1)$ and $(x'^A, y'^A) = (-1, -1)$ and $(x'^B, y'^B) = (-1, 1)$.*
- *If $2 < d \leq 2\sqrt{2}$, run Mechanism Two-Counting (O).*

Theorem 9. *For two-heterogeneous-obnoxious-facility location games with Manhattan distance, Mechanism 9 is a strategyproof and optimal mechanism.*

Euclidean Distance. We first show a lower bound of approximation ratio by the following theorem.

Theorem 10. *There is no strategyproof mechanism which can achieve the optimal social welfare for two-heterogeneous-obnoxious-facility location games with Euclidean distance.*

The proof of Theorem 10 utilizes the similar technique and construction with the previous proofs of lower bounds in Sect. 3. In the following, we give the upper bound of the approximation ratio of our mechanisms.

As for the upper bound of approximation ratio, we investigate in two cases, $d \in [0, 2]$ and $d \in (2, 2\sqrt{2})$, and design the approximate mechanism in each case. Finally, we summarize two cases into one mechanism.

Mechanism 10. *The mechanism always outputs two locations by the following rules:*

- *If $0 \leq d \leq 2$, run Mechanism Two-Fixed with $(x^A, y^A) = (0, -1)$ and $(x^B, y^B) = (0, 1)$.*
- *If $2 < d \leq 2\sqrt{2}$, run Mechanism Two-Fixed (O).*

Theorem 11. *For two-heterogeneous-obnoxious-facility location games with Euclidean distance, Mechanism 10 is a $(\sqrt{2} + 1)$-approximation strategyproof mechanism.*

5   Extensions

In this section, we briefly discuss how to extend our results to three related models.

5.1   Walking Distance

In this subsection, we first investigate the walking distance, where the agents first walk along the street to the boundary of the plaza (i.e., $(1, 0)$ or $(-1, 0)$) and then pass through the interior of the square to the facilities. Formally, if the facility's location is (x, y) and an agent i's true location is $(x_i, 0)$, then her cost (or utility) is

$$c_i(f(\mathbf{x}), x_i) = d_i = |-1 - x_i| + \sqrt{(x + 1)^2 + y^2}, \quad \text{or}$$
$$c_i(f(\mathbf{x}), x_i) = d_i = |x_i - 1| + \sqrt{(x - 1)^2 + y^2}$$

Note that the first part of the cost is independent of the facility's location, which means that under walking distance, the optimization problem is equivalent to deciding the facilities' locations when n_1 agents are at $(-1, 0)$ and n_2 agents are at $(1, 0)$ using the Euclidean distance. Interestingly, for all settings considered in this extension, we show that the optimal mechanisms are always strategyproof. Intuitively, when the agents are only at $(-1, 0)$ and $(1, 0)$, compared to the previous situations, the optimal locations depend on the number of agents at each point, rather than agents' exact locations. The problem is simplified to a special case of facility location games with Euclidean distance, which makes the strategyproof mechanisms with the optimal social cost (welfare) exist.

Theorem 12. *For both single- and two-facility location games with walking distance, no matter the facilities are homogeneous or heterogeneous, favorable or obnoxious, the optimal algorithms are always strategyproof.*

5.2 Facilities Can Be Placed Inside the Square

In this subsection, we consider a broader location space of facilities, i.e., the square and its interior. Mathematically, the location space is defined as $\bar{S} = \{(x, y)| -1 \leq x \leq 1, -1 \leq y \leq 1\}$. If the facilities are obnoxious, since the agents want to stay away from them, the optimal solutions should put the facilities on the upper and lower sides of the square (otherwise, we can always move the facilities vertically without breaking any constraint to obtain a better social welfare). It means that the expansion of location space does not affect the results of obnoxious facilities.

Next, we focus on the favorable case. For the single-facility or two homogeneous facility games, agents in fact care about the location of the closer facility, which implies that the optimal locations lie on the left and right sides of the square (otherwise we can move them horizontally to decrease the social cost), and thus our results carry through. It the following, it suffices to restrict our focus to the two-heterogeneous-favorable-facility location games.

When the distance metric is Manhattan, the description of the optimal locations is changed in some cases and a simplified version of our mechanism works and achieves the optimal social welfare. Recall that in Subsect. 4.1, we consider the optimal locations in four cases. However, as the facilities can be located inside the square, the first two cases can be combined into one simpler case. We simplify Mechanism 7 and obtain a concise optimal mechanism.

Theorem 13. *For two-heterogeneous-favorable-facility location games with Manhattan distance, when facilities can be located in the square, the mechanism outputting two locations by the following rules:*

- *If $0 \leq d \leq 2$, run Mechanism Two-Counting with $(x^A, y^A) = (-1, 0)$, $(x^B, y^B) = (-1 + d, 0)$ and $(x'^A, y'^A) = (1 - d, 0)$, $(x'^B, y'^B) = (1, 0)$.*
- *If $2 < d \leq 2\sqrt{2}$, run Mechanism Two-Counting (F).*

is strategyproof and can achieve the optimal social cost.

Once the distance metric is Euclidean, the characterization of the optimal locations becomes more complicated compared to the original model, since we cannot assume that they are located on the boundary to simplify the calculation. On the other hand, the expansion of location space may make the optimal social cost smaller, which could affect the approximation ratio. Nevertheless, we can still show that the approximation ratio of Mechanism Two-Counting (HtF) is 4, using a more general analysis.

Theorem 14. *For two-heterogeneous-favorable-facility location games with Euclidean distance, when facilities can be located inside the square, Mechanism Two-Counting (HtF) is a 4-approximation strategyproof mechanism.*

5.3 Locating Facilities in a Rectangle

When the shape of the central plaza is a rectangle, we use $(l, w), (-l, w), (-l, -w)$ and $(l, -w)$ to represent the four corners, where $w \geq 0$ and $l \geq 0$. We can use similar techniques as in the previous sections to achieve the guaranteed approximate ratios via strategyproof mechanisms. For example, the following theorem is straightforward extension of our result for squares.

We generalize Mechanism Two-Fixed (O) by disregarding the reports and outputting $(-l, -w)$ and (l, w). The generalization of Theorem 6 is:

Theorem 15. *Mechanism Two-Fixed (O) is a $((\sqrt{l^2 + w^2} + l)/w)$-approximation strategyproof mechanism for two-homogeneous-obnoxious-facility location games with Euclidean distance.*

Additionally, for single-facility location games, we also extend Mechanism Single-Fixed by setting $(x, y) = (0, w)$ to the setting of rectangle and still guarantee the constant approximation ratio (see the full version of this paper).

Since the techniques in the previous sections directly carry over here, we omit the detailed calculations of the approximation ratios for the other cases.

6 Conclusion and Future Direction

In this work, we design strategyproof mechanisms for locating facilities on a square where strategic agents are on a line crossing the square. We consider various settings regarding the types of facilities and the metrics of the distances. There are many future directions. One of the most obvious is to study the objective of egalitarian welfare by maximizing the minimum utility. In this work, we only considered one or two facilities, it is interesting to know whether the results can be extended to more than 2 facilities. Moreover, we can investigate how randomness can help improve the approximation ratio and the case when the plaza is of other shapes such as a circle. Finally, an underlying assumption in our work and most existing ones is that the facilities have unlimited supply or capacity. It is of both theoretical and practical importance to generalize our results to take the facilities' capacities into consideration.

Acknowledgements. This work is partially supported by the National Natural Science Foundation of China (No. 62102333) and the Guangdong Basic and Applied Basic Research Foundation (No. 2023A1515010592).

References

1. Anshelevich, E., Postl, J.: Randomized social choice functions under metric preferences. J. Artif. Intell. Res. **58**, 797–827 (2017)
2. Baur, C., Fekete, S.P.: Approximation of geometric dispersion problems. Algorithmica **30**(3), 451–470 (2001)
3. Ben-Moshe, B., Katz, M.J., Segal, M.: Obnoxious facility location: complete service with minimal harm. Int. J. Comput. Geom. Appl. **10**(6), 581–592 (2000)
4. Chan, H., Filos-Ratsikas, A., Li, B., Li, M., Wang, C.: Mechanism design for facility location problem: a survey. In: The 30th International Joint Conference on Artificial Intelligence (IJCAI 2021), pp. 1–17 (2021)

5. Chen, J., Li, B., Li, Y.: Efficient approximations for the online dispersion problem. SIAM J. Comput. **48**(2), 373–416 (2019)
6. Cheng, Y., Yu, W., Zhang, G.: Strategy-proof approximation mechanisms for an obnoxious facility game on networks. Theor. Comput. Sci. **497**, 154–163 (2013)
7. Dokow, E., Feldman, M., Meir, R., Nehama, I.: Mechanism design on discrete lines and cycles. In: EC, pp. 423–440. ACM (2012)
8. Duan, L., Li, B., Li, M., Xu, X.: Heterogeneous two-facility location games with minimum distance requirement. In: AAMAS, pp. 1461–1469. International Foundation for Autonomous Agents and Multiagent Systems (2019)
9. Feigenbaum, I., Sethuraman, J.: Strategyproof mechanisms for one-dimensional hybrid and obnoxious facility location models. In: Workshops at the Twenty-Ninth AAAI Conference on Artificial Intelligence (2015)
10. Feldman, M., Wilf, Y.: Strategyproof facility location and the least squares objective. In: EC, pp. 873–890. ACM (2013)
11. Filos-Ratsikas, A., Li, M., Zhang, J., Zhang, Q.: Facility location with double-peaked preferences. Auton. Agent. Multi-Agent Syst. **31**(6), 1209–1235 (2017). https://doi.org/10.1007/s10458-017-9361-0
12. Fong, C.K.K., Li, M., Lu, P., Todo, T., Yokoo, M.: Facility location games with fractional preferences. In: AAAI, pp. 1039–1046. AAAI Press (2018)
13. Fotakis, D.: On the competitive ratio for online facility location. Algorithmica **50**(1), 1–57 (2008)
14. Fotakis, D., Tzamos, C.: On the power of deterministic mechanisms for facility location games. In: Fomin, F.V., Freivalds, R., Kwiatkowska, M., Peleg, D. (eds.) ICALP 2013. LNCS, vol. 7965, pp. 449–460. Springer, Heidelberg (2013). https://doi.org/10.1007/978-3-642-39206-1_38
15. Fotakis, D., Tzamos, C.: Strategyproof facility location for concave cost functions. Algorithmica **76**(1), 143–167 (2016)
16. Kyropoulou, M., Ventre, C., Zhang, X.: Mechanism design for constrained heterogeneous facility location. In: Fotakis, D., Markakis, E. (eds.) SAGT 2019. LNCS, vol. 11801, pp. 63–76. Springer, Cham (2019). https://doi.org/10.1007/978-3-030-30473-7_5
17. Lu, P., Sun, X., Wang, Y., Zhu, Z.A.: Asymptotically optimal strategy-proof mechanisms for two-facility games. In: EC, pp. 315–324. ACM (2010)
18. Meir, R.: Strategyproof facility location for three agents on a circle. In: Fotakis, D., Markakis, E. (eds.) SAGT 2019. LNCS, vol. 11801, pp. 18–33. Springer, Cham (2019). https://doi.org/10.1007/978-3-030-30473-7_2
19. Meyerson, A.: Online facility location. In: FOCS, pp. 426–431. IEEE Computer Society (2001)
20. Moulin, H.: On strategy-proofness and single peakedness. Public Choice **35**(4), 437–455 (1980)
21. Nisan, N., Roughgarden, T., Tardos, É., Vazirani, V.V. (eds.): Algorithmic Game Theory. Cambridge University Press, Cambridge (2007)
22. Procaccia, A.D., Tennenholtz, M.: Approximate mechanism design without money. In: EC, pp. 177–186. ACM (2009)
23. Procaccia, A.D., Tennenholtz, M.: Approximate mechanism design without money. ACM Trans. Econ. Comput. **1**(4), 18:1–18:26 (2013)
24. Ravi, S.S., Rosenkrantz, D.J., Tayi, G.K.: Heuristic and special case algorithms for dispersion problems. Oper. Res. **42**(2), 299–310 (1994)
25. Schummer, J., Vohra, R.V.: Strategy-proof location on a network. J. Econ. Theory **104**(2), 405–428 (2002)

26. Xu, X., Li, B., Li, M., Duan, L.: Two-facility location games with minimum distance requirement. J. Artif. Intell. Res. **70**, 719–756 (2021)
27. Yuan, H., Wang, K., Fong, K.C., Zhang, Y., Li, M.: Facility location games with optional preference. In: ECAI 2016, pp. 1520–1527. IOS Press (2016)
28. Zou, S., Li, M.: Facility location games with dual preference. In: AAMAS, pp. 615–623. ACM (2015)

Author Index

The manufacturer's authorised representative in the EU is Springer
Nature Customer Service Centre GmbH, Europaplatz 3, 69115 Heidelberg,
Germany. If you have any concerns regarding our products, please
contact ProductSafety@springernature.com

Printed and bound by CPI Group (UK) Ltd, Croydon, CR0 4YY

29/04/2026

02099532-0006